'Provocative, enlightening ... Calder is the perfect guide around some of mankind's most substantial achievements, but never swerves away from asking hard questions' ***HERALD*, BOOKS OF THE YEAR**

'Superb' ***FINANCIAL TIMES***

'An engaging study ... It has something of the appeal of Jared Diamond's *Guns, Germs and Steel* ... Calder makes a simple and important point, often with engaging and unexpected detail: architecture is indeed made by energy, which makes crucial the next stage of its evolution'
ROWAN MOORE, *OBSERVER*

'An essential read: clarifying, alarming, but hopeful' ***ARCHITECTS' JOURNAL***

'Calder has written an energetic global history of architecture – energetic both in the vim he brings to a colossal subject, and in its particular focus ... For the general reader, it's an entertaining and original introduction to the history of architecture. For the architect, it helpfully sets the daunting challenges of our day in lively and inspiring context'
WILL WILES, *RIBA*

'Finally a book to replace Pevsner's standard history of architecture. Calder retells the story of architecture for the climate change generation'
DR JAMES W. P. CAMPBELL, HEAD OF DEPARTMENT OF ARCHITECTURE, UNIVERSITY OF CAMBRIDGE

'Brilliant ... a truly astonishing depth and breadth of research ... a new frame for architectural writing which frankly makes some of the previous architectural histories look at best parochial, or at worst irrelevant in the face of the global climate crisis'
PROFESSOR JEREMY TILL, *BUILDINGS AND CITIES*

'Arguably the most important new contribution to the field of architectural history in decades' **JAMES BENEDICT BROWN, *JOURNAL OF ARCHITECTURE***

'A brilliantly written and timely investigation into a fundamental truth that is often overlooked: energy, in particular the availability of certain types of fuel, is perhaps the single most important driver of architectural design' **FLORIAN URBAN, PROFESSOR OF ARCHITECTURAL HISTORY, GLASGOW SCHOOL OF ART**

'Fierce and elegantly written ... a fine a'
history, with a venomous sting in its ta
EXCELLENT ESSEX

BARNABAS CALDER

Architecture
From Prehistory to Climate Emergency

A PELICAN BOOK

PELICAN BOOKS

UK | USA | Canada | Ireland | Australia
India | New Zealand | South Africa

Penguin Books is part of the Penguin Random House group of companies whose addresses can be found at global.penguinrandomhouse.com

First published in hardback 2021
Published in paperback 2022

006

Book design by Matthew Young
Typeset by Matthew Young
Printed and bound in Great Britain by Clays Ltd, Elcograf S.p.A.

A CIP catalogue record for this book is available from the British Library

ISBN: 978-0-141-97820-8

To Helen, Charlotte and Harriet,
who make me happy.

Contents

Introduction

This book tells the story of how fossil fuels made the world a much better place for humans.

It is a disturbing story because an overwhelming scientific consensus makes it clear that unless we can get away from our all-embracing dependence on fossil fuels by 2050, we will make the planet apocalyptically horrible for humans and many other creatures.[1] The construction and running of buildings are currently responsible for 39 per cent of all human greenhouse gas emissions – the equivalent of some 9 billion tonnes of CO_2.[2] This figure needs to drop to net zero by 2050. It is the toughest challenge the world of architecture has ever faced.

This book is the first to ask how humanity's access to energy has shaped the world's buildings through history. The story of the changing relationship between architecture and energy over time shows us how much good fossil fuels have brought us, and how difficult it will be to give them up. But it also offers hope. The history of architecture is shot through with the remarkable ingenuity and adaptability shown by humans in meeting the challenges of their natural environment and in improving human lives in the most unpromising circumstances. If, in the past, the world has changed so much, so often and so fast, it only needs enough sincere effort for it to change again, this time towards sustainability.

Build Your Own Pyramid

The extent to which today's human energy use is historically abnormal can be seen by looking at one of the biggest buildings of the ancient world. Around 4,500 years ago, perhaps the most powerful person in the world was the Egyptian pharaoh Khufu. He commissioned for himself one of the heaviest buildings ever made: his monumental tomb, well over 5 million tonnes of sandstone, granite and limestone, was dragged up earth ramps on sledges by tens of thousands of his sweating subjects, until it formed a vast pyramid almost 147 metres high. The drawing of it here is to the same scale as the other line drawings of key buildings throughout the book, so you can see how much bigger it is than most of history's big buildings.

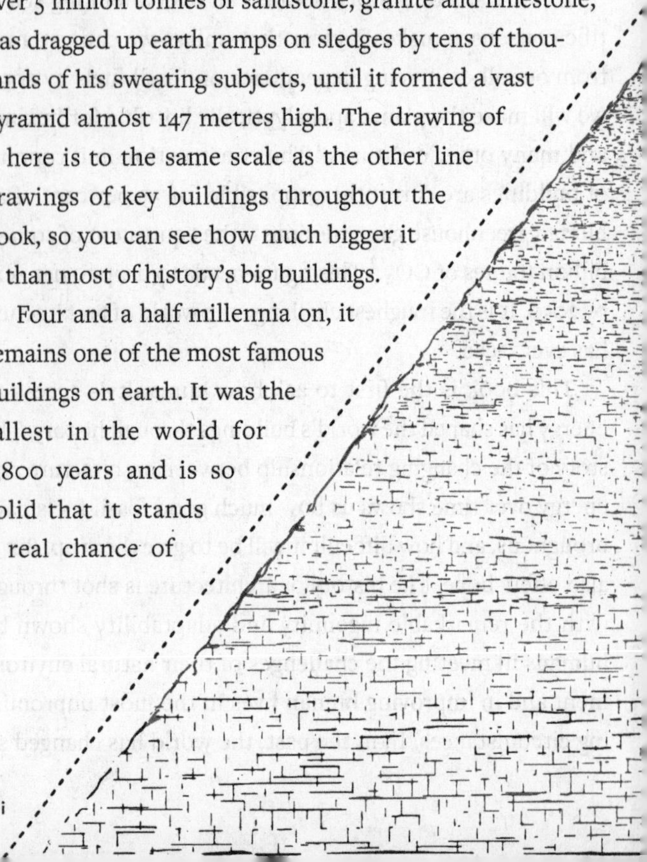

Four and a half millennia on, it remains one of the most famous buildings on earth. It was the tallest in the world for 3,800 years and is so solid that it stands a real chance of outlasting humanity on the planet – a kind of man-made hill. Estimates suggest that it cost around 78 million days of labour, spread over a force of tens of thousands of labourers, probably working for more than a decade.[3]

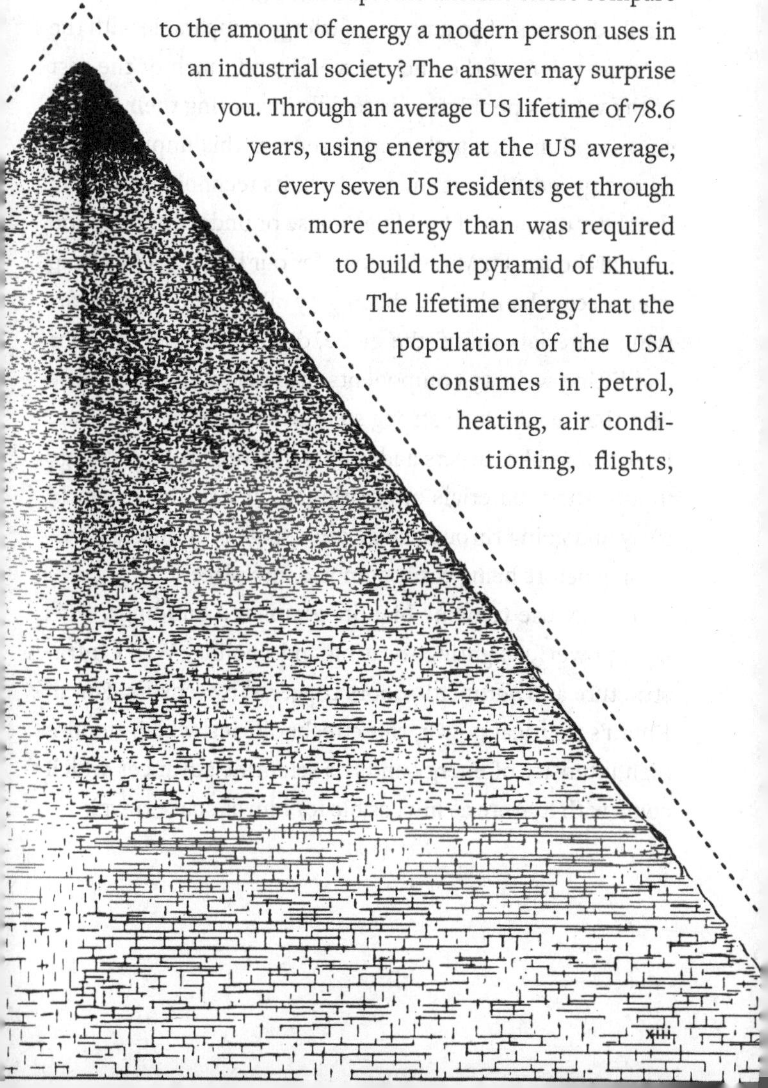

How does this supreme ancient effort compare to the amount of energy a modern person uses in an industrial society? The answer may surprise you. Through an average US lifetime of 78.6 years, using energy at the US average, every seven US residents get through more energy than was required to build the pyramid of Khufu. The lifetime energy that the population of the USA consumes in petrol, heating, air conditioning, flights, consumer goods, food and drink, and so on, is the equivalent of the amount it would take to build nearly fifty million pyramids the size of Khufu's. If you squeezed them in as tightly as you could they would take up more than a quarter of the entire landmass of the USA.[4] The average modern American family disposes of an amount of energy comparable with the greatest rulers of the ancient world, and much of the rest of the world's population is rapidly increasing their energy consumption. Our architecture reflects this unprecedented energy wealth in its materials, in its technologies and in the sheer quantity of buildings in use or under construction around the world. Miracles done for our buildings by cheap energy seem banal to us: the mighty oil-fuelled engines carrying materials round the globe, digging out foundations and lifting weighty components into place, or the intense heat that makes such strong steel and such large windows. Even things like carpets and ceiling tiles require high energy inputs, their materials originating in oil wells half a world away and going through stage after stage of industrial processing before being easily gummed into place.

Energy use remains high in completed buildings, with lifts, powerful lighting, heating and cooling, Wi-Fi infrastructure and automatic doors triggered by sensors. For all Khufu's political power, these technologies, which require high intensity of energy and the scientific advances which come with it, were as impossible for him as space travel.

Energy

Energy is the capacity to do anything. Without it nothing can be heated up or moved, nourished or destroyed. The sum total of human activity has always been constrained by the total amount of energy humanity could harness. Human actions have been shaped by the nature and limitations of their available energy, from the abundant food supplies for the pyramid builders to the coal heat that produced billions of bricks for English industrial cities.

Energy is the food we and our animals eat and the useful work done with those calories. It is also the heat from burning fuels, whether firewood, dried dung, coal, oil or natural gas. More recently, heat and muscle power have been joined by the mechanical power of steam engines and internal combustion engines, and hugely flexible electrical energy. Many hope that solar, wind and other renewables will soon replace fossil fuels.

Since the 1850s, scientists have known that all types of energy can be measured using the same units and converted from one form into another, however superficially different they may seem. The heat of a fire, the chemical energy in batteries and burgers, and the pushing power of water in a mountain stream can all be expressed in joules or calories or kilowatt hours. In theory, although the information is usually incomplete, the energy cost of any historical building could be expressed in these same units: the human and animal labour to procure materials and deploy them, and the heat required for making glass, metal or mortar.

The power of energy sources has grown sharply with

fossil fuel use. Human and animal muscle provided almost all mechanical force before the steam engine. Muscles are weak. An elite athlete at full sprint can achieve around 3.7kW* for a few seconds, but normal sustained effort in a day's labour of eight hours is more like 0.075kW.[5] This figure – the power output of someone labouring steadily – will be used at various points throughout this book to give a sense of how energy-hungry different building activities are. For comparison, many car engines are almost 3,000 times more powerful than a human's sustained effort.[6] Bulldozers can be three or four times more powerful again, replacing around 10,000 human labourers with just one driver and a mighty diesel engine.

Work can also be compared with heat. Throughout the history of architecture the two have indeed been related, even before steam engines allowed heat to become motion. In farming societies, finite amounts of land meant choosing between growing wood to burn for heat and food to feed people and animals.

Before fossil fuels, heat was expensive. A tonne of wood takes time to grow, locking up the land beneath it while it does so. When felled, transported and burned, new-cut wood can produce around 2,780kWh of heat energy per tonne. With fossil fuels, heat became cheap. A tonne of coal

---

* The kilowatt (kW) is a unit that expresses how much power a thing can bring to bear – the power at any one instant of an electrical current or of a horse pulling a cart. The kilowatt hour (kWh) is the amount of energy a force of one kilowatt would apply over a period of an hour, so it can be used to describe the amount of work done by the current, or the horse, or anything else, over a given period of time.

generates around three times as much energy as a tonne of wood. A tonne of oil is even easier to extract and transport, and holds more energy than coal – about 11,630kWh.[7] So a tonne of oil can replace over 150,000 hours of human labour – over 19,000 eight-hour shifts. At August 2020 prices, all that work can be had for just $332.[8]

With world oil consumption reaching 4.66 billion tonnes in 2018, it provides the equivalent energy bonus to human society of more than 400 billion extra people doing manual labour for an average US working year – the equivalent in pre-fossil fuel terms of more than fifty servants for every human being on the planet.[9] The other leading fossil fuels, coal and natural gas, provide the world with more than the same amount of extra energy again. Energy wealth is very unevenly distributed, but for good and ill our billions of fossil fuel servants shape every aspect of our world.

The different energy economies of different places at different times have structured what was affordable and desirable in the way of buildings. Even buildings which intentionally push the limits of what is possible have always done so in ways that reflect their energy context: we will see that the Romans prided themselves on transporting exotic building stones huge distances across their empire. It was ostentatiously difficult, but achievable with the resources they had in abundance: labour and shipping. They did not try to build in structural steel, which would have used amounts of firewood they simply could not spare. With modern, highly efficient steel production it costs around 5,000kWh of energy to make a tonne of steel. To obtain that much energy before fossil fuels would have required all of the wood felled in an area of 5,500 square metres of sustainably coppiced woodland to be burned into charcoal.[10] Worse – in lower-technology pre-industrial conditions, 5,000kWh of charcoal would have made far less than a tonne of steel.

Most of the architects and clients in this book would not have interpreted their own architecture in terms of energy. Clients and their builders often had an impressive practical grasp of the relationships between labour, fuel and construction, but before 1850 they certainly did not theorize them into a single notion of 'energy'. But however they understood energy, it acted on them. Energy is like gravity – you do not need to theorize it to be bound by its rules.

As the chapters that follow will show, producing buildings is a big physical challenge, and the type and quantity of energy available is a major determinant of what gets built, where and how. The history of architecture has a great deal to teach the present about how dependent we are on fossil fuels for our current standard of living and how comprehensively entangled our current architecture is with heavy use of fossil fuel energy.

The Energy Story of Architecture

Shifts in human access to energy have been a key determinant of architectural change, from the initial prehistoric impetus to create the first houses, cities, temples, palaces and monuments, through the revolutions in technical skill that produced the great temples of Greece and Rome, the towering pagodas of China and the delicate stone spider's webs of French Gothic cathedrals, and on into modernity. From the 1600s, fossil fuels came to revolutionize first Britain, then continental Europe and North America, and then other parts of the world, changing everything about buildings and cities.

The story of architecture and energy is full of human achievement and ingenuity, of ambition and optimism, but also of exploitation, inequality and cruelty. It is wider in its implications too: as one of humanity's most energy-hungry activities, construction forms a kind of illustrated history of the beliefs and capacities of powerful people over thousands of years. These beliefs and capacities themselves were and are shaped by our energy contexts, from Khufu's pyramid – perhaps the ultimate expression of the overwhelming dominance of one man over the ordinary farmers who created his wealth – through to the great railway stations that became the world-changing monuments of the coal-fired nineteenth century, and the social housing of the twentieth century, where entire energy-rich societies clubbed together to improve the living conditions of the worst-housed.

This book moves from place to place and period to period around the globe, stopping off to look in detail at some of the most remarkable and justly famous buildings ever produced, including the impeccably subtle Parthenon in Athens, the enormous Basilica of St Peter in Rome and the skyscrapers of New York that stole every record for height and construction speed. We will visit some of the world's most exciting cities, from Uruk, perhaps the first of them all, via ancient Rome and Georgian London, to contemporary Chinese megacities. We will look at the origins and characteristics of some of the world's great architectural moments, styles and ideas, many of which have gone on to shape today's architectural theory and practice. Beautiful materials will come and go: hand-carved marble and wood, well-laid brick, exposed concrete and expanded cork, each bringing its own requirements for artistry, craft and technological innovation to turn these raw ingredients into memorable, emotionally engaging works of creativity.

Along the way it will become clear how extraordinarily rich a tale humanity's buildings tell about human priorities, human views of the world and the relationships between individuals and groups. Khufu's pyramid reflects the cosmology of an elite who believed that their bodies would be physically resurrected in the afterlife. It is the antithesis of today's wicker coffins, which aim to allow humans to disappear without leaving yet more harmful traces.

The story of architecture and energy turns out to be the story of humanity. Family structure is reflected in homes, whether those of post-industrial families with separate bedrooms for everyone or the agrarian huts where entire extended families huddled for warmth with their animals, amid flea-ridden straw. Idealized political structures of different energy regimes are shown off by conspicuous energy expenditure on palaces or parliament buildings, royal tombs or law courts. Religious ideas and religious power are proclaimed grandly by temples, churches or mosques, using the very finest craft techniques of their place and moment. In the age of coal, the ambitions and dreams of industrialists were reflected in their factories and country houses. In the twentieth century, coal, oil and electricity brought unprecedented material improvements to daily life, but also ever greater wars and mightier weapons; the pitch of anxiety and excitement which these unstoppable changes provoked emerges in buildings which seem sometimes too clean, too smooth and too good for mere messy humans. Many in the architecture world today are trying hard to lead the difficult effort to end our fossil fuel use and avoid catastrophic climate change.

Beyond this tendency of architecture to mirror our fundamental social, economic and intellectual structures, there is something else too; something that means that builders and designers as far back as we can trace have pushed themselves beyond the basics. Whether in the carvings on some of the earliest known homes, the glorious ceramic work of Middle Eastern mosques or the extra attention given to the way steel meets glass in the best modernist towers, architecture seems to have its own momentum, and all over the world to produce buildings which can bring delight to the well-informed visitor.

Human societies have based themselves on three energy systems so far: foraging, which leaves few architectural traces; farming, which leaves a great many; and fossil fuels, which have grown rapidly in importance with each of the past three centuries, their use and influence spreading around the globe to power the worldwide building boom of the past hundred years.

The thousands of years of farming cultures covered in Part One represent much less change than the four fossil-fuelled centuries of Part Two, but they are discussed in detail to help adjust contemporary eyes to the metaphorical and literal darkness of the agrarian world. Today rapid technological and social changes seem so normal that we get irritated with the deficiencies of, say, smartphones that would have seemed like science fiction three decades ago. When farming was the main energy source things changed glacially slowly, and inequality of gender, class, ethnicity and age were universal and unchallenged norms. Great monuments were bought at the price of injurious poverty for most and enormous, insecure privilege for a small minority.

Part Two will take the story on through the past four centuries, where first coal and then oil, gas and nuclear energy changed architecture beyond recognition, setting many alight with excitement at the new possibilities, but filling others with anxiety and horror at the speed and scale of change, and its unforeseeable consequences. Architecture reflected both responses.

The relationship between architecture and energy changes as the story goes on. In Part One, the great effort invested in constructing a large building – what contemporary architects refer to as 'embodied energy' – was expensive and difficult to obtain. Buildings were built only when the need was pressing or the sponsors powerful. Once up, most buildings were designed to demand as little ongoing energy as possible – domestic open fires were much the biggest ongoing energy cost of architecture, providing warmth, but also cooking heat and light after sunset. With the rise of fossil fuels, the initial financial cost of materials and construction dropped sharply, even as the energy consumption of these processes rose. And by the later twentieth century the ongoing energy cost of running many buildings – what architects call the 'operational energy cost' – had risen considerably, heating being joined by air conditioning, abundant artificial lighting, lifts and so on.

As architecture changed technically, aesthetic fashions changed too. The working lifetime of a typical architect in the twentieth century went through a number of different styles that would have taken five centuries to play out in any period before the 1800s. These three were all designed by Basil Spence, and were completed in 1938, 1962, and 1976.

The subjects of much of the second half of the book will be disproportionately European and North American, because these areas industrialized first through the accidents of history and geology. Their architectural response to fossil fuels was often the earliest and was frequently, thanks to the economic and colonial power that accompanied it, influential on other parts of the world.

The examples I have chosen to focus on are sometimes world-leading: Uruk as perhaps the world's first city, Rome and Song-dynasty China as representatives of the biggest energy booms of the agrarian world, and the British and American Industrial Revolutions as successive early exploiters of fossil fuels. Other examples show patterns that recur with variation all over the planet: aesthetic conservatism; use of architecture to send messages or to reinforce privilege and power within an energy system; innovation rapidly adapting to new energy supplies; the best building designers of each period carrying the ideas and materials of their moment to an extraordinary pitch of beauty and skill.

Many of the buildings and architectural ideas in this book are parts of the 'canon' of architectural history that is used for discussion within many Western architectural contexts. I want to explore how the buildings so many in the West know and love emerged from their energy conditions. There is also abundant high-quality academic research already in existence on these buildings, making it possible to analyse them with greater certainty and precision. However, with the right research, any building or city, ancient or modern, could be approached using the same question that underlies this book: what role did the energy context have in shaping the architecture? Wherever you find yourself on the planet, you will be able to take the ideas of this book and apply them, looking for different forms of energy use in the stone, wood, concrete or steel of the building in front of you, and enriching your understanding of the past and the present through them.

I very much hope that this book helps you to enjoy and understand the architecture around you (not a jet-fuelled long-haul flight away), and to think differently about how and why we create buildings. In researching the book, I have resisted the temptation to fly to many of the sites discussed, writing about some of them only from published sources and from a wealth of photographs and drawings. Parts of the book might have been better with first-hand experience of the buildings, but in a world that can ill afford the carbon burden of jet-fuelled travel, the move towards sustainable energy consumption will involve bigger and tougher compromises than this.

PART ONE

CHAPTER 1

Life with Less Energy

Amid the colourless landscape of ice, snow and rock, the buildings would have stood out with shocking vividness. Visible from a distance on their raised ground, the nearer you got the more striking they would have become. Each structure was brightly coloured and bristled with protruding shapes like coral. Some were over six metres across; all were tall enough to stand up in. As you got closer, you might have recognized vast, curling tusks and gigantic bones. The structures were not just decorated with large pieces of skeleton, but made of them.[1]

At the base of each of the handful of buildings, a circle of twenty or so hefty mammoth skulls was partially buried in the earth. The natural openings of the skulls were used as sockets to hold thick, heavy leg bones, also from mammoths, which leaned in towards each other as they rose like the poles of a tepee. Animal hide was probably stretched over each robust structure to provide a weatherproof shelter, and then further bones were added for strength and decoration. Some or all of the bones were painted. In better weather, smoke might have drifted from vents at the top of the roof, but in colder times it probably only seeped out at the seams

between hides. Throughout most of history, smoky interiors were normal in cold climates; hypothermia kills far faster than lung disease.

Going inside, the air might well have been acrid with an intense smell like burning hair: remains seem to suggest that in this treeless landscape bones were burned as fuel.[2] Through the soot and darkness you might have made out by the firelight that the mammoth bones of the interior had been carved with chevrons and other recurrent patterns. The people who lived in these structures were modern humans in biological terms, but, as this was around 14,500 years ago, they had no knowledge of farming, no experience of cities, no metal tools, no pottery, no glass. As their homes showed clearly, they lived by hunting mammoths, the three-metre-high, woolly, ginger elephant cousins of the Ice Age north, and other animals that herded on these cold grasslands.

All that we know about these houses has been reconstructed by archaeologists from collapsed bones and other buried leftovers. A few such groups of remains have been found, mostly from around the same period, across a band of Ukraine and Russia. They are among the earliest buildings anywhere on earth to have left substantial evidence of their form and materials, their remains surviving better in the soil than those of less robust building materials.

Right back fourteen and a half millennia ago, with the earliest buildings to leave substantial remnants, we can see energy shaping architecture with a clarity sometimes obscured by the complexity of later societies. These houses were tools to allow humans to hunt the rich game of the icy northern grasslands. The houses were built to control energy, keeping in the fire's warmth. They were made of food energy by-products, and kept people warm as the bones left over after meat, fat and hide had gone were burned. Building the houses cost the residents energy directly – hard work moving big bones and stretching hides over them. Construction work also had an opportunity cost: time spent on construction was unavailable for hunting, fire-making, cooking or other activities that contributed to the useful energy of the group.

These are the essentials of architecture now as then: it costs us energy to build and to run, but in return it keeps us warm or cool, and houses many of our activities. In low-energy societies all or most indoor activities took place in the house, whereas with our vast fossil fuel energy wealth we have a bewildering range of specialist buildings for everything from medical treatment to industrial production to administration of our complex, populous world. Our range of well-housed leisure facilities might seem utterly bizarre to our low-energy ancestors, from airports processing millions of purely recreational travellers to indoor refrigerated snow slopes for skiing in warm countries. What would these ancient people, so expert in harvesting and conserving their precious supplies of energy, make of modern people driving to artificially lit and ventilated gyms to try and burn the too-easy calories of our heavily processed food on exercise bikes which themselves require energy input from the electricity grid?

Despite the gulf between our energy experiences and theirs, we can recognize architecture's tendency to go beyond the basics of its function. Even from the broken-down remnants of the mammoth huts, aesthetic handling seems detectable in the powerful regular zigzags of stacked jaw bones. Houses next door to each other use the same structural principles but have very different details: the mammoth-skull foundations are buried the other way up; tusks are sometimes removed to use the sockets to hold structural bones and are sometimes turned proudly outward to dramatize the exterior.[3] Our ancestors, even living in what we would see as threatening and marginal conditions, exhibited the same instinct for displaying their individuality that is shown in contemporary rich-world suburbs by house owners painting their walls and doors, growing distinctive plants or marking their own artistic identity with anything from plastic gnomes to modern sculpture.

Mammoth hunters, until the animals became scarce through over-exploitation, were privileged by historical standards in the quantity of protein and fat they could obtain as fuel for their muscles and brains. They built unusually robustly for hunter-gatherers, partly no doubt because they had tougher weather to endure. Most hunter-gatherer structures around the world may well have been more like the humbler ones found in numerous sites across the US state of Wyoming. Around two metres wide, these are little pits scraped down into the sandy soil. Many have signs of a fire having been made there, and in some of the comparatively elaborate ones food and waste were buried. A few have hearths, but otherwise their floors were of untreated sandy soil. Some had postholes where a superstructure of branches was pushed into the ground. The picture they conjure up seems unsophisticated: crude, improvised, small, humble.

In fact, though, simple architecture that cost little time and muscular effort to produce was probably part of a precise and complex series of adaptations by these builders to their specific energy context. Anthropologists have found that hunter-gatherer groups are remarkably good at determining who should do what to optimize their energy conditions.[4] Low-energy societies make highly rational decisions about how much effort to invest in building in proportion to the level of use they intend to make of the structure. Many hunter-gatherers move around throughout the year to exploit the best food energy sources of each season. For a very mobile group to spend extensive effort on obtaining and processing building materials like stone is a disproportionate use of energy, and out of twenty-nine societies in one study, none of the mobile foragers used stone or prepared timber (as opposed to branches). Instead these mobile groups favoured quick, low-effort structures which could be patched up and brought back into use when the group returned to the area. It was clearly a pattern that worked for the pre-Columbian culture of Wyoming, whose people seem to have used these sorts of structures from around 8,000 years ago through to the mid-eighteenth century. Groups who did not move area seasonally were much more likely to use stone and timber, avoiding the heavy energy costs of maintaining and replacing short-life materials for buildings they used a lot.[5]

The great differences between Wyoming's hunter-gatherer 'house pits' and the houses of the rich world today mask the fact that our ancestors were almost identical to us biologically. As this book will show, the enormous differences between how they lived and how we live arise from our massively greater ability to harness energy. In architecture, and in the rest of life, form follows fuel.

With their shelters, fires and cooked food, ancient hunter-gatherers were already a world away from the apes they had left behind in the tropics hundreds of thousands of years before. Humans had already called architecture into being with their big, clever brains. Yet we are far more distant from hunter-gatherers in our access to energy than hunter-gatherers were from apes. The energy it takes to fly a group of six friends on holiday from London to LA and back is around the equivalent of the entire life's food requirements of a long-lived hunter-gatherer.[6]

Our spectacular energy wealth shapes our world so completely that it is no longer obvious to us. The lower-energy world of ancient hunters, and our unfamiliarity with their lives, make their dependence on animal energy much more obvious. Already, at the very beginning of recorded architecture, we see the inextricable link between energy sources and buildings. From the fairly rapid extinction of the mammoth – and so many other large animals after humans arrived in their territory – we can also see a pattern of instability and imbalance in the relationship between humanity and our energy sources.

How Grass Supported Stone

Well over 1,000 kilometres south of our mammoth hunters, a world-changing energy revolution was just starting to take shape, from which mysterious new architectural forms were later to emerge. The rich soil and abundant natural springs of what is now south-east Turkey supported fruit, nuts and other edible plants, as well as diverse game. The area had grasses whose seeds could be used for beer-making and for calorific, easy-to-chew porridges and gruels. As the mammoth hunters were building their houses, their more southerly contemporaries were finding that, in the areas where they had gathered up grass seeds and carried them back to the camp, the seeds they had dropped had produced increasingly rich crops of grasses. The seeds of these could be stored for months as long as you could keep them dry, so the reduced food supply of the winter could be patched over by grain stored from the summer – something that was harder to achieve with seasonal abundance of fruit and meat. The grasses that these people were working with included the ancestors of modern wheat and modern barley. Over many centuries the importance of these grasses rose. This was where, around 12,000 years ago, farming first began.[7]

From this place and time is emerging a complex of buildings even more striking than the mammoth-bone houses. An exciting excavation at a site known today as Göbekli Tepe has uncovered some of the most impressive structures of the period: a group of substantial oval enclosures built of stone, with large stone columns around the edges and paired in the middle of each.[8] The columns may have supported a wooden roof made of long, thick timbers, perhaps thatched to keep those beneath warm and dry.[9] The most impressive features, though, are the columns, each of which is made of a single piece of stone, the tallest being as much as five metres high. The business of carving these from the bedrock must have been skilful and labour-intensive, but to raise them to their vertical position and to move them into place represented a substantial challenge in assembling and coordinating labour. The archaeologist who led the excavations argued that the level of effort was too great for housing and that he had found a religious site.[10]

Whether residential or exclusively religious in purpose, the buildings seem to suggest links between the natural and

supernatural worlds. The columns, in particular, are often carved. Some are made to resemble a giant stone human being and many have stone animals on them. Some of these are in shallow relief, barely emerging from the face of the stone, while others are fully three-dimensional, as if petrified while scurrying down the column.

These impressive buildings seem to show that, right from these earliest days when cereal crops were starting to emerge, the scale of the effort humans were prepared to put into buildings, both their construction and their decoration, rose.

The Rural Idyll

Farming brought with it a sort of energy trap: having been adapted to maximize its energy output, the land could support a larger population and higher birth rates than could the opportunistic food harvest of hunting and gathering. To furnish enough food to give a group of hunter-gatherers their 2,000 or 3,000 calories per day, the group typically needed a territory of more than two square kilometres per person. Even relatively unsophisticated slash-and-burn farming could support ten or twenty times more people on the same area of land.[11] Yet for many or most of the new individuals who owed their existence to cereals, the farming life was tougher than their ancestors' hunter-gatherer way of life: more hard physical work, much of it grindingly repetitive, a worse diet, increased levels of disease and earlier death.[12] Yet agriculture became the primary energy source for most of the human population for millennia.

In places where the climate and soil produce lower

levels of crop fertility, it is still possible today to see the architecture of farming at its clearest. Some of the world's least modernized farming villages are those of the Dogon in Mali. Many Dogon villages are spread along the Bandiagara Escarpment, a cliff rising several hundred metres out of the arid plains of inland Mali. The cliff retains invaluable water in the dry season and offers a good defensive position or hiding place when political troubles intrude.[13] The villages contain the essentials of farming architecture. Small, simple houses made of earth, straw and stones or potsherds, roughly rectangular with rounded corners, are interspersed with taller, narrower-based granary buildings, capped by pointy roofs made of dried leaves to keep the rain out of the precious millet seeds that provide this year's food and next year's crop to plant. Millet is more tolerant of Mali's unpredictable rainfall than higher-yielding but fussier grains like wheat. Men's granaries contain the millet seed and their farming tools

and other belongings. Women's granaries are independent, and contain valuables and food, in particular pulses and the spices that enliven a repetitious, cereal-heavy diet.[14]

These granaries are generally built or maintained by men towards the end of the barren dry season, in preparation for the farming year ahead. Labour-intensive wooden elements like the door and its frame are carefully conserved from earlier granaries for reuse. Parts of the earth walls of old granaries are broken up and pulped into the mud mix for the new granary, with an accompaniment of prayer and ritual in the hope of a good harvest. Dogon farmers negotiate tough conditions of drought, localized flooding, locusts, famine, nutritional deficiencies and disease.[15] In recent years, government clashes with Islamicist paramilitary groups have further worsened the situation for many villagers.

The economy with which Dogon villages are built shows that these farmers run only a modest energy surplus from their staple millet farming. Even so, they invest construction and maintenance effort on buildings which do not offer direct energy returns. There are extra houses, used only by women during menstruation – beliefs about religion or health playing out in the architecture even where so little building can be afforded. There is also a low, open-sided shelter consisting of a substantial square of thatch supported by timber, propped on simple stacked-stone columns. The function of this is as a meeting place in which the men discuss collective decisions of the village. The thatch keeps off the heat of the sun, and its low height means that if a disagreement becomes fierce the men, who are forced to sit, cannot stand up to fight.[16] Even in these technologically simple buildings, architecture

is being used to represent and to shape the nature of political discussions.

One house in some Dogon villages is grander, its façade often ornamented by baked-mud reliefs, and significantly higher than those around it. This is the house of the *hogon*, a senior man of the village who has a mixture of religious, judicial and diplomatic functions in promoting the well-being of the village.[17] The other farmers of the village produce enough calories each year to offer some of their own labour to ornament and aggrandize his house. The larger the food surplus produced by a farming society, the more other people they can support with food and labour, freeing these lucky individuals from the hard grind of food production to pursue other specialisms.

And with these specialist people has come specialist architecture. The *hogon* has his special house, with religious meanings to the art on its front wall. Larger, more energy-rich

societies have produced larger numbers of specialist non-farming people and specialist buildings, separating off into distinct building types the various functions of rule, religion, food storage, knowledge acquisition and so on.

The first half of this book will investigate the architecture of the fertile farming economies which for around twelve millennia produced the world's most complex and sophisticated societies, with buildings to match. In the labour-intensive craft and very large scale of their architecture, the Aztecs, the Mughal rulers of India and the Romans might all appear a world away from the Dogon villagers, yet the fundamentals of the Dogon system were the same: a farming economy producing enough food to support a clustered sedentary population and some non-farming activity. What different societies chose to do with that energy surplus has produced endless variation and brilliance.

CHAPTER 2

Farming, the City and Monumental Architecture

Uruk: The First City?

Amid miles of beige desert stands an equally beige hump, crumbly-edged and irregular. Around it for some distance are mounds, pits and ditches. Their profiles are messy and sloping-shouldered, but enough straight lines survive to betray that it was humans, not geology, making these disturbances in the desert. You are unlikely at present to visit these bleak leftovers, as they stand in the centre of Iraq. Since their first partial excavation at the start of the twentieth century, the archaeologists have been interrupted by a succession of wars fuelled by the modern dependence on petroleum. The site, abandoned once again in view of the current instability, still contains many mysteries.

Yet these are the sad remains of one of the most important locations in human history: Uruk, the first known city. This chapter will look at the ways in which, as farming yields increased with new techniques, the extra grain came to be used to support large populations of people who, for the first time, were not primarily occupied with producing food. Those who emerged as powerful in this new world commissioned the first monumental architecture – temples, palaces and other structures on a new larger scale. Already by around 4000 BCE there was some kind of special building with a central hall five metres wide and fifteen metres long. By 3200 BCE Uruk was a true city.

At Uruk we may be seeing the birth of specialism in architecture, with the appearance of workers who spent most of their time building rather than farming, hunting or gathering.

Not long before, almost every human had from early childhood been involved in food and fuel production as their primary activity, with a wide variety of supporting skills and knowledge.[1] With the birth of specialism, technologies became rapidly more complex. Uruk saw the beginnings of the earliest-known writing system, the first currency (long before coins, workers in Uruk were paid in energy itself: standard-sized bowlfuls of grain) and fast progress in bronze-casting.[2]

The food to support the new classes of specialists arose from productive new farming techniques. Animal-drawn ploughs, clay sickles and an ingenious system of irrigation which provided a reliable and nutritionally rich water supply may have increased by as much as five or ten times the amount of barley produced per farm labourer.[3] This enabled the religious authorities who implemented the new technologies on their farms to give barley to the other villagers to pay for farming labour, keeping what they needed to plant next year and what they needed to feed the animals that pulled the ploughs. After these expenses, paid in barley, the temple estates had around two thirds of their crop still left over as profit. With this they could buy other services from other people who wanted the calories – construction work, help with keeping track of the grain income and expenditure, and stone- or metalworking.[4]

The large non-farming population these new techniques could sustain produced this new form of civilization in only a few hundred years (precise dating eludes archaeologists in these last centuries before writing). Uruk, a city of some 250 hectares, may have housed around 40,000 people by 5,200 years ago.[5]

Earlier farming communities had showed little sign of social distinctions. One house was much like another, and in death too people were treated relatively equally, judging by the humble things they were buried with.[6] With specialization and the appearance of cities, society became harshly stratified. Suddenly a very small group of people had immense power, and a larger number had moderate power under these early leaders, or were fed and housed in return for serving their needs. Everyone else – most of the population – remained as relatively humble farmers. The grandeur of the elite depended on these farmers producing a bit more of the staple crop than their own family required to live, and handing it over to the rulers from piety, obedience or fear.

Architecture reflected the new social realities with its usual fidelity. The outlines can be seen even from the incomplete archaeology of Uruk, hampered by modern wars and numerous ancient rebuildings that have swept away all but the lowest parts of the early city. Whereas the mammoth-house settlement had only houses, no specialized buildings, Uruk by around 3500–3100 BCE had, alongside much ordinary housing, large formal buildings for religious or royal figures, an abundance of storerooms for grain and precious goods, and secure, barrack-like quarters for textile workers, who seem to have been held in something like slavery.[7]

A central cluster of particularly big buildings was raised physically above the majority of the city. Only the lower walls of these early monumental buildings at Uruk tend to survive, but enough remains to make clear that they were striking things. Mostly made of unfired brick, formed from clay and straw and dried in the sun, the buildings have thick walls, often vertically striped by patterns of projection and recession that probably acted as buttresses. The largest building is square, its sides fifty-seven metres long. Other impressive structures include an irrigated, walled garden-courtyard and open halls up to eleven metres wide which required substantial timbers to be brought hundreds of kilometres from the Levantine mountains, across the treeless plains of Mesopotamia.[8] The oldest surviving piece of literature, *The Epic of Gilgamesh*, describes a legendary king of Uruk travelling into wild mountains and fighting a monster to bring back such timbers. It also claims that he was the one who built the city's impressive circuit of walls.

Some of the grand buildings of Uruk's central precinct may well have stored barley or the treasures that surplus grain bought for the ruling elite. Others were almost certainly used for the kind of grandiose spectacles that have accompanied monarchs and religious leaders ever since.

Symmetry is a strong recurring theme, with most buildings having one axis of symmetry, some two. There are courtyards, perhaps as settings for larger gatherings or military displays, and streets were up to five metres wide.[9] Many of the buildings may have been painted, but at least one had a special treatment which must have both made it stand out and helped it to age better. Into its clay surface were pushed cones of coloured stone, making a mosaic pattern on the surface. These cones were extracted elsewhere and brought to Uruk in heavy loads. It may well be that the textile-like

patterns they formed, together with the undulation of the wall, evoked earlier religious structures made of fabric.[10]

The social structure of the region that Uruk dominated might have been recognizable four or five millennia later to medieval town dwellers over much of Eurasia: a large area of fields providing food, dotted with villages under the dominant influence of the city. The city itself was surrounded by a large wall, partly for reasons of military defence, but probably even at this early date also used for social and economic control: if there are only a few ways into the city you can mount an effective guard to make sure that taxes are paid and individuals regarded as dissident or criminal are kept out. In the centre of it all, the gods and the government were glorified by monumental buildings. It is a shape of city that proved highly durable in societies whose energy basis was the farming of barley and wheat.

Cities of Maize and Rice

Around the world, farming was independently invented several times, including rice farming in Asia and maize farming in Mesoamerica. City shapes co-evolved with each crop and the farming landscapes it produced.

The Classic-period (250–900 CE) Maya people of Guatemala, southern Mexico and areas of the modern states around these came to a different design of city based on their staple crops. The remains of their monumental buildings have always held a magnetic fascination for visitors from elsewhere – stepped pyramidal temples up to seventy metres high (around half the height of the pyramid of Khufu), poking up above the dense greenery of the rainforest. They still seemed sufficiently exotic and futuristic in the 1970s for George Lucas to adopt them as the alien moon Yavin IV in *Star Wars.*[11]

While sometimes associated with burials, the primary function of Mayan pyramids was as raised platforms on which to build temples. The obvious ostentation of height and the display of power involved in moving so much stone were augmented by layers of more subtle meaning. The number of steps on a pyramid, the orientation and the sculptural programmes have all been explained by archaeologists as deriving from Mayan cosmology and religion.[12]

The presence of these overwhelming monuments amid such apparently unpromising ground (rainforest soils tend to be shallow and are generally quickly exhausted by farming) has provoked theories ranging from alien construction teams through to religiously motivated visiting building parties from better farming land elsewhere.[13] In particular, archaeologists were long baffled by the absence of dense walled cities on the Mesopotamian model. Recent work using new satellite and airborne laser-scanning techniques has brought about the beginnings of a fascinating revolution in the understanding of ancient Mayan cities. It is now clear that the division between countryside and town was absent in the Mayan rainforest. Instead, the majority of people lived on the land as farmers, achieving high fertility from the maize and squashes grown on their small plots.[14] So productive was the range of farming techniques they employed that the overall population density of what was essentially continuous farmland could remain high enough to support powerful rulers and large-scale monumental construction activity. Rainforest soil that rapidly declines into near-desert when cleared of its trees was able, when farmed carefully as rainforest and fertilized with waste, to support dispersed cities that might in the largest cases have housed 100,000 people.[15] Long causeways allowed travel round these huge areas of farm/city, and farming techniques worked with the indigenous vegetation rather than fighting or replacing it. It provides an inspiring if remote dream to a generation today worrying about how to sustain dense urban populations without causing irreparable harm to the environment.

A comparable pattern was seen half a world away in medieval Cambodia, where still-conspicuous temple complexes, most famously Angkor Wat (built in the first half of the twelfth century CE), attract millions of visitors each year. The main complex at Angkor Wat, like the Mayan temples, is raised high above its surroundings, its dramatic silhouette communicating its importance to people near and far. As with Mayan temples, its design also embodies spiritual significance: its five central towers reflect the five peaks of Mount Meru, sacred to Hinduism and Buddhism, its surrounding moat echoes a mythological Sea of Milk surrounding the mountain, its vast programme of sculpture tells religious stories, and even the pillars supporting the galleries are carved to cast a shadow resembling the temple's towers on to its inner walls.[16]

Around Angkor Wat is a large rectangular moat. Lidar laser scanning from the air reveals that this water feature formed part of an extensive system of canals and reservoirs, which archaeologists now believe formed components of a geometrically regular, planned landscape of heavily populated farmland, making up a city of around thirty-five square kilometres.[17] The elaborate water system helped to keep the staple rice crop wet enough to maximize its fertility, but prevented floods from washing it away. The canals also provided transport for everything from rice itself to the tens of thousands of tonnes of stone from the hills some thirty-five kilometres away that built much of Angkor Wat.[18]

The peoples of Cambodia and Classic-period Maya lands had been out of contact for tens of millennia, since their distant hunter-gatherer ancestors parted ways, yet as they adopted comparable patterns of very high-yield farming from semi-flooded land, the rise of non-farming classes produced similar urban structures.[19] The basic needs of cities have remained the same through thousands of years of human history and are all related to energy supply: food, fuel, transport and architecture.

The Biggest Monument Ever Built

If the city patterns of agrarian societies (societies where farming provided the main energy source) were heavily influenced by the nature of their staple crop and how it was farmed, fertile farming cultures also shared a pronounced tendency to social hierarchy, with impressive monumental buildings constructed at the behest of the elite. The most extraordinary of all, at least in terms of sheer size, were those of ancient Egypt.

The pyramid of Khufu, discussed in the Introduction, came after a rapid sequence of smaller (though still very large), earlier pyramids. Before these, even the most powerful Egyptians had been interred in smaller mud-brick burial chambers – impressive tombs, but nothing close to the scale of the pyramids.

Until the 2600s BCE the Egyptians used very little stone in their buildings.[20] Instead, they built using the same mud that fed their fields, mixed with straw and baked into a robust brick by the very hot sun. Yet within a period of two or three lifetimes this suddenly escalated to building the pyramid of Khufu and inaugurating millennia of setting long-standing records for the scale and robustness of stone building.[21] Intense bursts of headlong change have been fairly common in the history of architecture since the advent of farming. Once a method of building is worked out, scaling it up can happen fairly fast, stopping only at the limit formed by the energy capabilities of the society: a village of a hundred could under no circumstances muster the labour energy to build the pyramid of Khufu, and even the Egyptians did not attempt to build a pyramid any larger than Khufu's – the two others built near it later are both slightly smaller. Because Egypt had so much good grain, its initial scaling-up was the most spectacular seen anywhere until the Industrial Revolution.

The ancient Greek historian Herodotus, writing nearly 2,500 years ago, long before the original limestone facing of the pyramid of Khufu was removed for another building project, records some fascinating and plausible details about the humble considerations which actually enabled the pharaoh to build on this scale. Herodotus recounts seeing an inscription on the base of the pyramid which recorded the prodigious quantity of radishes, onions and leeks bought for the workers building it. Over forty tonnes of silver was spent buying vast quantities of these cheap garnishes to cheer up the staple grain that fed the workers.[22] The inscription hints at the spectacular energy requirements of a workforce of tens of thousands doing hard physical labour over more than a decade. More than that, though, it suggests that the Egyptian administrators themselves were sufficiently proud of managing to feed and organize such an army of builders that they carved the most apparently humble details on the finished wall.

Herodotus was told that the first ten years of the project were spent building a causeway along which to transport the stone, the following twenty for the pyramid itself. This was probably somewhat longer than it took in reality: a decade or so of tens of thousands of people quarrying stone, dragging it on large wooden sledges to the building site, raising it via an ever-growing ramp (perhaps in the end as much as two thirds of the volume of the pyramid itself) to the height required, then placing it with precision and care according to the instructions of the project's astoundingly able surveyors.[23]

The pyramid was made mostly of sandstone, a medium-hard stone composed of grains of sand squashed together for millennia under layers of other soil and rock, and bound together by minerals deposited by water trickling slowly through it. Sandstone has proved to be one of the most used building stones ever since. It is available in many places, easy to cut from the quarry, often attractive in colour and good to carve. The best sandstones also stand up well to weathering. For the bulk of the pyramid, sandstone was the obvious choice, as it was available in large quantities very near the site.

This stepped sandstone core is what we see now on the surface of the pyramid. Originally, however, the builders placed glistening white limestone blocks, triangular in section, on to each step of the pyramid, producing a completely smooth, sloped, white surface. White is abundant in the world around us today, with glaring paints and plastics kept bright by powerful bleaches and cleaning products, everywhere. Yet before modern times, it was a special and unusual colour often given religious or aristocratic significance. Such a vast, perfect white triangle amid the blue of the Egyptian sky must have seemed more divine than human. The gold-covered top of the pyramid would have reflected the sun blindingly across miles of the green and beige world of the Nile Delta below.

Limestone and gold were not the only special materials used for Khufu's pyramid. At the heart of the pyramid too the material changes. Perhaps just to indicate how important it was, or perhaps to make sure the weight of the sandstone above could never crush it, Khufu's burial chamber at the centre was made of a much harder stone: granite. Granite is formed when the molten rock of the earth's outer core nears the surface and cools, making it very dense.

The level of technology the builders had was limited. Bronze tools were available for some work, including possibly quarrying the sandstone. However, because the unyielding, heavy granite that makes up the walls of the burial chamber was too hard for bronze, it would have had to be pounded from the unyielding bedrock by hundreds of low-status labourers using balls of dolerite stone up to five kilograms in weight. The workers' bodies would have been folded into a tiny work area alongside the flying stone chips and the dust, under the sweaty heat of the Egyptian sun.[24] In later periods at least, such work was a punishment for criminals and some absconded despite threats to their family.[25]

Lifting techniques were basic, without block and tackle systems to give the workers any help. Weak human muscle had to do all the work. What made the Egyptians able to build on such an astounding scale was their large population of well-fed workers whose labour was not constantly required for food production. The key to this was the fertility of the soil. Flooded with water and layered with rich mud from the Nile, it is thought to have provided more grain per hectare than anywhere else on earth, allowing the large farming population to generate the food surplus that supported perhaps the greatest concentration of people anywhere in the world who were freed from the protracted drudgery of food production.[26]

The Handiwork of the Gods

Although very big in total mass, the pyramid of Khufu was relatively modest in terms of another form of Egyptian construction prowess: the moving of very large blocks of stone. Almost all of the pyramid is made of blocks weighing from 2.5 to fifteen tonnes. This is a substantial weight, of course, particularly when dragging it up a slope to the top levels of the pyramid. Whereas two people can pull about a tonne on flat, wet sand, even a shallow slope of nine degrees requires nine people to shift the same weight. Nevertheless, it is fairly easy to imagine organizing 140 labourers to pull a fifteen-tonne block of sandstone up a ramp. At the heart of the pyramid, the tomb chamber itself was made of fifty-six much larger, heavier granite slabs, around fifty-four tonnes in weight and brought from considerably further away than the sandstone blocks that covered and surrounded them.

A painting on a tomb wall of around 1900 BCE shows a monolithic statue of around fifty-eight tonnes being pulled by 172 men, while another man stands at the front of the sledge pouring water into its path to keep the friction down. That used to be thought a symbolic gesture, but archaeologists have tested the technique and found it halves the number of people needed to pull the sledge.[27]

Fifty-four tonnes was, remarkably, far from the limit of what the Egyptians could quarry and move, even as early as Khufu's time, when blocks of up to 200 tonnes were used on occasion.[28] Later in Egyptian history, single-piece statues weighing many hundred tonnes were transported over great distances, and the famous obelisks that carried inscriptions and marked important places in royal and religious sites became increasingly vast over a number of centuries. A pair of granite obelisks around thirty metres in height and weighing 340 tonnes were quarried and carved at Aswan, transported over 200 kilometres, polished and erected at Karnak. One of them is carved with a boast that this was achieved in just seven years.[29]

To this day no one knows for sure how such heavy, unwieldy stones were raised to their final vertical position with the required precision and permanence, using only weak muscles and limited technology, while being surrounded by existing buildings that would have made large-scale temporary earthworks particularly challenging. The largest obelisk ever attempted never left the quarry – finally the Egyptians met the limit of their ability to escalate the same formula through bigger and bigger workforces. It was nearly forty-two metres long and, if fully extracted, would have weighed

a spectacular 1,160 tonnes. It seems to have been abandoned not because its engineers despaired of transporting or raising it, but because unsuspected faults in the rock started to crack as they worked on it.[30]

The Roman author Pliny the Elder reported that by his period Egyptians moved the largest stones by building a canal to the quarry from the river, placing boats half submerged by a cargo of heavy stones under the monument, then removing the stones until boats and block floated free and could be towed away.[31] Most of the distance was covered by boat on the Nile, and from the river to the building site a block was hauled. As impressive as the sheer weight moved for Khufu's pyramid is the precision of the setting out – almost exactly on the planetary north–south line, the four sides virtually identical.[32] The pyramid was almost perfectly square-based, but in the right light it is possible to see that each of the four sides is very slightly bent inwards to the middle line in the shallowest possible chevron, making it an eight-sided shape, and probably producing, when the pristine cladding stones were still in place, striking effects at particular times of day when the sun that the Egyptians worshipped reached certain angles. The surveying skills of which this was such a showpiece were honed by a landscape where seasonal flooding washed away field markers, meaning that a rapid and accurate re-establishment of land layout was required as the waters receded.[33] Since surveying was critical to Egypt's energy system, the best surveyors became remarkably skilled at it.

Ancient observers assumed intuitively that the client for such a phenomenal project must have taken ruthless, immoral measures to force so much work out of his people. Herodotus reported a folk tale that Khufu had closed the temples and forced his population into slave-like work by the hundred thousand in order to get the pyramid built. He also passed on a salacious rumour that Khufu prostituted his daughter to pay for the pyramid, and that she asked each of her visitors to contribute a stone to build her own little pyramid beside his – ancient sexual fantasy rather than architectural history.[34]

In fact, this miracle of aesthetics and surveying was achieved through fertile fields and sophisticated organization. Getting anything but chaos, corruption and civil disorder from the mobilization of such an army of workers required layer after layer of scribes recording every detail and stamping it with seals to confirm that they took personal responsibility for the honesty and accuracy of their accounting. While the workers themselves used crude but effective tools, the mathematical and administrative skills of the Egyptian bureaucracy were much more advanced, enabling any amount of labour to be kept in well-functioning order.[35]

The largest building in the world, then, was a triumph of micromanagement and accountancy. The pyramids were the achievement of the diligent middle manager, the rich Nile mud and the farmer.

One Species Worldwide

If the scale of Egyptian monuments was unusually big, other aspects of their pyramids are more closely echoed in other fertile agrarian societies. Across continents and millennia, societies whose farms achieved a substantial surplus often showed pronounced shared tendencies in what they chose to build as their civilization-defining projects. Similar technical skills and similar notions of beauty seem to emerge time and again around the world, wherever the human species had the spare energy to build for the glory of an elite.

The simplest shared tendency was size. Many or most agrarian societies worldwide expressed their hierarchical social structures by building substantially larger buildings for the elite than for everyone else. The mammoth-bone house described in Chapter 1 is perhaps on the larger side of pre-industrial housing for ordinary people, yet when it is drawn to the same scale as the monumental buildings in this book, it almost disappears. The agrarian peasantry lived in small, simple houses of a comparable size. They must have been even more struck by the scale and grandeur of the great monuments than today's visitors, jaded as we are by the ease of making very large buildings using today's abundant fossil fuels.

Most of the major buildings discussed in this book strike visitors as being big when they stand in front of them, but the line drawings showing them to the same scale as each other make clear that some societies' big buildings were very much bigger than others. Great monuments of farming societies tend to exhibit heavy labour in approximate proportion to the fertility of their land: the more people they could spare from the fields, the bigger the buildings constructed for the elite. The Parthenon in Athens is dwarfed by the pyramid of Khufu, just as the Nile-flooded farms of Egypt supported a much larger population, with a much greater surplus, than the hillier, stonier soils of the Athenian city-state.

Whether they were big or enormous, however, building

these monuments may well have had various functions beyond satisfying the egos of rulers. Large construction projects require discipline and muscular effort broadly comparable to those required in military undertakings. The great monuments may have had an element of war by proxy, like the rockets of the Space Race – the peaceful, scientific counterparts of nuclear missiles. The discipline of the monumental building site may often also have helped to instil habits of obedience and break down what agrarian rulers tended to see as unwelcome independent-mindedness.[36] The sultans of the Ottoman Empire, centred on modern-day Turkey, made extensive use of trainee soldiers and galley slaves in their construction schemes, and soldiers who did well on the building site could earn rapid promotion.[37]

On top of showing off quantity and organization of labour, the monuments of the farming millennia also tended to display high levels of skill in some of the crafts that were involved in producing them. The work of skilled craftspeople, as well as being a source of delight in itself, may also have sent signals to observers. First, the abilities of designers and surveyors could potentially be a proxy for those of military engineers: anyone who can build a pyramid with the speed and precision shown in Khufu's case can probably build a pretty good set of emergency defences, and possibly good siege machinery and other military aids. In Europe, the distinction between architects and military engineers only began to develop after the Renaissance, and even as late as the Second World War many trained architects were kept back from frontline fighting so that their skills could be turned to a huge range of military applications.[38]

A subtler requirement for nurturing sophisticated crafts is that the regime concerned should have a sufficiently stable energy surplus to support generations of craft tradition; complex crafts tend to arise from large bodies of craftspeople working over a long period. Thus the porcelain of China or the glass of medieval Syria struck wonder into observers from other countries for centuries because, however much money they threw at it, it was almost impossible for outsiders to compete with long-accumulated specialist knowledge and technical skill.

Methods of showing skill vary widely. Expert stone-carving and stone engineering have been among the most important to architecture. The other dominant world architectural material until the twentieth century was wood, plentiful in many areas and, until the 1800s, the best way to build a roof or an upper floor in most types of buildings. Beautiful woodworking traditions developed in cultures as diverse as Japan, Maori New Zealand and Scandinavia (like this twelfth-century stave church from Norway), often involving sophisticated handling of joints between wooden components, where puzzle-like interlocking pieces avoided the use of metal nails

which would have cost expensive heat energy to produce.

Metal, glass, mud brick, kiln-fired brick, glazed ceramic, paint, plaster and even paper were all raised to the highest architectural craft and art in one or more building traditions around the world, becoming the objects of intense competition between the people commissioning buildings and the craftspeople who produced them.

Thanks to competition, the employment of the most skilful craftspeople could demonstrate relative wealth too. Any Renaissance Italian state could afford artists, but to get the best they had to test their wealth against each other. Rulers paid big sums to attract the most admired artists and architects. The *Mona Lisa* is in France because Leonardo da Vinci died there, having accepted a substantial house and a large annual salary from the king.[39]

Size and craft, then, are the most fundamental shared aspects of the agrarian world's tendencies when building special structures. The similarities seem not to end there, though. Other tendencies also unite mud-brick temples, the palaces of the Classic-period Mayan elite, Gothic cathedrals and many other monuments from the agrarian age.

The first is a tendency to follow immediately visible rules, most often symmetry and repetition. If a building is nearly symmetrical but gets it slightly wrong, the human eye is quick to spot the mistake and easily irritated by it. The converse is also true: symmetry brings neurological satisfaction.[40]

For buildings made of naturally irregular trees, handmade bricks or hand-carved stone, achieving perfect symmetry is technically impressive enough to outweigh the potential staleness of cliché. For tens of thousands of workers to move millions of tonnes of stone into a pyramid, ending up with absolutely precise symmetry and regularity, shows the level of control the organizers had over the workforce, and even over the intractable bulk and natural variation of the rock itself.

The shared human enjoyment of symmetry may have been behind the recurrence for millennia of certain patterns and shapes, especially the geometrically perfect square and circle, in societies that had in some cases had no contact whatsoever since far before the invention of farming and monumental architecture. Egyptian and Mayan square-based pyramids have seemed similar enough for some to guess at transatlantic raft travel or alien-related nonsense. The plans of Cambodian temples, Italian Renaissance churches, Japanese pagodas and English castles in Wales revel in similar pleasures of four-sided symmetry, and buildings ranging from Stonehenge to the Dome of the Rock play with regular polygons and circles.

As for the functions of the great agrarian monuments, they were typically built to glorify the secular or religious powers that controlled the energy surplus. Palaces and other buildings for ruling powers, and places of worship and the other accommodation that went with them, make up the bulk of many traditional histories of architecture right down to 1800 CE.

These great buildings often made some appeal to cosmic order as understood by their society. All farming

Left to right, upper row first, plans of: Pyramid of Khufu; Maya pyramid temple; central building of Angkor Wat; Bramante's design for St Peter's; Nara Horyui Pagoda, Japan; Beaumaris Castle; Pantheon; Temple of Vesta, Tivoli; Stonehenge; Dome of the Rock

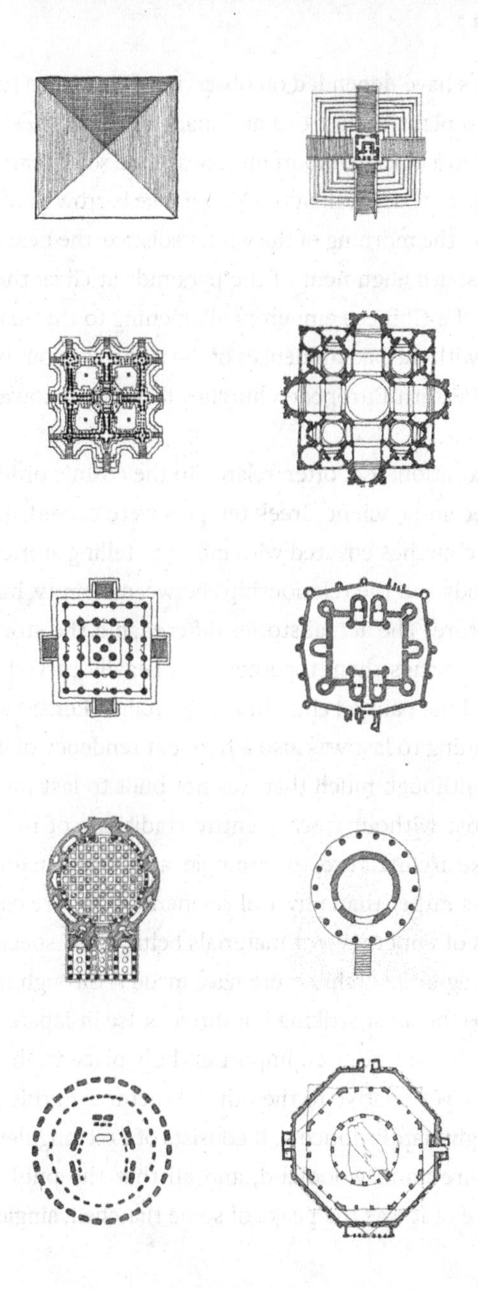

societies have depended on observing the stars to tell them when to plant their crops and many of the earliest monuments have links to important times of the solar year: the sun that hits the back wall of the Newgrange barrow in Ireland at dawn on the morning of the winter solstice; the near-perfect north–south alignment of the pyramids at Giza; the buildings of the Chinese emperors all opening to the auspicious south, with the bad influences of the north shut out by walls; most Western European churches facing east towards the rising sun.

Decoration, too, often related to the cosmic order: walls of Aztec and ancient Greek temples were carved, and Byzantine churches covered with mosaics, telling stories about their gods and the relationships between divinity, humanity and nature. The actual stories differed, but the storytelling instinct brings them together with heavily carved Angkor Wat and the painted churches of the Italian Renaissance.

Building to last was also a frequent tendency of agrarian elites, although much that was not built to last must have been lost without trace – entire traditions of impressive tent-like architecture, for example, are known mainly from pictures rather than physical remnants.[41] In rare cases, examples of impermanent materials being given special qualities in agrarian architecture have made it through to today. Perhaps the most striking is a shrine at Ise in Japan.

The Ise shrine is an important holy place in the Shinto religion, yet relative to the other structures in this chapter you might hardly notice it. It consists of a rectangular fenced enclosure amid woodland, and all that the public could ever see of it was the peaks of some thatched, single-storey

wooden buildings poking up above the fence. The woodwork is beautifully precise but basic in its techniques. Thatch and wood are not very durable, yet this pristine shrine dates, in a sense, from the seventh century CE. Because of the fragility of its materials, the shrine is rebuilt every twenty years, to precisely the same design. There are, in fact, two identical rectangular plots of land next to each other, each one left alternately vacant while the other takes its two-decade-long turn to house the newly reconstructed shrine. It is believed that this process has gone on with faithful precision for over 1,300 years.[42]

According to standard Western notions of what makes an 'authentic' old building – keeping as much of the original material as possible – the shrine is at most twenty years old, yet the design and the tradition here go back vastly further than most Western buildings. No single rebuilding of the Ise shrine comes close to the energy expenditure of assembling

Stonehenge, let alone that of building the pyramids. The message Ise sends instead is one of cultural stability: rather than a single episode of building frenzy hundreds of years ago, the imperial family renews and reaffirms its role in the state religion every twenty years, and has done so century after century.

With Ise, it seems as if the building process is itself an important part of the tradition. Could this also be the case with other agrarian monuments? To some extent, no doubt, the elites in agrarian societies built very large buildings because they could; grandiose buildings may be a natural side effect of the megalomania of great power. Obviously too, if you believe in the supreme power of a god or gods, it can make sense to build grand buildings in their honour.

Yet there may be another side to the spectacular quantity and scale of building seen in the most fertile ancient societies: building on a huge scale may in fact have been used as a sort of safety valve to let out excesses of energy surplus. Archaeologists have long been struck by the way that the ancient cities of Mesopotamia tended to demolish and rebuild their temples with a frequency that seems extraordinarily wasteful.[43] Ancient historians speculated that Khufu must have been brutal and megalomaniacal, yet the alternative for him might have been a very large number of healthy, strong, fighting-age men at a loose end. Keeping spare people busy on an inessential but harmless construction project might be less an act of pernicious self-obsession than an effective way of achieving unity, physical fitness and social control in a large population for whom food was available without them having to put all their effort into farm work.

In the context of the post-First World War European housing shortage, the visionary modernist architect Le Corbusier proposed that society had to choose between architecture and revolution.[44] It may be that Khufu faced a similar choice. If agrarian societies brought the constraints of low social status to most, they may also have imposed hard-to-escape patterns on the elite. Energy systems, then and now, bring with them their own powerful social, cultural and intellectual norms, and architecture, then as now, embodies and expresses them.

CHAPTER 3

Us and Them

The Parthenon and Parsa

This chapter looks at a cultural opposition that has often been seen as the originating moment of much in Western culture: the clash between the Greek city-states of around 500 BCE and the mighty empire of Persia. The differences between the two were ingrained in their farming patterns, which underpinned very different political systems.

The Achaemenid Persian Empire was the largest the world had yet seen – around five times bigger in land area than any earlier kingdom and home to by far the largest population ever to have been brought together under a single ruler.[1] In the fifth century BCE it extended from the borders of India on one side to the Balkans and Libya on the other. To the north it reached Kazakhstan and Georgia, and to the south, modern Sudan. Most Greek city-states controlled an area only tens of kilometres across.

The Persian Empire included what had up until then been some of the world's most powerful and fertile separate kingdoms: Babylonia, Assyria and Lydia (where the introduction of coinage had made King Croesus rich). At times the Persian emperors even managed to invade and hold Egypt. These well-developed economies, based on the fertility brought about by large-scale irrigation schemes, provided the staple cereal crops for the Persian king of kings' subjects. Perhaps the greatest challenge of such a large empire was to make its elites feel integrated, rather than the resentful victims of conquest, looking for the first opportunity to rebel. Architecture was brought into the service of this aim.

The energy basis of Athens and the other Greek city-states was very different: numerous smaller farms owned by equal citizens. Farms ranged from less than three hectares up to a few dozen, but were rarely bigger than that, and did not tend to feature large-scale irrigation schemes. Grain was the staple crop, with olives and grapes as a cash crop, tradable as valuable oil and wine. Many farms made the best they could of hilly land, and each was comparatively autonomous, not dependent on a central authority implementing and controlling the irrigation that made them viable. In the face of their enormous neighbour and enemy, the Greek city-states nurtured a sense of 'us and them' based on their identity as farm-owning citizens, which allowed separate polities to fight shoulder to shoulder against the common threat. Architecture contributed to this as clearly as it did to the Persian elite's search for a shared identity.

The two buildings in this chapter, the Parthenon in Athens (largely built 447–438 BCE, with work on the sculptures continuing until 432 BCE) and the palace of Parsa (or Persepolis) in Iran (built in successive campaigns from around 520 BCE), were among the most celebrated monuments of a period that has been widely regarded in the West as a high point for philosophy, the arts, history-writing and democratic government. Parsa and the Parthenon show how architecture was used to display and entrench the differences between societies that had different energy bases and a shared eagerness to bring together their people in the face of war and division. The two projects display how these great enemies chose to show themselves and their world what they cared about, what they believed in and who they thought they were, or wanted to be.

Classicism: The Columns That Would Not Die

In the case of the Greek city-states, part of their collective identity arose from their shared architectural language, known today as 'classicism'. Classical architectural motifs have proved enduring and adaptable, and are still common sights in today's European-influenced cities around the world. Existing over 2,500 years ago, ancient Greece is remote both in time and in culture, yet the Parthenon has often been treated as a foundational contribution to a 'Western tradition' of architecture.

It is relatively easy to tell the story of the Parthenon as an early example of the way we do things now: it is the earliest building featured in this book to be designed by people who used the title 'architect' (although at the time *architekton* meant a chief craftsman of boats as well as of buildings, and equivalent positions by other names must have existed at least since the monumental construction sites of Uruk).[2] It is the earliest building in this book whose budgetary records partially survive and where we know about the committee processes that gave rise to it. It had a sort of overseeing board (as new public buildings usually do today) and, under them, a sculptor and two or three leading architects. Even the administration of its construction seems familiar: what we would now call subcontractors signed up to contracts for specific parcels of work, complete with penalty clauses for substandard or late delivery.[3]

The designers of the temple too fitted into patterns that are still widespread in architecture: two of the leading architects involved wrote a book about their design ideas, just as so many architects since have written to justify controversial work or to advertise their services to future clients.[4] In another timeless pattern, the sculptor who oversaw the entire design was a personal friend of the powerful politician who pushed through the commissioning of the Parthenon.[5]

Yet the apparent modernity of the project may not be because the Greeks thought up these innovations for the first time, but simply because they left extensive records which have been very carefully studied by archaeologists. If equivalent records survived from Uruk, we might find that some of these patterns had already been taking shape from the dawn of monumental construction. Perhaps they are the obvious way of going about a large, complex project.

Thanks to the remarkably long life of the Graeco-Roman classical styles of architecture, not only the process but the Parthenon's architecture itself also looks oddly timeless to those brought up feeling that the British Museum or the Lincoln Memorial represent their culture.

Classical column designs and other signature motifs of classicism will keep cropping up throughout this book. They have appeared on everything from Roman amphitheatres and temples, via Renaissance palaces, to twentieth-century banks and the incongruous porches of suburban houses in contemporary England. Pennsylvania Station in New York

City (1904–10), for example, was designed for steam trains yet was a close architectural copy of Roman public baths. Classical columns are still widely used by film-makers today in the West as shorthand for specialness, dignity or privilege. Even today buildings using classical motifs are occasionally still constructed, and many contemporary Western-educated architects who do not explicitly quote classical architecture in their work still see its language as a formative influence and a touchstone in their design thinking.

By the twentieth century, classical columns were made and placed with the aid of mechanical stone-cutting, hidden steel structures and big diesel-engined lorries and cranes. This makes it easy to underestimate the achievement of the earliest builders of stone temples in the classical tradition. The accounts of the Parthenon show that the process of construction was reasonably quick, but very expensive. The marble – all 22,000 tonnes of it – was quarried on Mount Pentelikon, sixteen kilometres away, and slid down a smooth stone slipway to the plain. It is a beautiful marble but prone to faults, so these had to be avoided by skilled quarry workers. Extracted in pieces of up to fifteen tonnes, it was then dragged to Athens by substantial teams of draught animals borrowed from the farms around and hoisted up the final few metres on to the Acropolis itself. Once approximately shaped, it was lifted into position on pine scaffolding, the stone still rough so that it could later be carved in place into its perfect final form. The bottom few centimetres of each column were fluted, but the remainder was left rough. The outer row of columns – the 'peristyle', derived from the ancient Greek for 'columns around' – was built first, so all the rest of the material needed to come past before the final surfaces could at last be carved, safe in the knowledge that they would then not be damaged by continuing building works.[6] These most famous of classical columns were a showpiece of precision and perfection, but themselves were only a cautious step on from those of other, earlier Greek temples.

Buildings seem always, as far back as one can trace, to have got their designs from copying and adapting something else. We have already seen that Uruk's mud-brick temples may have imitated earlier fabric structures for worship. Similarly, some of the most characteristic features of Egyptian stone buildings appear to have been carried across from earlier mud-brick architectural ideas: very stocky columns (mud brick is weaker than stone), a big protective overhang at the top of the wall, called a cornice, to run rainwater off the vulnerable mud face of the building, and walls much thicker at the bottom than the top to strengthen the brittle mud. These

result in over-engineered stonework, but humans are cautious about change.

Even the pyramids found their inevitable-looking final form only by progressive building up of earlier ideas. Earlier grand burials were under mud-brick platforms now known as *mastabas*, initially laid out like a house as a dwelling for the dead ruler or aristocrat in the afterlife. These then became stone for the grandest burials, because it is more weather-proof in the very long term. The first pharaoh to have something like a pyramid essentially had a multi-storey pile of mastabas; only subsequently did designers smooth the stepped volume of stacked mastabas into the familiar pyramid shape of Khufu's burial place.[7]

As with the architectural styles of Uruk and Egypt, however close we get to the origins of classicism it seems to be born out of copying earlier buildings. These first steps in classical architecture's development took place in the area we now call Greece, early in the first millennium BCE. The move from wood to stone may have been a sign of admiration for the durability and magnificence of Egypt's great monuments. However, in terms of the architectural details, the Greeks seem to have based their stone buildings not on Egypt's but on their own earlier wooden buildings erected for similar purposes. Archaeologists have found the remains of a large wooden hall from around 1000 BCE, which seems to get us as near as anything yet to the origins of classicism.

Rectangular in shape, with a U-shaped apse at the western end and an entrance to the east, the hall is at Lefkandi, on the Greek island of Euboea. It was crowned with a thatched roof, whose outermost wooden supports landed in a line outside the wall, creating a veranda-like structure. The roof itself would have been pitched, with a gable at the entrance end of the building.[8]

This is our strongest surviving evidence for the wooden prototypes of the classic Greek temple: a walled room surrounded by a circuit of columns. The stone versions continued to show some details that are purely decorative in stone but probably started out in the wooden versions as important structural components.

In the most robust and ancient of the styles of classical architecture, the columns always have above them a motif which is probably a stylized depiction of a timber beam end. Beneath it a row of protrusions probably recalls the wooden pegs that once held the beam securely in place. The gable end becomes, in Greek temples and twenty-five later centuries of classicism, a triangular decorative element called a pediment.

From its roots, then, classical architecture was looking backwards even as it improved upon and refined the revered earlier ideas and motifs. That was to remain the case right through to the twentieth century, with classical architects looking back to imagined golden ages even as the new fuels and energy technologies of the Industrial Revolution allowed them to far surpass the technical and organizational achievements of the earlier designers they so admired.

This tendency to assume that the golden age was in the past, and that change was very likely to be threatening and harmful, is particularly strongly associated with societies that depended on farming as their main energy source. Without a scientific understanding of weather, crop fertility and disease, the annual variation in crop outputs tended to be understood as being controlled by a god or gods. As with athletes wearing the same pair of lucky socks in which they set their personal best, farming societies tended to repeat formulas that had worked in previous years: we made sacrifices in the temple and had a good year, so if we interfere with either the sacrifices or the temple, how are we to know that our fortunes will not change for the worse?[9]

In every one of the great innovations described in the first half of this book the architectural counter-voice can be detected, reassuring a cautious and conservative world that it has deep historical roots and will not represent too radical a change, whether that is in the tent-like style of Uruk's first brick temples, the wooden details in the Parthenon or the meticulous precision with which the shrine at Ise is rebuilt identically every twenty years.

Even with such widespread conservatism in farming societies, classicism lasted a remarkably long time, continuing with this partially backward-looking progress. It certainly profited from a measure of luck in the military successes of the cultures which adopted it, but perhaps its greatest strength lay in its balance of reassuring rules and invigorating freedoms. On the face of it, classical architecture was prescriptive and potentially artistically narrowing, yet, as we will see in this chapter's investigation of the most famous of all Greek temples, even the core rules of classical architecture included contradictions that forced designers to fudge, compromise and invent. There was always choice, even from early on, with three column types and associated decoration – known as 'orders' – produced in different parts of Greece under specific local influences. These three, Doric, Ionic and Corinthian, started to offer a menu to architects even as early as the fifth century BCE.[10]

Classical motifs have offered millennia of familiarity and reassurance to some, and sets of perceived rules have allowed mediocre designers to produce something adequate – witness the countless houses designed through the eighteenth and nineteenth centuries for or by British aristocrats who, having been on the grand tour to Italy, came back and rebuilt their family seats using the charismatically illustrated treatise on architecture by Palladio, the great architect of the Veneto (see Chapter 6).

Yet the language of the classical orders has also proved sufficiently flexible to inspire rather than constrain the remarkable aesthetic experiments of the most talented architects, figures like Michelangelo, whose St Peter's Basilica we will see in Chapter 6. In Paris, under Louis XIV, Jules Hardouin-Mansart took a similar dome idea and the classical orders for his church of Les Invalides, and managed through rhythm and composition to produce for the military retirement home a sort of trumpet fanfare of stone and gilding – stirring, beautiful and unforgettable. Just years later in London, another fanatical architect, Nicholas Hawksmoor, was to hybridize classical columns with the chunky beauty of medieval cathedral fronts in the church of St Mary Woolnoth. These masterpieces are each strikingly original and individual – a world away from the repetitive stodge of the typical Palladian country house – yet to those who feel included by the classical language they also offer the atavistic pleasures of familiar architectural details.

Classicism with a Twist

The Parthenon shows both these tendencies: careful obedience to and refinement of conventions, and brilliant, self-confident variations on them.

The Doric order contains a great many subtle complexities which fascinate some people – they clearly absorbed the original temple architects – and leave others feeling baffled. I am going to explain one of them to demonstrate, but if you are not an enthusiast for spatial puzzles do not struggle too hard with this one: you do not need to understand it to get the fundamental point that Greek architecture set the designer very complex design challenges.

Looking at the Parthenon from the corner where ancient visitors would first have had a clear view of it, it seems perfectly regular.[11] Yet as you stare at it more, small eccentricities appear. In particular, the columns at each corner are visibly a bit closer to their neighbours than the rest of the evenly spaced row. This is often the case with Greek temples in the Doric order and comes about because two rival 'rules' of Doric architecture disagree with each other. One rule insists that each column should have a triglyph (the decorative element derived from the beam ends of earlier wooden temples) on the stonework directly above its centre. But another rule dictates that there have to be triglyphs right up at the ends of the frieze, meeting each other at the corner. You cannot obey both rules at once.[12] The solution followed at the Parthenon, pushing the column closer to its neighbours, produces considerable visual strength, turning an awkward fudge into a powerful modification.

To make the problem-solving even trickier, the architects voluntarily submitted themselves to additional complications.

Throughout the long history of architecture many designers worldwide have derived satisfaction from imposing on their buildings systems of mathematically related proportions. Many have believed – and many still do – that there is something intrinsically beautiful or perfect in establishing these numerical relationships between parts of a building. The Parthenon's ground plan and short-end elevations have the same proportions: height to width ratios of 4:9. The long sides are related to this by having a ratio of $4^2:9^2$. The columns are spaced using the same numbers: the thickness of the columns is in a 4:9 ratio to their spacing.[13] The implications of these voluntary restraints are complex for the designers to resolve.

Combine these proportions with the tough rules of the Doric order and the exercise of designing a temple somewhat resembled a giant three-dimensional sudoku puzzle. And, as explained, it was not in fact fully soluble: there were contradictions within the rules which meant every temple designer had to choose where and how to compromise and distort the perfect order. Definitely a sudoku you would rather solve in changeable pencil than permanent stone.

As if all this were too simple, the architects of the Parthenon then chose to introduce a set of complications so mind-bending that it is hard even to describe them, let alone build them. Taking its lead from an older temple that had been under construction on the same site until it was destroyed by the Persians in the previous war, the Parthenon is built not on a flat base but on a base that bends upwards towards its middle, as if it were made of rubber and stretched over a very large spherical surface. This distortion then radiates upwards through the building, with every horizontal line or surface from base to roof curving in precise magnification of the base's curve.[14]

The complications do not end there. The main verticals of the building, the columns, were also distorted. One of these refinements is obvious to the naked eye, and familiar from classical architecture before and since: the columns are not perfect cylinders, but swell slightly near the bottom and taper gently towards the top, in a distortion known as entasis. The second distortion of the columns is far less conspicuous: they are all tilted in very slightly, so that if you continued a line up from the centre of every column the lines would all meet about 2,200 metres above the top of the temple's side walls.[15]

There is a third adjustment too: the four columns at the corners of the building are one fiftieth wider than the rest of the columns, possibly to compensate for the fact that they will be seen against the bright light of the sky rather than the dark shade behind the other columns. The level of precision required to follow through such delicate tweaks is considerable, and sure enough the Parthenon was built to astoundingly low tolerances, with a difference of just 2.5 millimetres in length between the two seventy-metre sides.[16]

These distortions are very subtle. The corner columns, for example, are only four centimetres wider than the others. And the biggest adjustment is high on the side walls, where the curve of all horizontals makes the centre just 11.25 centimetres higher than the ends. Most people would probably never even have noticed them by eye alone. So why are they there? One popular explanation, originally recorded by the

Roman architect Vitruvius, was that the distortions corrected optical illusions that would otherwise make the temple seem less than perfect: long straight lines, he felt, can seem to sag when seen from below.[17] Another explanation, which I prefer, is that it was an expression of the artistic preoccupation of Athenian sculptors at the time with movement and liveliness. In contrast with the dignified monumentality of perfectly symmetrical, static Egyptian sculptures of pharaohs, Greek sculptors preferred to pose their figures in more lifelike, mobile-looking stances, their bodies achieving the beauty of imbalanced liveliness. The word that Vitruvius used for the curving of columns, entasis, is the Greek for tension or strain.[18]

Energy, Architecture and Audience

In a sense, the reasons behind the distortions or corrections in the Parthenon are less important than their subtlety. Going to so much effort throughout design and construction to produce bends and angles that are imperceptible to the normal observer tells us something important about whom the Parthenon's designers were talking to: it must have been people who would tell each other about the extraordinary achievement. If they had wanted to impress strangers from all over the world, building it less precisely but bigger might have been a better use of the resources available. To appreciate the subtleties of the Parthenon not only do you need to be familiar with other Greek temples but, even more significantly, you need to be told

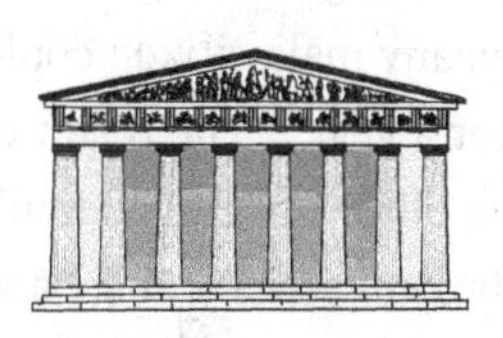

about them. The Parthenon is a temple built to impress expert eyes, and to impress in the conversation and debate that gave rise to the rich intellectual atmosphere for which ancient Athens is so famous in the West.

The roots of this culture of public debate and intellectualism lay in Athenian patterns of energy production, most importantly the pattern of landownership. The pharaohs of Khufu's time claimed to own all the land in Egypt, and the scale of labour they were able to command suggests that this was more than just a legal fiction.[19] Greek landownership, by sharp contrast, was widely dispersed among the male full citizens of each city-state. There were around 30,000 citizens in Athens at the time the Parthenon was built, of whom perhaps around 80 per cent were middle-income, with enough land to support a family and to keep a few enslaved people who would do the bulk of the farm work.[20]

This shared control of the city's energy supply supported an unusual political system – democracy – in which citizens had equal rights and equal political status. Rich and poor citizens could debate without the poorer having to give way out of deference or fear. So keen were the Athenians to avoid any citizen rising to king-like dominance that they had an annual opportunity to vote to 'ostracize' any citizen, no matter how powerful, and send him into exile for as long as ten years. Their confidence in the equality of their citizenship was so great that they chose their most important governing council entirely randomly through a lottery: any male citizen could end up as part of the ruling body for a year.[21] The result of this legally equal, wide landowning elite was a proud culture of competitive education, lively intellectual debate and competition between the male citizenry of cities. The Parthenon's hidden subtleties were part of this rich culture.

The great threat facing Athens was the Persian Empire, under the overarching rule of a single king of kings. With its constituent parts so sophisticated in their energy economies and related politics, the Persian kings adopted a policy of allowing the existing local patterns of rule across their vast lands to remain largely as they had been before conquest. The results were complex but overwhelmingly hierarchical: the Persian monarch and his family owned a large amount of good farmland, and also controlled important canals and reservoirs.[22] They dominated a series of subrulers, who in turn were the leading figures in their large region, with a local elite under them. Farmers owed dues of physical labour and grain to those above them, in proportion to their landholdings. Farmers across millions of square kilometres surrendered much of their agricultural production to people further up the hierarchy, who in turn could use the food to support soldiers, craftspeople, administrators and other non-farming specialists.[23] A system of tribute – food, fuel, animals, fabric and other goods given to the ruling elite by conquered peoples – formed another important component in the elite's exploitation of their empire.[24]

The two contrasting cultures, the Greek city-states with their messy citizen-equality and the Persian Empire with its single king of kings atop a steeply hierarchical social pyramid, show clear correlations with the types of food production that dominated in their lands. Even today, dependence on large irrigation systems like those of the Persian Empire has been found to correlate with more dictatorial central authorities.[25] The farming of Athens was conducive to a broad elite of more independent citizens, and to thriving water-based trade that encouraged wealth and confidence among a swathe of the male population.[26]

Persia and the Greek city-states had been intermittently at war for decades before the construction of the Parthenon. Persia had invaded Greece in 492 BCE, but was unexpectedly defeated at the Battle of Marathon in 490 BCE. The previous temple on the site of the Parthenon was being built to celebrate this triumph when, in 480 BCE, the Persians returned, taking Athens and massacring those who had stayed behind to fight after the rest of the population evacuated the doomed city, intentionally burning and smashing the religious buildings and statues of the Acropolis. Another unexpected victory by the Greeks, achieved through a superior navy beating a much larger land army, turned the tide and sent the Persians home defeated.

For thirty years after these traumatic events the Acropolis was left in blackened ruins, reminding Athenians of the ongoing Persian threat and of their need to maintain a powerful navy and an alliance with other Greek city-states and islands against the next invasion, which was expected any year. It appears that only when the Athenian-led Greek alliance and Persia signed a peace deal around 449 BCE did the Athenians feel secure enough to start rebuilding the Acropolis in grand style. The Parthenon was its crowning glory.[27]

Esperanto Architecture

The two Persian kings who invaded Greece, Darius and his son and successor, Xerxes, also built magnificently, including a very large new palace near the centre of their huge empire. The building work took place in two major campaigns stretching from around 520 to 450 BCE, spanning both their reigns. The final work on the palace was under Xerxes's successor. Starting with a large outcrop of rock at the base of a range of hills, they levelled it by lowering parts and raising others to make a platform around 300 metres deep and 450 metres wide, rising 12 metres above the plain in front of it. Up to it they built a symmetrical double staircase 7 metres wide, with 110 broad, shallow steps.[28] On the top of the platform they laid out a series of large buildings, courtyards, gateways, storehouses and other accommodation.

Even the way the Persian kings pursued their palace project is characteristic of the energy manipulations of agrarian rulers: it is located in an area that is naturally too arid to reliably produce thriving food crops. Initial archaeological surveys have revealed signs of a twenty-kilometre canal and other systematic interventions to bring water from the range of hills against which the palace stands down to the plain, to irrigate the farmland which – accounts preserved in the ruins of the palace indicate – supported building operations and the ongoing personnel of the palace.[29]

The Persians called their palace, and the settlement that grew up on the plain below the terrace to serve the palace, Parsa. It is now better known in the West by the name the Greeks gave it, Persepolis, or by its modern Iranian name, Takht-e-Jamshid. The Parthenon and Parsa embody in stone the differences that arose in the politics and societies of Athens and the Persian Empire as a result of their very different energy distribution.

The first great contrast lies in whom the buildings seem to speak to. As we have seen, the Parthenon was Greek architecture for those who spoke fluent Greek – both the literal and the architectural languages. By contrast, Parsa's most fundamental statements could not be missed by anyone from any culture: height, size, symmetry, order, ornateness, gilding and bronze-clad doors twelve metres tall to make visitors feel like crushable ants next to the superhuman might of the

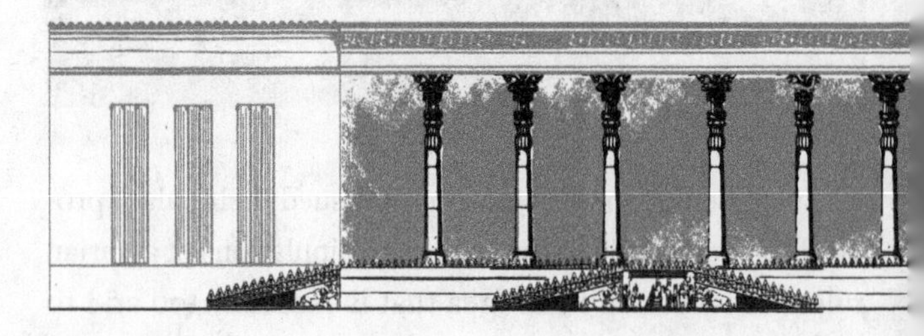

king.[30] The layouts of the Athenian Acropolis and Parsa are sharply contrasting. The Acropolis looks somewhat higgledy-piggledy; no two important buildings quite aligned or parallel. Parsa maintains a thumpingly consistent geometry throughout.

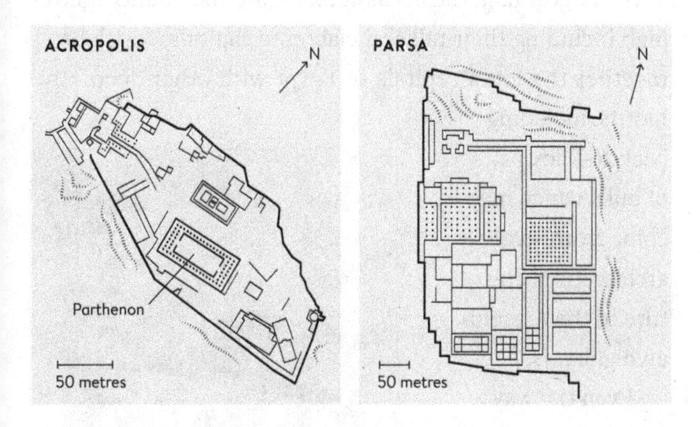

The Acropolis, although rebuilt after its destruction, was not neatly replanned, but carefully reflected the locations of lost buildings and the centuries of Athenian traditions and history they embodied. If the Acropolis is the architecture of the first democracy, Parsa is hierarchical architecture. It tells visitors from anywhere on earth that the hand which commanded this palace into existence was vastly powerful and well organized, and its orderly thoughts became realities.

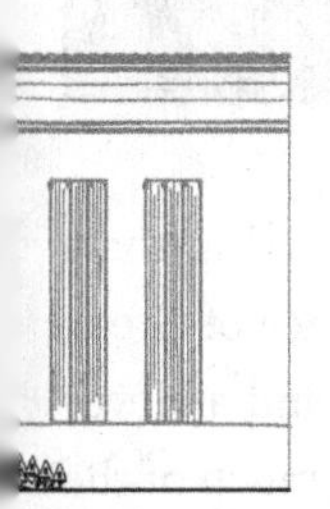

Of course there is an architectural language at Persepolis too, but it is as intentionally hybrid as the Parthenon's is not: a kind of architectural Esperanto blended from architectural ideas and motifs from all over the empire. The artificial terrace beneath echoes Mesopotamian palaces, as do the vast statues of winged bulls which guard the gates. The images of the Zoroastrian god whom the Persians worshipped may be derived from Assyrian art. The columns in the largest hall, the *apadana*, were around twenty metres high including their tall and elaborate capitals, which bring together the lotus capitals of Egypt with other decorative motifs, including back-to-back pairs of bulls which may come from an older architectural culture of the Armenian highlands.[31]

Even the way Persepolis was built shows influences from around the Persian Empire. The stone platform on which the palace stands uses an ancient masonry style employed in Armenia and elsewhere, while the stone of the buildings was worked using tools and techniques that seem to indicate the presence of Greek masons: for example, toothed bronze chisels and a clever technique that involved slightly hollowing out much of the touching faces of the blocks so that the remaining margin of stone where the surfaces met fitted perfectly.[32] The fluting on the columns also reminds the Western observer of Greece, but the curvy bases on which the tall, slim columns stand are unfamiliar.

The vast complexity of the Persian Empire's relationship with its constituent kingdoms required just such a palace: not clearly allying with one region against another, but offering courteous inclusion to the elites of all regions, while producing something distinctive that was the new architecture of the Persian Empire. Even the inscriptions at Parsa tend to be in several parallel languages, the king of kings showing his legitimacy as ruler of his subkingdoms.

If the architectural languages used at Parsa and the Parthenon embrace the difference between the hierarchical energy regime of the Persian Empire and the flatter dispersal of the control of energy among the citizenry of Athens, so do their functions. Parsa was built as the dwelling, treasury and audience hall of a supreme king. The largest room in the palace, the magnificent *apadana*, is thought to have been an audience chamber for the king to meet his courtiers, or to receive delegations from his provinces or beyond the empire. It was a gigantic square in plan, with sixty-metre sides, its roof made of magnificent cedar timbers, doubtless heavily carved and coloured, towering twenty metres above the humans below. It could have held as many as 10,000 people.[33] The purpose of the other large audience hall is unknown, but some have proposed that it was associated with the king's elite military force – a display, if so, of his direct personal power.[34]

The Parthenon's columns are only just over half the height of those in the *apadana*, and its interior was far smaller, yet it too was a home and an audience hall. The Parthenon housed the goddess Athena, represented by a spectacular gold and ivory statue 11.5 metres high. The Parthenon was never intended as a place of worship in the way of most churches or mosques, where all the worshippers would crowd in for a collective experience. Instead, Athenian public religious festivities were held outside, where a much larger crowd could be present, worshipping at an outdoor altar and an outdoor statue of the goddess.

Parsa also incorporated its divinity. Darius, under whom much of the building took place, had an inscription carved saying that 'by the grace of Ahuramazda I built this fortress. And Ahuramazda was of such a mind, together with all the divine beings, that this fortress [should] be built. And [so] I built it. And I built it secure and beautiful and adequate, just as I was intending to.'[35] The god only gets credit for supporting Darius, however: Darius himself is clearly the man who actually got it built.

Authorship was considered equally important by the Athenians, but in the opposite respect: the Parthenon had to be built collectively by the citizens of Athens, not by any powerful individual. The historian Plutarch, writing around five centuries later, claimed that Perikles, the politician who wielded most influence at this period in Athens, was attacked for the vast expense of the Parthenon and, to shut up his critics, offered to pay for it himself and call it after himself. The Athenians were horrified, and agreed to continue funding it collectively.[36]

In these two buildings sculpture was integral. Not only was there much of it carved into or attached to the buildings, but in the case of the Parthenon we are told that the sculptor in charge of the statue of the goddess also had general authority over the building that would house his sculpture.[37]

The many architectural sculptures at Parsa and the Parthenon clearly reflect the different political set-ups of Persia and Athens. The Parthenon had more narrative sculpture than any earlier Greek temple, all of it painted, and some of it enlivened by the addition of sparkling bronze horse tack and weaponry.[38] Yet the Parthenon's hundreds of figures included no living or recent historical people. There are not even representations of the heroic battles against the Persian invaders. Instead almost all its many sculptures show mythological figures, and the frieze that many believe shows a religious procession in contemporary Athens depicts a vast mass of unidentifiable Athenians as a collective political entity, not specific individuals.[39] The spirit of Athenian collectivity is clear, and individual prominence is entirely absent.

In Parsa, by contrast, the sculptures are obsessively centred on the figure of the king. He appears again and again, sometimes symbolically defeating terrifying beasts, sometimes in more literal-looking scenes, sitting on thrones held aloft by courtiers and receiving tribute from twenty-three regions of the empire. These scenes have been treated as fascinating evidence of the clothing and weaponry of many of the empire's peoples, and the luxury items coming from different areas (Ethiopian ivory, Indian gold and so on). The most striking thing about the reliefs, however, is that every one of the groups bringing tribute is bringing energy

supplies: beehives full of calorific honey, meat driven to the royal supply rooms still on the hoof, beasts of burden to work the royal fields and a huge range of jars, baskets and pots that were presumably recognizable as the distinctive packaging of the crops and delicacies of different regions.[40] This direct contribution of energy to the central authority of the Persian Empire was a crucial support to the king's political power.[41]

Parsa and the Parthenon, then, display at its starkest the contrast between these two very different energy systems: the palace of the Persian kings is one of the great peaks of the agrarian tendency to hierarchy; the Acropolis demonstrates the anti-tyrannical obsession of Greece's strange, disruptive new experiment in avoiding it.

The Athenians played up this difference as an important defining feature in their own view of themselves. The later reputation of the Achaemenid Persians has done badly out of it, as the Greeks left behind a great quantity of written material from their perspective, but little or no equivalent material survives for the Persian Empire at this period. The result is that for the Western world the Greeks have enjoyed 2,500 years of being the heroes of their own narrative, with the Persian kings represented by Greek authors as growling tyrannically, 'The guilty and the innocent will alike bear the yoke of slavery.'[42]

The Athenians felt that, apart from the king, no one else in the entire vast Persian Empire was as free as an Athenian citizen. Every man in Persia apart from the king had to defer to his superiors and ultimately to the king in the way that an Athenian wife deferred to her husband; an Athenian citizen deferred (they claimed) to no one. As if to heighten the comparison, Persian male costume – long, all-covering robes and abundant personal jewellery to show wealth – somewhat resembled female fashions in Greece.[43]

However, the aggressive opposition that Athenian culture showed towards the Persian Empire has concealed similarities which are more profound than the differences.

The Inescapable Shape of Agrarian Societies

If we take the wider Acropolis rather than just the Parthenon, the similarities to Persepolis are striking: each is raised haughtily above its city on a part-natural, part-modified rock platform; each is a self-evidently impressive, hard-to-build, big complex of buildings; the main structures of each dwarf ordinary dwellings; each seems in its original state to have been a brightly coloured feast for the senses, the impact of which was almost certainly heightened for special occasions with perfumes and music; each was the centre of great processional festivities which dramatized the social and political order of the society, with the stairs at Parsa set wide enough and shallow enough for important people to ride up on horseback.[44] You might manage to ride a horse up the Acropolis too, but getting it back down the rather steeper stairs would be nerve-racking.

Both the Acropolis and Parsa stored the impressive wealth of their states: gloating displays of plunder taken from defeated enemies (the nine-metre-high outdoor statue of Athena was made of melted-down bronze weaponry dropped by the routed Persian army as they fled), mounds of precious tribute given with greater or lesser willingness by visitors and subjects, and abundant coinage or coinable precious metals. Even the statue of Athena was clad in gold panels, which were intentionally designed to be removable when future wars or emergencies left no alternative, to be repaid with interest after future victories. Meanwhile, the treasure at Parsa is said to have taken 10,000 mules and 500 camels to remove when the palace eventually fell.[45]

Perhaps the most surprising similarity lies in the huge sculptural friezes in Parsa and the Parthenon. In both, groups of people are shown coming with tributes of food, drink, animals and precious goods. Historians have puzzled over why the Athenians would do something so similar to the art in the heart of the hated Persian regime.[46] The answer may lie in the fundamental, inescapable similarity between agrarian energy systems: ultimately, for an elite population to enjoy leisure and sophistication, a larger population of others has to work the land.

In the case of Persia, the picture is familiar from Uruk

and Egypt: a hierarchy with a broad base of farmers supporting a steep, narrow pyramid of craftspeople, traders and elite. In the case of classical Athens, the picture appears very different, until you take a closer look. The Athenians tried to be as different as they could from the Persian model of tyrannical hierarchy, having instead a relatively broad elite group, with relatively small differences in status and wealth between members of the elite, protected by ostracism and their constant neurotic caution about individual power. Even the Greeks, however, depended on an underclass larger than the citizenry, and gave very limited rights to women, even those of high social status. For between 30,000 and 40,000 male citizens to live in political equality, educating their male children, selling surplus crops and buying the basics and luxuries of life from traders and artisans, a large population of people needed to work as slaves in the fields. More than this, to rebuild the Parthenon and the rest of the Acropolis on a large scale and to a very high standard required not only Athens's own surplus energy and the money into which that surplus could be converted, but also extensive grain imports from their own colonial settlements on other shores, and, perhaps most interestingly, some of the surplus of other city-states.[47]

By the time work began on rebuilding the Parthenon in 447 BCE, Athens had for three decades been the leading city in a coalition set up to keep the Persians out of Greece. At some point in that period Athens crossed the fuzzy line from being the most powerful member of a cooperative alliance to profiteering from a protection racket that extorted money and grain from weaker neighbours under the threat of their mighty navy. In 454 BCE, the Athenians moved the collective fighting fund from its agreed home on Delos to their own city, perhaps later even storing it within the Parthenon's treasury. Plutarch tells us that even within Athens many saw it as a tipping point, and associated the removal of the money with the construction of the Parthenon and its neighbouring buildings: 'The Greeks are surely insulted by a grave dishonour and subjected to open tyranny when they see that we are gilding and ornamenting our city with their enforced contributions for the war just like a prostitute, wearing costly stones and statues and thousand-talent temples.'[48]

Most strikingly, they could only control the fundamental hierarchical tendencies of agrarian energy economies by giving the position at the top of the hierarchy to a useful fiction: the goddess Athena. The desire to show off wealth and skill, to concentrate food and treasure in the hands of the head of the society and to live off the labour of a large body of farm workers were all there in classical Athens, but partially disguised by dedicating the glorious new building to a goddess rather than a mortal human being. The phenomenal burst of creativity and ingenuity promoted by the large citizen body of classical Athens could get them away from the rule of kings, but even they could not escape the irresistible pressure that farming exerted towards hierarchy.

The Athenian experiment with empire was not long to outlast the construction of the Parthenon. The years of work decorating the completed structure finally came to an end in 432 BCE and the following year Athens went to war with their Greek neighbours the Spartans and never regained their power.[49] Even so, the Parthenon survived for almost two millennia in surprisingly good condition, converted eventually into a Christian church and later again into a mosque. The level of damage the structure shows now is not from weather and time, but from a single devastating incident in 1687, when the army of the Ottoman Empire was using the Acropolis as a fortress and the Parthenon as an ammunition store. The Venetian army attacking it managed to land a shell on the roof, causing the gunpowder inside to explode, wiping out the interior and roof, and doing considerable harm to many of the outer columns too. Around half of the statues which survived were then removed at the turn of the nineteenth century by a Scottish aristocrat, who sold them to the British Museum, where they remain. The debate on the rights and wrongs of the removal of what the British call 'the Elgin marbles', and whether they should be returned to Greece, shows no signs of dying down. Nevertheless, the fact that the Parthenon lasted almost 2,000 years in largely original condition is remarkable.[50] At the time it was built it must have seemed very likely that it would suffer the fate of its predecessor – destruction at the hands of a further Persian invasion. In fact the tables were, astoundingly, soon to be turned.

In 330 BCE, 4,000 kilometres from Athens and protected by perhaps the largest armies ever mustered, Parsa was taken, looted and destroyed by Greeks. Less than two centuries after building works had begun on the palace at Parsa, Alexander the Great, ruler of a marginal northern Greek region, managed to defeat the Persian army and overrun its vast kingdom. Alexander admired the achievements of the Persian kings and emulated many of them. However, at Persepolis, egged on by stories of the long-ago destruction of the Athenian Acropolis and by the goading of some of his companions, he set light to the great palace, burning its roof and leaving it a ruin, never again to be inhabited.

CHAPTER 4

Energy Booms

The Roman Empire and Song-dynasty China

Imperial Rome in the early centuries CE and Song-dynasty China (960–1279 CE) might superficially seem like odd chapter mates. Their cultural traditions were different and they were separated by most of a millennium and thousands of kilometres of steppe and desert. Their histories were very different too. Rome had, by the 50s BCE, become the dominant military power of the Mediterranean basin, and its empire continued to expand through an aggressive campaign of invasions and alliances for another 150 years. Song-dynasty China was far from being the triumphant military force of its region. It was under consistent pressure from powerful northern neighbours and in 1127 suffered a catastrophic defeat at the hands of the Jurchen, who captured the capital, the emperor and a large portion of the dynasty's northern lands. The rump Song regime withdrew to a temporary capital in the south until their defeat by the Mongol Yuan dynasty in 1279.

Their architecture was very different as well. The grand public buildings of the Roman Empire are lastingly famous for their exceptional robustness: hefty stone, brick and concrete used in vast quantities, forming huge arches which were built with such solidity that many have outlived their original functions by millennia.

Classical Chinese monumental architecture, by contrast, used wood as its chief structural material. Comparatively slender columns sprouted from bases of rammed earth, brick or stone. At the top, the wooden columns branched out through bracket sets of elaborate carpentry to support a wide roof of fired-clay tiles. The roof's substantial overhang protected the woodwork from the rain and also showed off the skill of the carpenters, their slender woodwork supporting the heavy fired clay above.

With such divergent histories and architecture, why does this chapter bring together these disparate societies? The answer lies in their one great similarity: imperial Rome and Song-dynasty China saw two of the most rapid and spectacular energy booms in the whole of agrarian history. The shared experience of rapid and substantial increase in available food energy, it turns out, overcame many of the differences between them sufficiently to promote comparable architectural responses.

Antony, Cleopatra and Zhenzhong

The Battle of Actium in 31 BCE saw the decisive defeat of a celebrated couple: the Roman general Mark Antony and the last Egyptian pharaoh, Cleopatra, both of whom committed suicide in its wake. The victor became the first emperor of the Roman Empire, Augustus (dominant in Rome from 31 BCE to 14 CE), and gained control of Egypt as a personal property. Suddenly the vast grain supplies of Egypt were available to feed Rome.[1] Thousands of shiploads per year were taken from the Nile to the River Tiber, which flows through Rome, in order to support a population which historians estimate grew from 250,000 in the 120s BCE to a million under Augustus – by far the biggest city ever to exist in patchily fertile Europe to that date. Indeed, it remained the biggest city ever seen in Europe for well over 1,500 years after Rome lost Egypt's grain supply and shrank to the extent that farm animals grazed between the mutilated monuments of its centre.[2]

The decades of peak monumental construction in Rome, and of its widest military expansion, coincided with a period that climate historians report as having seen exceptionally reliable and fertile Nile floods, offering unusually good crop yields even by Egyptian standards.[3] Not only Egypt but the Roman Empire as a whole enjoyed uncommonly good climatic conditions for farming during these years, and architecture and city growth boomed with unprecedented vigour, run from and culturally centred on the great city of Rome.

In Rome, many authors wrote polemically of the imperial period as one of decadence and decline after the military expansion and intense political discussion of the Republic that preceded it. In architectural terms, however, everything you probably think of when you call ancient Roman buildings to mind was built under the emperors: the Colosseum, the Pantheon, the Forum, the great bathhouses, temples and large mausolea, most of the surviving city walls of Rome and so on. Remarkably little survives of Republican Rome. Augustus, profiting from vast Egyptian grain imports, is said to have boasted that he found Rome a city of brick and left it a city of marble.[4]

Rome's size – perhaps the biggest population of non-farming people concentrated in one city anywhere up to that date – produced the richness of culture, politics, commerce and social structure that helped make Rome so lastingly famous. It was, however, a smaller urban boom than that of the militarily unsuccessful Northern Song dynasty. Their chief city, Bianjing (today's Kaifeng), in what is now eastern China, rose to somewhere in the region of 1.5–1.9 million people, the sheer weight of economic growth and population smashing down the internal walls with which earlier Chinese dynasties had sought to segregate and contain their urban populations.[5] Bianjing and other cities burst out into an unprecedented flowering of popular and elite culture, shopping, nightlife and food. Bianjing was a 'city of towers'.[6] Above the single-storey houses that made up the bulk of any Chinese city rose apothecaries' shops advertising their services with the nearest equivalent they could manage to a 1970s Las Vegas casino sign. Multi-storey restaurants were even more prominent, draped in brightly coloured silk, with bridges leaping from balcony to balcony.[7] The grand entrance gates of the silk shops, offering the supreme Chinese luxury product, were to be found in the specialist high-end shopping areas, surrounded by gold shops and jewellers catering to large commercial and political elites with money to burn.[8]

As night fell, designated entertainment districts continued to sell acrobatic or theatrical entertainment, food, drink and sex, right through to the morning.[9] New low-denomination iron coins enabled even relatively poor people to join in this money economy rather than bartering for smaller-value transactions. Right through the crowded centre of the city ran a 200-metre-wide open space, in a straight line from the south gate of the city walls to the south central gate of the palace.[10] What better demonstration of power and wealth could a ruler seek than empty space in a city with rocketing land values? A similar route was later to run south from the gate of Beijing's Forbidden City, leaving its mark on the Chinese capital even today.

This scale of urban boom came from a vast energy revolution, but here it was not based on conquest and importation as in Rome. The Song dynasty's government under Emperor Zhenzhong (997–1022) had engineered their energy boom quite deliberately in response to a drought in the important rice-growing areas around the Yangtze and Huai rivers.[11] Fear of northern invasion had driven many farmers southwards, exacerbating the challenge of feeding not only the civilian population but also the army. By imperial order, 30,000 bushels of a new variety of rice were distributed to farmers, along with written instructions which enabled local officials to give advice on how to plant it.

The new rice was brought from Champa in what is now central Vietnam. A cross between Indian and Chinese varieties, it ripened around 100 days after it was planted, as opposed to the usual 180 days. Champa rice did not mind what time of year it was sown either, whereas the normal Chinese crop had to be grown in its habitual season.[12] One benefit was the potential for early planting and early harvest, which reduced the risk of summer drought wiping out the crop. More than this, though, across a large area of the best growing land the farmers could suddenly grow and harvest two annual crops (either two rice crops or winter wheat and summer rice) in each field. Initially farmers were cautious, fearing that yields would not be good enough to repay the considerable extra labour of a second entire planting and cropping cycle. Landlords too showed resistance, fearing that the intensity of exploitation would exhaust the land. Experience soon showed that these fears were unjustified and farmers threw themselves into the new farming pattern, producing the biggest surpluses in Chinese history.[13] With tax incentives and low-interest government loans to support expanding production, together with improved networks of irrigation and transport, large tracts of land were newly brought into cultivation and high-yielding farms spread into areas that had previously been considerably less productive.

Some farmers moved away from rice and wheat, specializing in cash crops like silk and sugar, then buying food cheaply for their families instead of growing it.[14] The cash economy exploded.

With unparalleled new energy resources at their disposal, in the shape of food and the human and animal labour it sustained, what did the Roman and the Song-dynasty emperors choose to build and why?

Feeding the Workforce

The first answer was infrastructure: large interventions that tended to repay the energy they cost through improving the functioning of the city's energy systems. Because these were built in hope of a substantial return, they could on occasion be much bigger than buildings whose purpose was less materially functional. So, if the 22,000 tonnes of marble of the Parthenon demanded a lot of animal and human labour, this was dwarfed by the material and labour required for a project like the expansion of Rome's port facilities under the early emperors.

In 42 CE when the emperor Claudius came to power, Rome used as its harbour a natural river mouth at Ostia, thirty kilometres away as the crow flies. It was too shallow and narrow for the substantial ships which were bringing grain and other supplies from the rest of the empire, so small boats had to go out to the ships anchored offshore and the freight transferred between rocking vessels of different heights, making the all-important grain cargoes vulnerable to the waves and the elements.[15]

Artificial harbours had been around for millennia by this time – the earliest known one was built around the early 2500s BCE on the Red Sea coast of Egypt and used by Khufu.[16] At Ostia, Claudius initiated the construction of a much larger new harbour, 800 metres wide and considerably more than that in length, with a depth of seven metres. A pair of protective arms made of Roman concrete was built out into the sea, sheltering hundreds of moored ships.[17] Cargo could then be transferred on to river barges to go up to Rome, or taken into vast new granaries alongside the docks. Thanks to this hefty infrastructure, Claudius was able to ensure Rome's core food supply in an attempt to avoid the political instability that came with fear of hunger.

This large harbour was repeatedly expanded and improved over the following decades.[18] The emperor Trajan, best known now for the monumental thirty-metre-high column showing his military victories, and for being the ruler who took the empire to its greatest extent, built a polygonal extension to the harbour (still a prominent landscape feature near Fiumicino airport). He chose to show it off as a proud achievement on one of his coins, demonstrating the political as well as practical importance of his magnificent contribution to the security of Rome's energy supply.[19]

The Song dynasty also conducted substantial harbour-building projects to service the growing volume of international trade as their energy and money economies boomed.[20] Perhaps even more important to the Song, however, were canals. Rome also built canals, but Song-dynasty China went further with them, using them both for transport and irrigation, increasing crop yields and allowing the resultant surplus to be carried affordably to cities. The capital Bianjing was located near a meeting of major canals. The amount of construction activity on canals in this period in China was so great that it led engineers to develop the first locks, which allowed canals to go up and down with the landscape rather than having to follow a single contour over their entire length.[21]

The Song showed the full range of benefits possible from these energy-hungry investments in future energy supplies: the set of new canals they built near the disputed northern borders of their territory irrigated the land to provide the crops to feed a larger army for longer periods near the border, and accelerated the movement of armies and military equipment towards the front line. The positioning of the canals also made them a helpfully defensible obstacle to the enemy land army. With the speed of modern warfare it might seem an unfeasibly long game to hand-dig hundreds of kilometres of canal, but the Song canal-building project was clearly threatening to the Liao, the northern neighbours targeted by the programme. When their repeated attempts to invade pre-emptively and disrupt the building campaign were unsuccessful they struck a peace deal with the Song.[22]

Both Rome and Song-dynasty China made other infrastructural energy investments, including substantial road networks with numerous robust bridges. Roman and Song bridges still survive in daily use in a number of places.[23] Both empires built sewers and drains in their cities to deal with floods and to take away unhygienic human and animal waste. Again, some of these systems are still in operation today, as are a few of the Roman aqueducts which bring clean water from the neighbouring hills to the city centre, freeing the population from dependence on the lethally polluted waters of the Tiber.

Whatever humane considerations were at play in these sanitation measures, the reality remained throughout the agrarian era that a healthy population was the main power source available to rulers. Plague caused, among other things, a drop in the supply of the muscle power that converted crops into work.

With their large labour forces, the regimes of Song-dynasty China and imperial Rome transformed both city and countryside in a self-reinforcing circle, with more food supporting more specialism, which in turn generated further improvements to energy systems.

Water Power

The age-old dependence on human muscle was a major limitation on agrarian energy systems, with repetitive jobs like grinding indigestible grains into usable flour taking up large amounts of time and effort. As the population rose, grinding went from a family affair to an industry, with donkey-powered mills and eventually large watermills using the powerful weight of water to turn the millstone.[24]

By the later Roman Empire special aqueducts were sometimes built to service watermills, like that at Barbegal in modern France, where two staircases each of eight mills must between them have extracted from the descending water the maximum possible energy under the conditions of the time.[25]

Even at the enormous Baths of Caracalla in Rome, a complex of giant pools (mostly heated) and other facilities for washing and leisure, the huge energy consumption needed to keep everything functioning was no reason not to find efficiencies where possible. The

outflow of dirty water from the baths ran down through a small mill.[26]

If the hard work of grinding corn could be entrusted to the unending power of streams, so could the cutting of building stone, particularly marble. By the third century CE, saws driven by waterwheels were being used on large blocks of stone.[27] As we shall see, these precisely cut claddings were to be a major feature of Roman imperial monumental architecture.

The watermill technology in Song-dynasty China was even more impressive. Mills were used for grinding cereals, but also for textile manufacture and to power bellows that fanned furnaces for metallurgy.[28]

Standardization and division of labour

The ingenuity and innovation that come with substantial populations of non-farming specialists brought improvements to all areas of Roman and Song building activity. Workers who do a bit of building, a lot of farming and a bit of cloth manufacture are never likely to build up the speed, quality and expertise of a worker who spends all day every day just on one task, whether that is carving stones, laying bricks or shaping wood.

To pursue these efficiency improvements, both the Roman and the Song building industries implemented programmes of standardization. In the Roman Empire, the construction techniques used for major projects evolved rapidly as the energy boom brought a corresponding building boom. The *opus incertum*, or irregular work, that characterized the structural walls of earlier centuries involved skilled masons patching together similar-sized but randomly shaped stones like crazy paving, as facing on a concrete and rubble core. Already by the time of the first emperors this had come to be largely replaced by *opus reticulatum*, in which standard-sized square-faced blocks of local stone were laid at forty-five degrees to the horizontal, again as facing to a concrete and rubble core.[29] The preparation of the outer stones for *opus reticulatum* was more labour-intensive, but laying them into a robust facing was presumably easier and quicker than the individual artistry and experience required for good *opus incertum*. If the twentieth-century experience of standardization is a fair analogy, the possibility of a wall being spoiled by an *opus incertum* worker having a bad day may also have been reduced by the standardization brought in by *opus reticulatum*.[30]

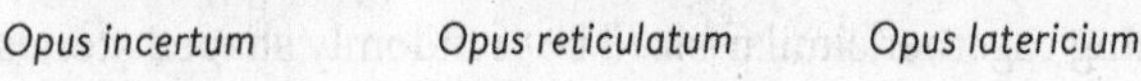

Opus incertum *Opus reticulatum* *Opus latericium*

The standardization of production techniques tends to make it possible to train new labourers (whether slave or free) in their part of the process considerably more quickly than they could have been apprenticed for older, higher-skilled techniques.[31] Huge workforces of the urban poor could therefore be given gainful employment with little training on the vast projects of the grain-rich emperors.[32]

The new material of the emperors, however, was yet another step towards the industrialization of building: *opus latericium* consisted of broad, shallow, triangular tile-like bricks laid in an overlapping pattern or 'bond', not unlike a brick wall now, their points receding back into the concrete like the points of the *opus reticulatum* stones. Both *opus reticulatum* and *opus latericium* tended to cost considerably less labour than *opus incertum* for a given quantity of wall construction.[33] *Opus latericium* was so successful that it not only became the standard technique for grand buildings, but also replaced mud brick as the predominant building technique for all permanent structures in Rome, producing a much more robust architecture.[34] When visiting Rome today, it is these gigantic craggy walls of pink Roman brick that most strike the eye. Rome's great building boom was largely composed of these brick-faced concrete and rubble walls, built at very large scale and considerable speed, often with several major imperial projects going ahead concurrently within a few hundred metres of each other, alongside many private building operations, big and small, all over the city.

Opus latericium departed from the earlier facings in using fired brick rather than stone. There are some indications that the ability to grow enough plant-based fuel for Rome's substantial requirements, including for brick kilns, may have involved the deliberate introduction of managed forestry on land where the pressure to grow maximum food calories was relaxed by the imported grain from Egypt and elsewhere.[35] The surface of these concrete and brick structures was then covered, for prestigious buildings, in decorative stonework, which we will return to below.

The huge energy wealth of the early emperors supported standardization initiatives not only of brick dimensions but of many building components, from roof tiles to sewage pipes.[36] Some scholars even believe that standard dimensions were used for the gigantic Greek-influenced columns that fronted temples and other grand public buildings. For major prestigious buildings, each column shaft was generally a single large piece of stone. Many of the biggest monoliths came from the same quarries in Egypt where the obelisks and other Egyptian granite monoliths had been carved for millennia, as we saw in Chapter 2.

The process of quarrying granite remained very slow, and the transportation of columns to Rome added considerably to the length, complexity and riskiness of the process. Rome is 1,700 kilometres from the Aswan quarries as the crow flies, but by water it was far further, winding along the Nile on a barge, then, after transferring to an ocean-going ship, hugging the coast, scurrying into harbour at any sign of abrupt and challenging storms, and struggling against the prevailing wind.[37] Some ships carrying columns sank in storms, or were delayed by wind or weather. Even transmitting an order for a set of columns from Rome to Aswan would have taken sixty or so days, the order perhaps transmitted by a skilled worker who went in person to the quarry to oversee the choice of stone and its roughing out into transportable blocks of appropriate size and shape.[38]

Some scholars think that, to speed up this process, column shafts were sometimes produced in standardized sizes on spec, for stockpiling in Rome against future demand, in sizes up to fifty Roman feet (14.8 m), a granite Corinthian column shaft of which size weighed around 100 tonnes.[39] Stone for revetment (thin surface cladding) was almost certainly produced on such a basis.[40]

While the standardization of columns and other large stone elements is debated, the standardization of smaller components and techniques is clear, and seems almost certain to have needed a push from the emperor's circle, as standardization can weaken the economic power of individual actors in the building industry (a proprietorial brick size, for example, could keep the client from buying from a cheaper rival brick kiln, whereas standardization pushes down prices and pushes up competition on quality). It would be fascinating to know how the Roman emperors' bureaucracies intervened in the construction industry with such impressive effectiveness. The indications which survive include an increasing control of brickworks in the area around Rome, first by senators and later by the emperors themselves.[41]

The role of the imperial civil service in achieving standardization in Song-dynasty China is much better documented. A book commissioned by the emperor in the late eleventh century survives in several copies of two editions. Called *Yingzao Fashi* (*State Building Standards*), it was written by

Li Jie, Superintendent of State Buildings. Completed in 1100, it was adopted by the emperor in 1103.[42]

Yingzao Fashi's thirty-four chapters impose the sort of standardization seen in the archaeological record in Rome: bricks and tiles were to be made in seven different sizes each, with special slope-edged bricks for city walls and other battered walls (walls that narrow as they rise from a broad base). Woodwork too was standardized into eight sizes, each expressive of the level of cultural importance of the building as well as its actual dimensions.[43]

To modern Western readers, *Yingzao Fashi* has about it an echo of the most famous architectural treatise of European antiquity, *Ten Books on Architecture* ('book' meaning 'chapter' in modern terms), which was written around 28–25 BCE by Vitruvius. Though a millennium apart, both the Roman and the Chinese books provide guidance to the government official commissioning architecture and to the designer. Vitruvius is best known among Western architects for his useful summing-up of the key virtues of a good building: *utilitas*, *firmitas* and *venustas*, often given in English in their famous early translation 'commodity, firmness and delight'.[44] Each is full of practical advice and accumulated experience, but at the same time both authors write about architecture as a topic of intellectual and literary seriousness, seeming to want to lift it above the lower-class sweat and dust of the building site. Each reveals the complexity and sophistication of the discussion of architecture in his period and the ways in which building designers and makers used metaphors from nature. Li speaks of the beautiful complexity of the bracket sets that branch out from each column to support the roof in botanical terms: 'branches', 'leaves', 'flowers' and 'petals.'[45] Vitruvius relates architectural beauty to the proportions of idealized bodies.

Both authors also put into writing a mass of experience which prior to that point had predominantly been passed down orally. They did this partly to make it easier to access and partly to give it the intellectual lustre possessed by the written word in eras when literacy was rarer and more prestigious than it is today.

Li's *Yingzao Fashi* was commissioned by the Song emperor Shenzong (1067–85) in the 1070s.[46] The leading civil servant, Wang Anshi (1021–86), was at the same time attempting to conduct a major reform of the imperial civil service. It was aimed at fighting corruption, in particular the improper promotion of family members irrespective of their competence. Wang wanted to revise the process of selection and training for civil servants, making it less of an abstract ability test and more of a pragmatic training in the specific disciplines needed by the government.[47]

Yingzao Fashi can be seen as part of this same reform project, and the extent to which the emperor really wished to use it to audit building projects is clear from the fate of Li's first draft: his decades of work produced a text which was rejected by the emperor because it did not give numbers for the amount of labour and materials required for different projects. Without these, the emperor said, 'it could not be enforced and thus became meaningless words.'[48]

This gives an idea of the level of ambition and courage in the Song government's plans for the building industry. Just as they had stepped in to revolutionize the energy basis of the empire and the workings of the civil service, so they planned to change the building world. *Yingzao Fashi* was an attempt not only to standardize the dimensions and techniques of the empire's builders, but also, it seems clear, to professionalize the industry and introduce rigorous cost controls.

The sophisticated building crafts of China had for centuries been passed down from father to son within each trade in the form of long poems. These hereditary carpenters, masons, ceramic workers and so on had long maintained something like a monopoly over architectural production. The secrecy and the element of heredity may on occasion have allowed second-rate carpenters to thrive monopolistically, but even if standards remained high, the supply of construction specialists was inelastic. A limited supply of skilled workers potentially restricted the capacity of the industry to deal with the great energy-fuelled building boom of the Song – a boom so spectacular that they worried about exhausting their timber supplies.[49]

Wang Anshi's reforms sought to avoid nepotism in the civil service. *Yingzao Fashi* seems to have set out to do the same for the building industry, recording huge quantities of information that had never been available to those outside the guild before (only 8 per cent of the book seems to have ever been written down before; the rest came from quizzing practising construction workers), breaking the monopoly of the trade families and allowing the rapid training of new workers.[50]

Even the lowest end of the building trade, the unskilled labourers who dug the earth and moved the stone for infrastructure and construction projects, was brought into the great reforms of Wang Anshi, who moved away from the tradition of compulsory unpaid labour for such tasks towards paying labourers in cash, further increasing the liquidity and reach of the money economy.[51] This extension of the monetized, competitive economy into the building industry is likely to have contributed to the quality and quantity of building conducted by both government and private clients under the Song dynasty. Experts in Chinese architecture comment on the refinement which characterized the great buildings of the Song.[52] The Sage Mother Hall (1038–87) at the Jin shrines, Xishan, is one of the most elaborate Song buildings to survive. Its careful distortions are reminiscent of those seen at the Parthenon – columns slightly curving inwards at top and bottom, with each leaning inwards slightly more as you move away from the centre of the façade.[53] The roof ridge is curved and the eaves sweep up like the wings of a bird, full of life and grace.

The standardization of *Yingzao Fashi* was not only technical, but also aesthetic. The same architectural motifs and decorations were to be used across the entire empire – a convention going back hundreds of years in Chinese imperial history.[54] Equally standardized were rules about who could build what: certain colours and designs of roof tile, for instance, were reserved for imperial buildings and some decorations were forbidden on ordinary houses.[55] Thus someone from the extreme south could go to the capital or indeed the extreme north and recognize an indisputable Chinese character and familiarity in the architecture and city planning.

The level of interest that the emperor and his court took in architecture and infrastructure is clear from the role that these projects assumed in court art. The court culture of the Song included a considerable emphasis on painting by leading courtiers. Buildings, fortifications, bridges and so on feature heavily not only as the backdrops to many scenes, but also as an entire subdivision of painting known as *jiehua* (ruled-line drawings).[56] One important surviving example of this idiom is a beautiful image of a waterwheel: architecture, energy and art coming together for the pleasure and education of the emperor and his closest advisers.[57]

The Leading Folly of the Day: Ornamental Stone in Imperial Rome

In the Roman Empire too the elite took a close interest in architectural affairs and one of the ways they sought to compete was bedecking the buildings they promoted with decorative stone. Prodigious amounts of marble, granite, porphyry and other beautiful stones were extracted and transported during the first three centuries CE. Contemporary commentators were staggered at the scale of a stone market that provided little but luxurious aesthetic pleasure. Pliny the Elder complained that it was an assault on nature's mountains and called it 'the leading folly of the day'.[58]

The human effort it represented was indeed very substantial. A fluted column shaft of twenty Roman feet (5.9m, with a diameter at the thickest point of 74cm) in marble took 123 days of skilled stonemasons' time to carve. One the same size in a stone as hard as granite, even without fluting, took at least 684 days. A two-metre-high Corinthian capital in hard marble might take 3,600 hours of skilled carving.[59]

To pay for this time-consuming skilled labour and expensive transport, huge sums of money were spent on decorative stonework. The 75,000 *denarii* left by a man called Flavius Catullus to provide marble revetments for the baths of Mandeure in modern France would have bought a year's grain for more than 3,000 people in Rome.[60]

Only a minority of the many tonnes of ornamental building stone shipped round the Roman Empire was for imperial projects.[61] Yet when emperors built, they pushed the stonemasons and transportation teams to their absolute heights of achievement. Two of the most famous projects of Rome itself, the Pantheon and the Baths of Caracalla, show the scope of what they could achieve in concrete, *opus latericium* and ornamental stone.

The Pantheon was a temple and imperial audience chamber begun late in the reign of Trajan, around 112–14 CE, after its predecessor had burned down in a lightning strike, and was completed soon after the accession of his successor,

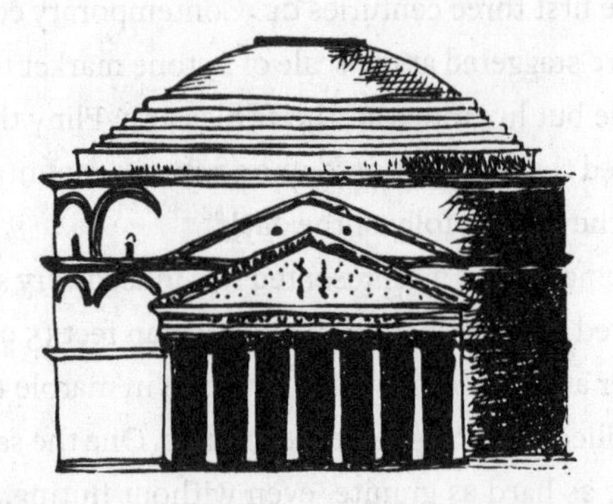

Hadrian, c. 117–21 CE.[62] Any visitor to the emperor in the completed building would have been in little doubt that they were meeting the most powerful man on earth. The Greek-looking column shafts of the portico (columned porch) were monolithic granite, forty Roman feet high (11.84m). These impressive pieces of stone are lasting monuments to the power of the emperor. Their original raising was a striking piece of theatre. When the emperor Vespasian (69–79 CE) was offered a clever new technique to reduce the workforce needed for raising columns, he refused it. The quantity of sweating, muscular bodies pulling together in prominent central Roman sites was presumably a welcome spectacle for the emperor, a dramatic symbolic sample of his power to command people and to get things done.[63] Materials for imperial construction projects were the only exception to the ban on freight transport on Rome's congested streets during daylight.[64] The importance of the public display involved in raising very large single-piece column shafts is underlined by the fact that only the largest columns were systematically monolithic. Smaller columns, which would have been easy enough to make and move in one piece, were often built like classical Greek ones, of more than one drum, stacked.[65]

The ceiling of the Pantheon's portico was clad in darkly shimmering gilded bronze. Beneath this the entrance to the building is through seven-metre-high bronze doors. The considerable impact of the interior on visitors today is nothing compared with how it must have shocked people in the second century. As the visitors' eyes adjusted to the shaded interior of the Pantheon, they would have seen the widest uninterrupted indoor space that had ever been constructed. The interior was (and is) cylindrical up to half its height, a massive drum 150 Roman feet (43.3m) across, clad in rich-coloured stone revetments. Above the drum a hemispherical dome rises a further seventy-five Roman feet to an open circle of twenty-eight Roman feet (8.2m), through which the sun pours, casting a spectacular spotlight on to whichever part of the wall or dome the hour and day dictates. When it

rains the centre of the floor gets wet, but a drain in the middle of the marble pavement discreetly carries the water away.

The roof of the Pantheon was covered in gilded bronze tiles, and the interior of its dome may also have been covered in richly decorated golden bronze, long since removed if so.

Even after almost two millennia, the Pantheon is outstandingly impressive. It echoes the achievements of other agrarian societies in precise surveying and in moving and raising heavy weights. Yet the Pantheon's engineering represents a new level of human technical achievement. Its breadth of column-free roof was not to be surpassed anywhere on earth until the later nineteenth century and beyond when iron trusses and, in more recent times, steel-reinforced concrete were added to the engineer's structural options.

The dome of the Pantheon is made of Roman concrete. The volcanic ash *pozzolana*, found by lucky chance in volcanic rock beneath Rome itself, was mixed with lime and water to make a sludge which would rapidly set hard and soon be as strong as stone. Modern concrete is poured into moulds, but Roman concrete was applied with trowels, plastering it on in layers around pieces of stone or brick which would make the concrete go further and add their strength to the finished building. The *opus latericium* that surrounded this concrete and rubble core was laid in courses as the building rose.

In order to make the Pantheon's 4.5-metre foundations and walls strong enough to take the world's biggest dome, the builders laid rough courses of robust, heavy travertine stone into the concrete. Within the dome itself, by contrast, they wished to keep the weight down, so instead of the extra-strong, heavy stone, they used brick and a lighter stone called tuff for the lower parts of the dome. As they got up towards the top, where the concrete was holding only its own weight rather than the weight of a lot more dome above it, they introduced to the mix very light, air-bubble-filled volcanic stone, not very strong but vastly less heavy.[66] The fact that it has survived for 1,900 years, including through occasional moderate earthquakes, shows that the experience-based engineering of the unknown Roman designers was sound.

The remarkable dome of the Pantheon was the greatest of all implementations of the engineering idea which the Romans exploited more impressively than any earlier civilization: the arch. In ancient Egypt, any large interior was full of columns. This was necessary because to build a roof, every spanning stone needed to stand at each end on either the outside wall or a column. This simple structural system, known as trabeation, is very stable, but stone is at its weakest when pulled and twisted, as it is in a lintel, so the pieces needed to be thick and not span too far.

To get lengths of stone of several metres, thick enough not to break under the load they would need to bear, required quarries of good stone without natural weak points, and a huge force of strong, well-nourished workers or animals to drag the stones from quarry to building site. These massive blocks then had to be raised to roof height. The labour required was vast, and the amount of material involved was very large, because of all the extra columns and the thickness of the stone lintels. Strong stone like granite is also around a fifth heavier than weaker sandstone. And after all that effort, the resulting interiors were compromised by forests of columns.

While Parsa and many other great buildings exploited lighter timber, with proportionately better tensile strength, for their roofs, this too required large trees to be sourced and transported. And, as Parsa showed when Alexander the Great arrived there, wooden structures burn all too easily.

By contrast, the arch represented a substantial improvement over any existing spanning technology. In an arch, small pieces of stone or brick, arranged into the correct curve, can span a considerable width by leaning on each other for support until the weight is carried down to a wall or pier. Each piece of stone or brick can be small enough to be lifted by a single building worker and carried up a wooden scaffolding rather than requiring the earth ramps needed for Egypt's massive building blocks. The stone or brick in an arch is only compressed, not pulled and twisted as it is in an Egyptian stone beam. Compression plays to the strengths of stone, meaning that a sturdy, long-lasting arch can be made with weaker materials that are much easier to come by than strong, faultless granite, and are also considerably quicker and easier to carve. The resulting masonry arch does not burn like wood and can be very robust.

There are two limitations to how big you can make an arch. The first is that, since each block pushes sideways as well as downwards on the block below it, a large arch produces very substantial pressure outwards on the side walls. Strong, heavy buttresses of some sort are needed to keep the arch from breaking the walls outwards and falling in.

The second limitation on arch size is that to build one it is necessary to have a temporary structure in place first, in the shape of the arch. Until all the blocks are in place an arch cannot stand up, so the 'falsework' structure, generally of wood, holds the individual blocks during construction and can then be removed when the arch is finished. For a very wide-spanning arch, therefore, a large quantity of strong wood is required. It can, however, be reused once the arch is completed, and studies have shown that in projects like aqueducts, where arches were built in their hundreds, standardized falsework was sometimes moved from arch to arch as the construction progressed.[67]

If you can get the wood and are willing to have sufficiently thick and robust outer walls, there is little limit to how big an arch you can make. A dome is essentially an arch spun on its centre. The wooden scaffold which must have supported the vast, heavy dome of the Pantheon during construction is not recorded in any ancient image or description, but must have been an extraordinary sight. There may have been those who felt nervous as the supporting wood was taken down, wondering whether the largest concrete span yet built might collapse – it is half as big again as the next largest Roman dome known.[68]

Emperors tended to be declared gods after their death. The Pantheon must have seemed an exhibition of almost godlike powers. The reality is simpler: such feats of engineering arise naturally from repetition. When there is enough energy to support a lot of building activity, the best designers and their teams have a lot of experience and vie to achieve ever greater wonders.

For all the achievements of its designers, the Pantheon has some significant oddities. The portico has above it a higher pediment on the wall of the main part of the building, as if they had originally meant it to be that height. It seems that the builders had already begun work on the main cylinder of the building when the design changed from very large column shafts fifty Roman feet high, weighing around 100 tonnes, to shafts of forty Roman feet, which weighed only half as much and were proportionately easier to extract and transport. In much classical architecture each part is related in its proportions to the others; the smaller shafts meant shrinking every detail of the portico, which as a result no longer fits the architectural layout of the rest of the building.[69] With evidence that the portico might have been completed under Hadrian, it seems possible that he decided that the huge cost and technical challenge of the larger column size were more than he could justify.[70] These very large monolithic stones were pushing the limits of what even the Roman energy system could support. A letter has been discovered from Roman Egypt, dated during the construction of the Pantheon, begging urgently for more supplies of grain to feed a team of draught animals transporting a column shaft of fifty Roman feet.[71] In Rome itself, however, the labour requirements of the Pantheon were perhaps lower than you might guess to look at it: perhaps never more than 240 people working on the site at any time, and often far fewer, over a nine-year construction campaign.[72]

When Roman emperors built at their largest scale, the labour requirements were much higher. The cheaply abundant form of energy in Rome and most other fertile agrarian economies was labour. A large population made it actively desirable to give extensive low-skilled employment to very large numbers of the urban poor, so designing to require workforces of thousands of labourers was a good idea. The economic historian Walter Scheidel has demonstrated that unskilled labourers were paid fairly comparable amounts (when converted into the litres of grain they could buy) across societies from 1800 BCE to the medieval period.

Typically pay was inadequate to support a family to a level of 'bare bones subsistence', unless the family's women and children also generated supporting income.[73] Despite Rome's vast energy wealth, the pay for unskilled labourers seems not to have been at the upper end of the normal historical range and may even have been towards the lower end.[74] The superabundance of labour in a city as heavily populated as Rome is indicated by the fairly modest extra pay for skilled labourers: only twice the day rate of unskilled labourers. Even those whose artistic prowess was crucial for the quality of the work (mosaicists, for example) were paid only 20 per cent more than normal skilled labourers.[75] The hierarchical tendencies of agrarian societies ensured, in other words, that however rich the city as a whole, the majority remained poor.

The leading specialist in Roman construction, Janet De-Laine, has conducted a remarkable study of one of the largest single buildings known from an agrarian energy regime, the Baths of Caracalla (built c. 212–16 CE). She has shown that the labour requirements of the rapid construction of such a big central block were truly immense, with an average workforce of 7,200 men involved in materials production and construction, and a further 1,800 men plus oxen involved in transport of materials in

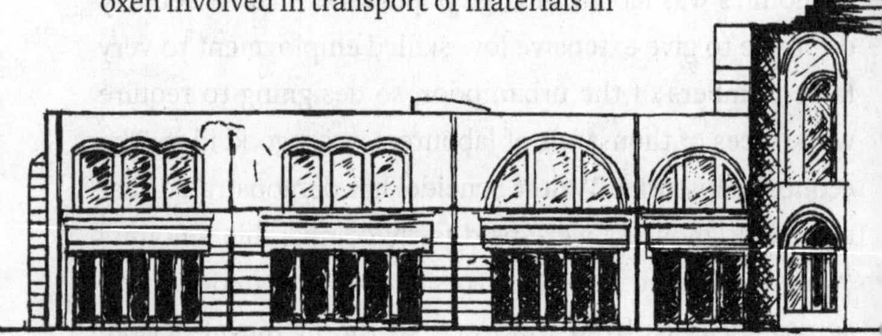

and around Rome. At peak times these figures could rise as high as 13,100 men involved in building the central block of the baths, with more working on the surrounding buildings in the complex.[76] Thanks to the relatively modest level of training required to get new workers contributing usefully to *opus latericium*, plus abundant roles in carrying, mixing and so on, the majority of this huge workforce could be recruited rapidly from the Roman poor, generating substantial employment.

The resulting building looked, in ancient times, like part of the 'city of marble' boasted of by Augustus. Its columns were impressive monoliths, its entablatures (the decorative group of mouldings that always go on top of classical columns) were substantial solid blocks of exotic stones and its surface had a revetment of marble. Yet this opulent appearance was subject to hidden economies. While long-distance transport of beautiful stones made for a desirable show of power – and conspicuous consumption of human and animal energy and of shipping capacity – the amount of these stones used in Roman imperial architecture was as small as possible consistent with maximum aesthetic effect. In the case of the Baths of Caracalla, marble formed less than 0.5 per cent of the volume of

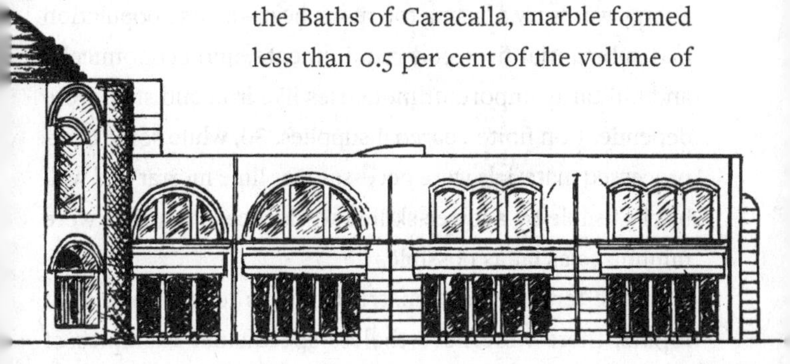

the central block – a highly visible but insubstantial veneer to chunky structural walls of *opus latericium*.[77] Prominent features like columns and entablatures could be seen in the round, so needed to be solid stone, but much else was either done in skinny revetment or, higher on the building, in painted plaster.

If long-distance bulk transport was minimized (a form of energy which was in limited supply and could be used for other purposes capable of generating greater economic return), so too was heat energy, and for similar reasons. Lime, produced by heating limestone to over 900 degrees Celsius, was needed for the concrete, but in relatively small proportions, making up around 3.2 per cent of the volume of the building. Bricks too made up a very small proportion of the building's volume – 2.7 per cent. Like lime, bricks required kiln-firing at around 850 degrees Celsius.[78] Most of the building (76 per cent) was made of stones and other materials that not only required no heat to process, but were also sourced within twenty kilometres of the building site.[79]

This is quintessentially the architecture of an agrarian society. Even a farming economy with abundant imported energy was wary of using too much heat – a large population had substantial firewood requirements, and economically and militarily important industries like iron and steel were dependent on finite charcoal supplies. So, while some heat-processed materials were necessary for lime mortar and wall bricks usable by a lower-skilled workforce, quantities were minimized as far as possible.

If the tendency of construction was to minimize heat inputs, there were nevertheless spectacular examples of heat being conspicuously consumed in Rome. The baths themselves, their hottest rooms heated to sweating point by hot gases circulated from furnaces in the basement, guzzled fuel. The operation of bathhouses in every Roman city has been suggested as a significant contributor to deforestation around the Mediterranean under the Roman Empire. To retain the heat of both sun and furnace, baths increasingly used energy-hungry glass panes up to seventy by forty centimetres in size. Nevertheless, the total consumption of firewood for bathhouses in a big city may have amounted to 8,000 twenty-five-metre trees each year.[80]

Pagodas

Despite comparable anxieties about wood supplies in the context of the Song-dynasty population and construction boom, embodied heat was used more freely in some important Song-dynasty buildings. Built in 1049, the so-called Iron Pagoda – actually constructed from brick, but with the sheen and colour of iron when seen from a distance – is the sole surviving part of a large monastery just outside the capital, Bianjing. The rest of the monastery, built of wood in the customary fashion, was swept away by a flood in the nineteenth century. The pagoda, having been rebuilt in brick after its short-lived wooden predecessor was struck by lightning and burned down, survived that flood as well as others, and has since withstood tens of earthquakes over the course of nearly a millennium.[81]

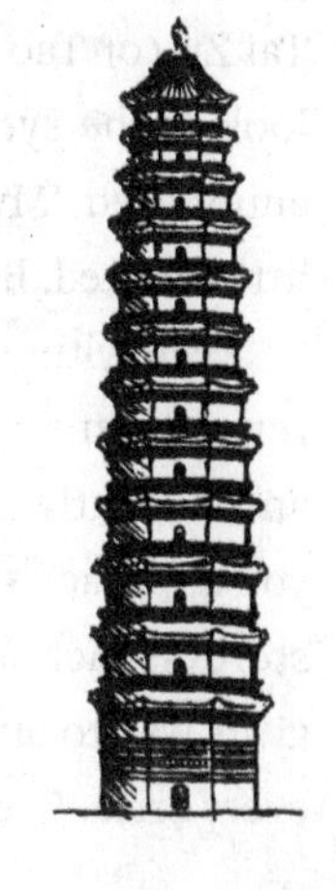

Originally, the pagoda was the tallest structure in a complex of Buddhist monastic buildings that included halls to house hundreds of monks, large open-air assembly spaces, buildings for worship and so on. The monastery was used for the great examination through which people could enter the civil service – an exam so important that the emperor himself marked the essays of the handful of best candidates.[82]

Monasteries in the Song dynasty enjoyed the same sort of spectacular scaling-up seen elsewhere across their society thanks to the energy wealth of the period: the need for a Thousand-Monk Pavilion at Mount Jing Monastery, Lin'an, in 1140 hints at the scale of the major temples of China at the height of the Song dynasty.

The layout of major religious and royal building complexes in imperial China flaunted the luxury of open space – a spectacular contrast with farming and urban landscapes that were characterized by very high densities of population and building. Order too was used as a means to show imperial power. Decades before the arrival of Champa rice, Emperor Tai Zu (or Tao Zu, 960–76 CE) summoned his courtiers to look at the symmetry and clarity of his palace's plan and announced, 'My heart is as straightforward as all this, and as little twisted. Be you likewise!'[83]

Even without its original surroundings, the Iron Pagoda remains an impressive structure, taller than the Pantheon and elegantly slim in relation to its considerable height. As you approach the pagoda, the orderly clarity of its thirteen storeys, each a slightly smaller version of the one below, gives way to an enthralling richness of decorative detail on every brick. Even the apparent sobriety of the coloured glaze

on its facing turns out on closer view to break into bright bursts of reddish brown, yellow and green.

These glazed ceramics were a technological challenge to produce. *Yingzao Fashi* describes an extraordinarily elaborate set of procedures for achieving the level of waterproofing that would make them durable. For tiles, clay was moulded on to a large, textile-covered, bulbous mould, spun like a potter's wheel to produce a cylinder. They would then be cut off in quarters or halves for different positions on the roof and scraped down by hand with a piece of stone or tile to produce a flawlessly smooth surface, removing the pattern of the fabric against which they had been moulded. The tiles were then coated with either talcum powder or white clay before firing. Even after firing there was more work to be done, cutting and trimming the tiles and, for the semi-cylindrical ones, rocking them in a specially made mould to check that they were geometrically perfect.[84]

All of this labour was in service to the main event – the firing itself. To fire clay into ceramic requires a sustained temperature of around 560 degrees Celsius. To get this level of heat throughout a kiln requires an intense fire for which wooden logs are inefficient. *Yingzao Fashi* details the fuels used for tile and brick kilns: brushwood twigs and wheat straws, for example, burn fast and hot because their surface area is high in relation to their volume, meaning lots of contact with the oxygen in the air. These quick-burning fuels were needed in very substantial volumes to run a kiln at 600 degrees Celsius for a day and kilns competed with other needs for heat, including cooking and warmth (the average daily temperature in winter in Kaifeng today is around freezing).

For the bricks in the Iron Pagoda another stage was required: they were glazed. This required a second firing at even higher temperatures (around 700 degrees Celsius) in order to melt a thin layer of powdered coloured glass over the surface, making the finished brick or tile exceptionally resilient to time and weather, and giving it a distinctive and visibly special appearance.

Chinese ceramic workers had a remarkable level of knowledge of how to exploit a dazzling range of chemical processes to produce diverse and high-quality results. For a grey brick they poured water into the kiln as it cooled, producing steam that reduced the oxygen content in the kiln, keeping the iron in the bricks grey rather than the red that resulted without steam. They knew the proportions of lead monoxide, quartz powder, copper powder and iron needed to produce specific browns and yellows, greens and reds, and could come up with stand-ins for key materials when they were in short supply.[85]

The level of specialist technical knowledge of the Chinese ceramicists came from centuries of accretive experience, intuition and trial and error. The knowledge in *Yingzao Fashi* could only arise from a very long period of sufficient food and fuel supplies in the hands of the elite to support a large and continuous demand for specialist pottery workers.

If the Iron Pagoda was an impressive display of energy consumption and craft skill, another group of Song-dynasty pagodas were to compete with it on both. One, from 1061, survives in decent condition at Yuquan Temple, Dangyang, Hubei. It is made entirely of iron. Iron production requires very high temperatures for smelting (over 1,250 degrees Celsius) and even higher ones for melting for casting (around 1,500 degrees Celsius). It is hardly surprising that the proud iron founders of the Yuquan pagoda should have recorded not only the day and time that they embarked on perhaps the most challenging and exciting pour of their careers, but also the exact weight of iron used in the eighteen-metre structure – 38.3 tonnes.[86] This may well have cost more than 200,000kWh of energy to smelt and cast. If this had been done using charcoal made from firewood, the wood demands would have been exorbitant

– even making the initial pig iron, without the hotter process of casting, might have cost more than all the trees in two square kilometres of forest.[87]

In fact, however, this apparent profligacy was probably the outcome of another revolution in Song-dynasty energy supplies. Over the course of the eleventh century, Song industry came to make increasingly heavy use of a rival fuel that burned hotter even than charcoal and which cost almost no land to procure: coal.

Coal, the remains of ancient forest swamps, crushed for millions of years under layers of rock, had been known about in China, Britain and elsewhere for centuries, but had not tended to replace wood on any major scale. It is harder to light; its gases are rank and dirty, polluting food; it burns problematically hot if you are used to wood, but with a shorter flame, and releases chemicals into furnaces that can corrupt industrial processes until complicated redesign is undertaken. Nevertheless, when some of these difficulties were overcome, coal represented a huge supplement to the Song dynasty's ability to conduct heat-hungry activity. Once Chinese metallurgists had found ways of coal-smelting iron, iron tools produced further reinforcements to agricultural and industrial productivity, iron coins enlarged the cash economy and iron pagodas became a now largely forgotten wonder of the medieval world.[88] Once again, with growing energy use came growing energy efficiency, and Song-dynasty ceramic kilns became more and more effective in extracting as much as possible of the useful heat from the fuel rather than pouring a large proportion of it fruitlessly out of a chimney.[89]

It is one of history's great puzzles that Song-dynasty China developed most of the key technologies of the British Industrial Revolution seven centuries earlier, but these advances did not develop into a prolonged self-reinforcing pattern like the growing worldwide exploitation of fossil fuels of the past two centuries. The fall of the Song dynasty to the Mongols in 1279 may have been a sufficient dislocation to derail the remarkable energy developments of the previous centuries.

Decline and Demolition

If Song-dynasty China's political collapse affected its energy systems, the opposite may have been the case in the Roman Empire, where a cooling and increasingly erratic climate seems to have reduced crop fertility and destabilized politics and economics to the point where centralized Roman rule collapsed over most of the European part of the empire in the fifth century CE. The contrast between the energy economies of Rome at its height, complete with substantial Egyptian grain imports, and Rome fending for itself agriculturally with a worse climate is nowhere clearer than in architecture. The great monuments of ancient Rome were pillaged in medieval and Renaissance times for every scrap of resource that could be recovered from them. This included modifying old buildings rather than making new ones. The Arch of Constantine on Rome's Via Triumphalis is one example of a monument that survived because it was built into a medieval fortified house to save on construction labour and materials, and many other ancient buildings like the Colosseum still show the beam holes that once supported later insertions of wooden floors which turned ancient corridors into medieval housing, farm buildings or storerooms.[90] Medieval Romans removed stone and brick from disused buildings to save on quarrying or firing more. They also took the metals. In the early centuries after the fall of the Western Roman Empire the solid bronze roof tiles of the great temples were removed by rulers hungry for this valuable metal, and lead was stripped from the roofs of disused buildings all over the empire.[91] Medieval Romans even mined from deep in the stone walls the lead-covered iron ties that Roman masons included as standard reinforcements to masonry.[92] By the Middle Ages, the energy that had long ago smelted these small pieces of metal was sufficiently valuable to justify digging out mere morsels of iron and lead.

Imperial Rome and Song-dynasty China, so different in their political histories, both burned brightly when their energy supplies were large and rising, their building booms improving both the technical proficiency and the efficiency of their construction industries. Rome's boom was to prove the more transitory, and the next chapter will go on to look at the history of the Mediterranean over the following centuries, where rulers haunted by the ruins of Rome's great buildings vied to make their own eternal monuments.

CHAPTER 5

'A proportional indicator of power'?

Tradition, Energy and Mosques

This chapter looks at a new building type that emerged in the seventh and eighth centuries CE to serve the needs of a new religious and political presence, Islam. The Muslim conquests were among the largest and fastest political and cultural shifts in agrarian history: a new religion, a transformed political map and shifting networks of trade and cultural exchange. This chapter will look at a handful of major mosques built by Muslim rulers in different places with distinct agricultural, economic and cultural settings. They offer us the chance to see the new building type being adapted to local conditions across some of the wide diversity of cultural and energy contexts touched by Islam in its first millennium.

Compared with the expansion of the Roman Empire over several centuries, the Islamic conquests of the seventh and eighth centuries CE were remarkably fast. In under 130 years Muslim armies originating initially in the Arabian Peninsula took control of an area from Yemen up to the Caucasus, and from the borders of India to Morocco and Portugal, a territory that spanned more than 8,000 kilometres from one end to the other. Islamic faith contributed to a sense of unity and confidence for this sensational campaign, which, among numerous other conquests, took over the Sassanian Persian Empire and cut the Byzantine Empire – the Roman Empire's continuation centred on Constantinople – down to a fraction of its former size.

The features which typically unite mosques may have originated within the lifetime of Muhammad, and were certainly coalescing into a building type with its own regular characteristics, present throughout the newly conquered lands, within very few decades of his death.[1] Mosques had, and continue to have, a crucial role in Islamic spiritual life. The five daily prayer sessions that are a core requirement of Islam can take place anywhere clean, but for the main Friday prayers the early Islamic aspiration was for the Muslim population, particularly the men, of a city to gather together in one place to pray as a single body, with differences of wealth and rank subsumed into a common religious identity and a shared spiritual experience. The reality was to become considerably more complex, but the buildings discussed in this chapter are all substantial congregational mosques aimed at large Friday prayer gatherings.

The fundamental functional requirements of congregational mosques are not especially restrictive in architectural terms: clean, special prayer space, facing towards the holy city of Mecca. Ideally, worshippers are divided as little as possible by walls, to allow the congregation to pray shoulder to shoulder. Generally there is a fountain, as Muslims are required to wash before prayer.[2] This relatively open brief enabled different typologies of mosque to emerge in different cultures as they adopted Islam.

The Making of the Mosque: The Great Mosque of Damascus

Our first building in this chapter was part of an ambitious project, somewhat in the mould of Parsa or the imperial buildings of Song-dynasty China, of developing and solidifying a cultural identity over a large land mass. The sudden acquisition of political and economic power over a vast area – over perhaps as much as a third of the world's population – brought threats to the identity of the conquerors as well as to the conquered: men who found themselves rapidly elevated to the ranks of the most powerful and richest rulers on earth had to forge a new identity that was at once faithful to the Qur'an's teachings and capable of stabilizing their political hold over diverse regions. Once again, architecture proved a powerful tool in shaping the new idea of a mighty Islamic caliphate.

The first caliph (religious and secular leader) of the Umayyad dynasty, Muawiya I (ruled 661–80 CE), chose as his capital not one of the great metropolises of his new lands, but a less important regional city within the territory recently conquered from Byzantium: Damascus. Whereas the Muslim garrisons of many other conquered cities lived in semi-separate forts, Damascus was relatively stable, with conquered and conquerors living and worshipping in the same streets and public spaces.[3] In addition, the city was well set up for comfort, with a pleasant climate, good drinking water, excellent fruit and famous public baths.[4] It was well located to keep in contact with an empire of such reach, while not being as vulnerable as a coastal city to sudden attack from Byzantine fleets on the Mediterranean.

Ruling from Damascus, Abd al-Malik ibn Marwan (caliph 685–705 CE) was a radical innovator and reformer. Previously the Arab Muslim conquerors had ruled with a light touch, allowing local laws, religions and political structures to remain substantially as they had been before the conquest, on the condition that subjects paid tax to their new rulers.[5]

Abd al-Malik's reforms brought the caliphate a more centralized structure of rule, especially through one great administrative reform, making Arabic the language of government across all his domains, and one financial reform, the introduction of his own system of coinage.[6] He and his successors increased the agricultural productivity of Syria and Palestine through ambitious new irrigation canals and the widespread introduction of good new crops.[7] In towns this improving energy base was reflected in provision for growing commerce: shopping space was built into existing city streets and whole new market streets were built.[8]

In architectural terms too Abd al-Malik stamped his presence firmly on important cities, building the Dome of the Rock in Jerusalem as a spectacular marker of Muslim power on the most sacred site of a holy city shared and sometimes contested between Jewish, Christian and Muslim believers. The decoration of the Dome of the Rock has been changed significantly in later renovations, but the beautiful core goes back to Abd al-Malik's time: a dome standing on columns and arches around the exposed sacred bedrock, with an outer ring of columns and piers unobtrusively converting

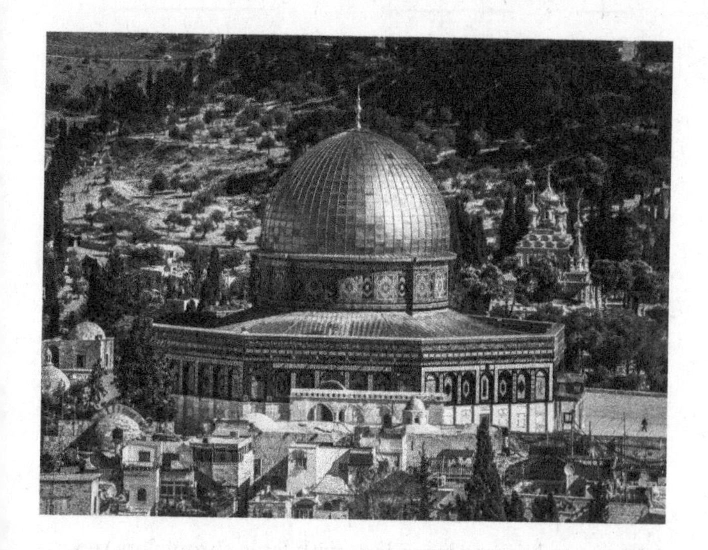

the circle of the dome into the regular octagon of the outer walls. The geometry of the plan is deeply satisfying, a sophisticated overlapping of rotated squares setting out the positions of columns, walls and dome. The simplicity and clarity of the architectural form provide a stunning counterpoint to rich decorative surface treatments of stone, mosaic and tile, added and reworked over subsequent centuries.

Abd al-Malik's son and successor, al-Walid I (caliph 705–15), built on the agricultural, administrative and architectural programmes of his father. His dominion took in many of the greatest cities and buildings of the ancient world, and it's possible that al-Walid wanted to show that the Muslim empire too could not only conquer, but also commission great monuments as reflections of the faith. He built magnificently at Islamic sacred sites in Medina and Jerusalem, and in his capital at Damascus he built a prodigious new sacred space for Friday prayers: the Great Mosque.[9]

Before al-Walid, Friday prayers in Damascus took place in an existing open area within the city: the courtyard around the Christian Cathedral of St John the Baptist, which still accommodated Christian worship, but fortuitously had a long wall correctly oriented to face towards Mecca.[10]

Al-Walid demolished the cathedral, returning the city's confiscated churches to Christian use in compensation. He turned the whole courtyard into a new mosque with a prayer hall considerably bigger than the church had been: a clear statement of confidence and ambition at the political heart of the caliphate. The size, complexity and expense of the project was to become legendary, with later chroniclers reporting in awed tones that even to move the receipts generated by the job was a major undertaking: eighteen camels were required to carry them all.[11]

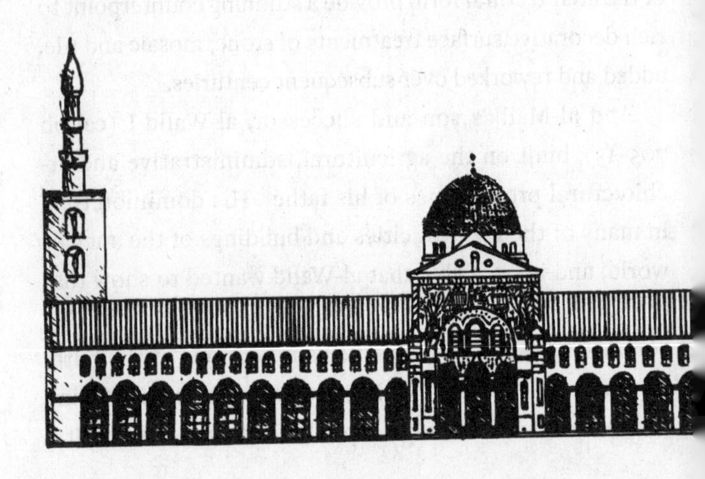

Al-Walid was just the latest in a series of rulers to change the religious meaning of the site. The first known use of the precinct had been as a temple to a Syrian god of storms, but the Roman authorities had replaced this with a temple to Jupiter, enclosing their building with an imposing outer courtyard wall, much of which survives in the present mosque. As the Roman Empire Christianized, the new Church authorities tore down the temple to build the cathedral, which in turn gave way to al-Walid's mosque.[12] This multiple layering of activity by different religious traditions on a single site is a story that recurs all over the world, from Aztec temples replaced by Catholic churches to pagan, Jewish, Christian and Muslim use of sacred places in the Middle East.

The new mosque made clever use of its site. Opening up the space by demolishing the cathedral, the designers created a large courtyard surrounded by a colonnade and leading into a broad, covered prayer hall. Amid the narrow streets of the densely built-up city, the feeling of space on entering the mosque was and still is a piece of spectacular theatre.

The vast investment of money and skilled effort that the building represented was to make it a yardstick for important mosques for centuries to come, and the format of courtyard and prayer hall that it explored so effectively was to prove lastingly influential.[13] It accommodated itself well to the large and growing Muslim population of a great city, producing on a new scale the thrilling experience of a unified crowd sharing a sacred ritual. It also succeeded in being clearly a new building type for the powerful new empire; it occupied the remaining walls of the Roman temple precinct, yet you could not mistake the Great Mosque of Damascus for the church or temples that went before it. This was an original and distinctively Islamic type of building.

Yet if the shape of the building was a new departure, the engineering and decorative techniques employed the most sophisticated craft traditions already present in the region. The arches, domes, mosaic and marble of the new mosque are related to Byzantine crafts. Mosaicists and marble workers of Byzantine Syria had inherited techniques and artistic skills from a well-networked Roman Empire. Their work was beautiful and durable. In designing new structures, the leading local masons were highly proficient in making slim, elegant arches stand atop tall columns. The centre of the prayer hall exploits similar structural principles to the gable of

Byzantine Syrian churches, complete with a dome above. The rest of the prayer hall, extending to each side, is structurally comparable to an exceptionally long Byzantine church.[14] Its columns are Roman monolithic column shafts with Roman capitals, taken from older buildings in the city to support and adorn the new mosque, just as was happening in the city of Rome at the same period in the construction of Christian churches.[15]

The craft that went into the Great Mosque was exceptional in both quality and quantity – so much so that it was an ever-present fixture of many rival lists of Wonders of the Islamic World. In one it even featured twice, its dazzling mosaic decoration counting as separate from the rest of the building.[16] Lower on the façade is marble revetment, almost certainly removed from Roman buildings for the purpose, used in the mosque in careful continuation of the Roman craft of placing slender veneers in such a way that the grain of the stone produces beautiful, orderly patterns. Marble window grilles in the new building follow geometrical patterns of beauty and mathematical ingenuity, carved into the brittle stone. To sculpt such delicate bars in marble was technically challenging, and the repetitive geometry left nowhere to hide if a slip, or a fault in the stone, broke a piece off.

It is unsurprising that al-Walid used craftspeople who had inherited longstanding local traditions. Construction seems to have been the second-largest economic sector after agriculture itself in many agrarian societies, and complex supply chains of stone, brick, building timber and lime for mortar developed gradually over centuries. If you wanted to build impressively in eighth-century Syria, drawing on Byzantine materials, engineering and decorative crafts was the obvious way to do it.

The Great Mosque, fundamentally new and revolutionary in its form, but with strong continuity in detail, turns out to echo closely the political and economic reforms of al-Walid and his father. There too the changes were ostentatious, with spiritual and political significance, and a strong role in asserting a new Islamic identity. Yet continuity was also much more normal than change. Agrarian societies respond slowly, and often reluctantly, to change and the many political powers that had risen and fallen in the Middle East since Uruk had presided over great continuity of local hierarchy and farming practice. Like other empires which had gone before, the political unity of the caliphate proved impermanent. Some of the great Islamic powers that emerged in subsequent centuries echoed the borders and cultures of pre-Islamic empires in the same regions: Iran retained the Persian language, while the Ottomans ended up ruling much of the Byzantine Empire, whose capital, Constantinople, they took in 1453. As we will see, their religious architecture in each case responded to local architectural traditions.

The World's Richest Man

As Islam spread, continuity of local building traditions is seen across a range of different energy contexts. Mansa Musa (c. 1280–c. 1337), the emperor of Mali, whose military, economic and political power dominated West Africa, built a new Friday mosque, Djinguereber, in Timbuktu in the 1320s. He gave the order for the new mosque to be built on his journey home from Mecca, where he had been on the hajj in the most spectacular fashion, flaunting his fabulous wealth in gold, with perhaps a hundred camel-loads of it, and taking with him a retinue of tens of thousands.[17] Mansa Musa brought back with him from the Hijaz – the region that contains the holy cities of Mecca and Medina – a scholar and artist originally from Muslim Spain called al-Saheli, whom Musa commissioned to oversee the design of the mosque.[18] Bringing back a scholar from his hajj may imply that Mansa Musa wanted to imbue his new building with the authority of the holiest places in Islam.

This mosque was similar in its fundamentals to the Great Mosque of Damascus, with a walled courtyard, a minaret

and a covered prayer hall. Stone arches are used in part of the mosque, which may represent a gesture towards the construction technologies of the Mediterranean and the Red Sea. However, most of the building was to be erected using more typical local materials and techniques.[19] The structural walls and piers were built largely of *banco* – local clay mixed with straw to make a sun-baked brick. Palm-tree trunks and branches provided the roof structure, over which palm matting was laid.[20] Local families of masons, who still play a significant role in the refurbishment of Timbuktu's

mosques, appear to have a long history in the city, performing animist rituals in association with construction projects. Supernatural powers are ascribed locally to the masons, who retained most of the craft specialism required for construction projects.[21]

Lower dependence on subspecialization in the Timbuktu building trades (as against Damascus, for example, where several different types of specialized trade were active just within the stonework) accords with its energy context. Large areas of the Middle East and the Mediterranean had patterns of climate and soil which permitted reliably high-yield farming over very long periods, offering a very good return on investment for complex irrigation systems, terracing of land and other labour-intensive but productive interventions. The cities supported by the resulting harvests could feed large populations of non-food-producing people, encouraging specialization and subspecialization of many different activities, including architectural production.

West Africa, by contrast, has for the most part relatively challenging annual patterns of rainfall for grain farming, encouraging the use of more reliable but lower-yielding millet and sorghum rather than high-yielding but easily drought-affected cereals like wheat. In addition, West Africa has for tens of thousands of years had a cycle of changing climate where dry periods have seen the Sahara Desert expand southwards by hundreds of kilometres, then retreat northwards in wetter centuries.[22] The combination of relatively low-yielding crops and longer-term climate instability discouraged the development of long-lasting, large cities across much of West Africa. Cities sprang up repeatedly, but then shrank or vanished, probably when the climate rendered them unsustainable or surplus to requirements.[23]

Mansa Musa's spectacular wealth came, unusually in this book, from an arid, less fertile period in West African history. Political borders have often been related to ecological borders in West Africa. Grassland pastoralists from the north were unable to travel into the wetter southern regions because the tsetse flies that thrive there were lethal to their cattle and horses, upon which they depended. This natural frontier protected rainforest populations from military domination by northerners.[24] The dry climate of the thirteenth century allowed Malian cavalry to reach further south, closer to the goldfields that had previously lain deep behind the natural defences of tsetse flies and rainforest.[25] The gold trade between the southern parts of West Africa and the gold-hungry Mediterranean grew very important, and Mansa Musa controlled it. His wealth in gold was such that some modern historians maintain that his spending and donations when on pilgrimage to Mecca were so lavish that they depressed gold prices in Egypt for the following twelve years.[26]

For all his wealth, however, Mansa Musa presided over land whose long-term energy instability was a powerful disincentive to over-investment in agricultural improvements and population expansion. What was seen by later European invaders as a failure to develop high-energy forms of complexity is much better understood as a successful long-term adaptation of West African society, allowing individuals and groups to remain more mobile and making them less dependent on a fixed metropolis than on slowly moving ecological boundaries.[27]

In this context, the limited body of specialists required for the construction of Djinguereber was advantageous. The supporting bulk labour could be provided seasonally by people whose primary occupation was food production, so they would not be vulnerable to a decline in the building market. *Banco* is very good for houses, the essential building type of any city, under the hot sun of Timbuktu: thick *banco* walls with few openings retain some of the cool of the night through the heat of the day, then radiate the heat back out through the colder night.

Banco is considerably less durable than marble or fired brick, with both rain and wind-blown sand abrading its surface substantially in a matter of months. In the case of Djinguereber, this limitation has been counteracted by a combination of occasional major reconstructions by pious rulers, and steady maintenance in between by the people of Timbuktu. At present the cycle of maintenance is typically over two years, with the imam of Djinguereber determining when restoration is needed. He solicits contributions from the congregation, consisting of paid work whose proceeds are then donated to meet rebuilding costs, or direct contributions of materials (windows, doors, drainpipes or palm-tree trunks). Local transport cooperatives provide free transport for the *banco* and other materials that will be needed. The project itself takes place on a Sunday, with everyone getting involved – those who do not volunteer to do so are sought out and covered in *banco* for shirking their religious and social duty. The leading mason, wearing an incense-holder filled with magical ingredients, climbs the minaret to the religious chanting of the crowd and applies the first load of *banco*. Unskilled volunteers fetch and carry *banco*, water and other materials, and at the close of the day's repairs there are more prayers and appropriately celebratory feasting centred on the masonic families.[28]

Feet of Clay?

The level and stability of agricultural productivity in his empire helped to shape the buildings of Mansa Musa, but energy context also exerted pressures on rulers in higher-fertility regions. Complexity and scale increase fairly easily in building traditions, but are hard to reverse, perhaps thanks to a kind of long-term peer pressure. An English writer of the twelfth century commented with unmistakable disapproval on the trend for very large buildings that had come with the country's new Norman kings and which 'now almost all men emulate at great expense'.[29]

When Shah 'Abbās I (1571–1629), the Safavid ruler of Persia, commissioned a new mosque in Isfahan in 1611, he was bound into an Iranian tradition of large, technically complex, expensive construction. As the great Tunisian historian Ibn Khaldun (1332–1406) had written: 'Monuments are a proportional indicator of power, and dynasties are remembered for their monuments. The monuments exist by the power which founded the dynasty, and the vestiges left by that dynasty are proportional to its power. The monuments of any dynasty are its building projects and principal edifices.'[30]

'Abbās had both military and cultural rivalries with the Ottoman Empire to his west and with a range of powers on his eastern borders. He moved the main base of his rule to Isfahan, safely remote from the military threats on each side.

Isfahan was an oasis city on a plain more than 1,500 metres above sea level, most of which was too dry to support high-yielding agriculture.[31] To feed the city's planned growth, 'Abbās embarked on a major irrigation scheme to bring water and fertility – a pattern reminiscent of the supporting irrigation programme for Parsa a millennium earlier and 320 kilometres south. Part of 'Abbās's scheme came about, but a considerably larger plan to divert a second river into the one that watered the plain proved too challenging and was abandoned.[32]

Nevertheless, the new irrigated farmland was enough to support an increased scale of economic and political activity in the city. To convert some of the new food supply into profitable luxury goods for export, 'Abbās had families of skilled silk workers moved from Armenia.[33] To house his court and aristocracy, he built an urban extension on the grandest lines. A broad ceremonial route through the city was flanked

by the houses and gardens of wealthy courtiers. The road crossed the river on a large, handsome bridge.[34]

The new urban area centred on a public square 160 metres wide and 560 metres long called, with sweeping ambition, 'the square on the plan of the world' (Maidān-e Naghsh-e Jahān).[35] Around it were a very large palace, a royal mosque at which the court could worship daily (accessed from the palace by a secure private tunnel under the square) and a bazaar for trade and shopping. The square is bounded on all sides by a two-storey arcade, its orderliness and continuity a display of royal power amid the winding narrow streets of the city. The only interruptions to the regularity of the arcading are for the entrances to the great royal buildings, where beautifully vivid blue and turquoise tiles shine out amid the beige of the arcades around. The great square hosted festivals, markets, public executions and polo matches (the original marble goalposts still stand).[36]

Dominating the square from one of its short ends were the dome and entrance gateway of a new mosque for communal Friday prayers. The mosque faces Mecca, of course, but since the square was oriented at a different angle, the dome and minarets of the mosque make a more dramatic and enigmatic appearance above the wall surrounding the square.

The entrance gateway of the mosque is the most impressive of all the gates of the square. The arcade steps

back to make a polygonal open-sided courtyard funnelling the faithful into the gate. Every surface is richly decorated in bright tilework and two minarets rise above. Each is inscribed with religious texts and, paired with the two minarets within the courtyard of the mosque, they contribute to a conspicuously impressive skyline.

Above the doorway is a far larger porch, shaped like a domed room that has been cut in half – a structure known as an *iwan*. The *iwans*, domes and courtyard of the Shah Mosque are related to earlier mosques in the region, which, in turn, may have partially derived these motifs from similar features used in pre-Islamic Persian palaces like that at Ardashir, built in the early third century CE.[37]

The entrance *iwan*'s half-dome is decorated with a series of stalactite-like hanging indentations, a decorative treatment known as *muqarnas*. The complexity involved in calculating the geometries of these decorations on a curved surface, followed by the craft skill to produce and attach the tilework for them, are spectacular displays of the greatness of Safavid Iran's architectural tradition.

Muqarnas was a distinctively Islamic decorative form seen from Iran to Spain, though there is considerable uncertainty over where it originated.[38] It proved a pleasing way to resolve the unavoidable complexity of making round domes sit on square or polygonal bases – the transition between the two is always structurally and aesthetically difficult and partial smaller domes and arches in each corner were one way of making it work. The result was so beautiful that it was frequently continued across entire vaults.

Inside this stupendous gateway, a narrow passage ingeniously turns the visitor through the necessary angle to reach the Mecca-oriented main courtyard of the mosque. Even after the great square this courtyard is impressive. The two-storey arcade of the square continues here, but now covered all over with colourful tiles. There are large *iwans* in the centre of each side of the courtyard. Their large, complex, curved surfaces provide opportunities for further tiled decoration, virtuosic in geometry and craft. Thanks to the small scale of the ornamental motifs and the simple clarity of the overall shape of the building, it is muscular in its architectural effect despite the delicacy and ubiquity of the decoration.

Within the prayer halls the vaults and dome are again covered in blue and yellow tiles, their patterns inspired by plants and geometry, accompanied in key locations by beautiful calligraphy – forms of decoration which respected the Islamic tendency to avoid religious imagery that depicted animals and people. The display of energy wealth represented by so much kiln-fired brick and tile would have been much more legible to its original audience than it is to modern eyes, accustomed as they are to ubiquitous, cheap, high-quality ceramics produced by low-cost fossil fuels.

The magnificence of 'Abbās's building programme at Isfahan might suggest that he was boundlessly wealthy, yet there are indications that it in fact stretched his resources problematically. He was sufficiently worried to ban exports of coinage in an attempt to bolster a struggling economy.[39] Similarly he quietly economized on his mosque, using weaker foundations than it needed and compromising on the quality of the tiles.[40] The mosque has had to undergo very substantial restoration in the 1930s and since.[41]

Keeping up with the Romans

If Shah ʻAbbās was pushed by his context to compete with the ancient buildings and cities of Persia, the monumental architecture of much of the Mediterranean found itself for centuries under the long shadow of the great ancient Roman energy boom, with its optimal climate and abundant imports from a particularly fertile moment in Egyptian history. The first centuries of the New Rome, that is Byzantine Constantinople, saw monumental construction on a scale that seemed to make it a worthy successor to the former imperial capital. The most spectacular Byzantine project was the magnificent cathedral of Hagia Sophia, built under the emperor Justinian in 532–7 CE, when his empire still included Egypt and many other fertile lands, though with a less ideal farming climate than the earlier empire had enjoyed. The span of its magnificent central dome is smaller than the Pantheon's at around thirty-one metres, but Hagia Sophia is higher, and its more delicate-looking structure admits more light through more windows. It had a much bigger internal volume than the Pantheon. Hagia Sophia is different from the buildings of imperial Rome, but still very ambitious and technically impressive.

Even Justinian, however, seems to have been haunted by the prodigious Roman buildings that dotted urban and rural landscapes alike. He sent out an order to strip marble and other beautiful cladding-stones from pagan temples all over his reduced empire in order to furnish materials for Hagia Sophia.[42]

For centuries afterwards, rulers across the former Roman Empire busied their more modest building forces with stripping the best stones from Roman ruins and using them to clad new projects. These reused stones, known to historians as *spolia*, appear in projects from the early churches of the city of Rome and in Sicily, where column shafts of rare porphyry were recarved into decorative roundels and even sarcophagi, to the twelfth-century cathedral of Pisa; and from the Great Mosque of Damascus to the mosques and churches of Spain, France and the coast of North Africa.[43]

Ruins which were accessible by water were preferred, for removal by ship of their marble revetments and column

shafts up to about three metres. Substantial ships capable of carrying the largest monolithic shafts seem to have been very rare in the Middle Ages, but smaller columns could be craned aboard more modest vessels, like the spare wooden masts that they often carried.[44] When an army led by Venice captured Constantinople in 1204 they took away marble and other valuable stones that in some cases may already have been taken from Egypt to Rome and from Rome to Constantinople. By the sixteenth century the Ottoman rulers who had conquered Constantinople and renamed it Istanbul were able to obtain, transport and erect three columns of thirty Roman feet (8.8 m) for the magnificent Süleymaniye Mosque. One of the giant shafts came from Istanbul's own palace stone stores, one from Baalbek in modern-day Lebanon and one from Alexandria, brought on a barge.[45] The three columns were joined in the new mosque by a fourth, originally even larger, which was cut down to match the other three. An eyewitness account of the process of demounting this giant from its previous location in Istanbul makes clear how exciting a process this was, with numerous timbers as big as the masts of ships being erected all round the column, which was then dislodged by a workforce of 'thousands' of galley slaves and trainee soldiers, some pulling on ropes, some marching on treadmills, to tighten ropes as thick as a man's body. When the shaft reached its tipping point and its weight, perhaps twenty tonnes, yanked on the restraining cables, 'sparks streamed forth from the [iron] pulleys like a thunderbolt' and fibres from the straining ropes flew off in clouds of fluff.[46] It was the spectacle of construction at its most operatic.

The mix of cultural preference and energy-saving expedience in such practices of reuse is impossible to untangle. Clearly it was generally less effort and required fewer types of skill to reuse existing well-made marble cladding than to quarry and saw it from scratch. But associations with the glories of the past were also felt strongly by observers like the tenth-century Kurdish geographer Ibn Hawqal, writing about Alexandria with understandable admiration of:

> a host of antiquities and the authentic monuments of her erstwhile inhabitants, eloquent testimony of royalty and power, which proclaim her domination over other countries, her grandeur, and her glorious superiority . . . immense columns and all sorts of marble slabs, any one of which could be moved only by thousands of workmen, and which are hoisted between earth and sky [. . .] The whole is decorated with astonishing effects and prodigious colours. Such remains represent the past.[47]

One source claimed that porphyry *spolia* were worth more than their weight in gold in the fierce Mediterranean market for the purple stone, which was now available only from Roman ruins after the closure of the Roman quarries amid the political and economic turbulence fostered by the cooling climate in the fifth century.[48] The Ottomans were sufficiently aware of the finite quantity and preciousness of their Roman stones to feel it was worth making it illegal from 1577 to sell marble or porphyry to non-Muslims.[49]

In Istanbul, with its prodigious pre-Islamic architecture, a succession of Ottoman sultans and their families built impressive palaces and mosques, their architecture responding to and adapting the Byzantine masterpieces they had inherited. In particular, the scale, engineering and amazing interior space of Hagia Sophia rapidly became an inspiration to the mosque builders of the city, and the work of Ottoman architects repeatedly transformed and adapted the precedent with profound sophistication, in a rich dialogue with the aristocratic and royal men and women who commissioned numerous projects.[50]

Building a great mosque was both a spiritual good work and a route to lasting admiration. The result was a period of intense mosque building concentrated in the capital. The greatest of the Ottoman mosques, the Süleymaniye (1550–57), designed by the royal architect Mimar Sinan (c. 1490–1588), had a workforce that averaged up to 3,000 people during the three seasons when construction work was at its peak, over a period of eight years.[51] *Spolia* came from all over the empire for the project and other materials were also shipped into the city: brick came more than 140 kilometres from Hoşköy and Gallipoli, timber from 250 kilometres along the coast and iron 470 kilometres as the crow flies from Samokov in modern Bulgaria.[52] Ottoman royal officials compiled lists of workers and their skills to aid in rapid recruitment of good workforces and also catalogues of building stones and column shafts available around the empire, complete with samples.[53]

Some of the results of this scale of project were positive: professor of Islamic art Gülru Necipoğlu argues that the complex networks involved in securing the labour and materials for grand buildings not only represented a valuable display of imperial strength, but also aided political integration across provinces.[54] Royal decrees that a proper price should be paid for materials suggest a concern that the scale of building activity might damage local economies or cause problematic ill feeling. The same tenderness did not apply to the workforce: masons and carpenters could be brought to the site under guard, perhaps because conditions or pay rates on other jobs were sufficiently tempting to make workers consider deserting major royal construction sites that paid fixed wages.[55]

Whatever the economic reality, contemporaries were concerned that cultural and religious incentives to build large mosques were sapping the energy of the city and the empire. An influential theologian proposed in 1581 that only sultans who had secured substantial military triumphs over non-Muslim armies should build grand mosques in the capital.[56] They ought to be a celebration of enrichment to the empire and the faith, rather than a drain on the substantial but finite resources of the existing realm.

At the start of the seventeenth century the young Ottoman sultan Ahmet I (1590–1617) went against this injunction. He was anxious to demonstrate that, militarily unproven though he was, he was already a worthy successor to ancestors who had ruled Asia Minor, much of the Middle East and Egypt, and had captured Byzantine Constantinople, taking armies as far north and west as Vienna. To demonstrate his piety, power and competence, Ahmet embarked on a very large new mosque in a prominent location which would dazzle visitors entering Istanbul by boat along the Bosporus.

Ahmet's limited military success was not his only disadvantage: it appears that the energy base of the empire was less healthy by the time his project began than it had been

On the left, the Sultanahmet Mosque,
on the right the former Byzantine cathedral of Hagia Sophia

half a century earlier when the Süleymaniye Mosque was going up. Relations with many regions of the empire were strained, and there are signs of a declining rural population and the abandonment of some settlements in the plains.[57]

The Sultanahmet Mosque reflects these problems in its less ambitious sourcing of materials: stone from the shores of the adjacent Sea of Marmara, marble quarried by enslaved people on an island in that sea and roof tiles made in the immediate vicinity of the city.[58] Ahmet may have been pushing exploitation of his local resources harder in the context of declining energy levels across the empire. Nevertheless, his architectural achievement at the Sultanahmet Mosque indicates clearly that increasing limitations on his political and military power did not cause a sudden collapse in the building expertise available to him. A struggling empire with a long-standing habit of very substantial and high-quality construction projects can still muster extensive supplies of skilled labour and specialist knowledge. Some observers, both Ottoman and Western, described the Sultanahmet Mosque as the most beautiful in Istanbul – an impressive claim in a city with such fine competitors.[59]

Often called the Blue Mosque because of its bright internal tiles, the Sultanahmet Mosque was another variant on the familiar layout seen in early form in the Great Mosque of Damascus: courtyard and prayer hall, minarets and dome, all oriented so that the praying faithful would face Mecca.

Ahmet courted controversy by giving his mosque six minarets – four was the normal number even for an important mosque.[60] The only other mosque with six at the time was the mosque of the Ka'aba in Mecca, the holiest site of Islam. During construction too Ahmet took every opportunity to establish the exceptional importance of his building, holding numerous religious ceremonies to mark special dates and momentous stages during its construction.[61]

At the Sultanahmet Mosque, as seen so often in this book, glamorously profligate use of space is an important display of wealth and grandeur. Despite the density with which most of

Istanbul was built up, with narrow streets and every square metre of land valued and exploited, the leading mosques of the centre adopted an opulently spacious landscaping, standing in open space within a walled garden.

The location of the Sultanahmet Mosque within Istanbul had Byzantine resonances in its dominance over the ancient horse-racing track, which had become an important public space in the life of Constantinople. The location was clearly important to Ahmet, who paid a lot for the site, and was criticized for building a capacious mosque in an area with so few residents and the huge Hagia Sophia church, by then a mosque, already catering for them. Elsewhere in the Ottoman Empire far larger populations crowded in to worship in smaller, less impressive mosques.[62]

The relationship of the Sultanahmet Mosque to its city surroundings is strikingly different from that of the Shah

Mosque to its surroundings in Isfahan. Architects today discuss this under the heading 'figure/ground'. In Istanbul the mosque is the figure, the object of interest, standing in an essentially leftover open space that acts as background to its architecture. In Isfahan the square and the mosque courtyard are two open spaces that are the object of interest against a background of continuous dense-grained city.

As the examples in this and other chapters have shown, innovation and change in farming societies tended to be intricately connected not only with the practical limitations and strengths conferred by energy context, but also with the cultural preferences for continuity and admiration for the past that were common features of the agrarian world. Even within the world of architecture today many clients and many architects have pronounced attachment to stylistic motifs that originated in earlier energy settings. The next chapter will look at the ways in which the architecture of Rome was to influence hundreds of years of buildings in western Europe, as the region's access to energy fell and rose again.

CHAPTER 6

Plague and Prosperity

Medieval and Early Modern Europe

Early Medieval Europe

Compared with the Muslim Mediterranean, the north-western provinces of the former Roman Empire suffered a profound economic and energetic contraction that must have seemed catastrophic to many. By the time the Great Mosque of Damascus was under construction, the ordinary people of much of Europe spoke what a Roman would have seen as messy, impure dialects of Latin which were to become French, Spanish and so on. Further north Latin had disappeared as an everyday language. It clung on as a specialist language associated with the Christian Church, which after a period of retreat was expanding once again among the pagan tribes that had conquered the former Roman territories.

The Roman cities of the northern provinces were abandoned, or were shadows of their former selves. Huge areas of once-productive farmland had fallen out of use and been reclaimed by wild forest. Things were sufficiently rough in England for a church rule-book to specify what ought to happen if your pig or your chicken ate from a human corpse.[1] Rome's overarching political control had been succeeded by a patchwork of local warlords whose power rose and fell with their individual charisma and competence in an unending sequence of battles between neighbours.

In England, very few buildings from this period survive in anything like their original state. Those which do are small, simple and often constructed by reusing basic materials (brick and ordinary building stone) taken from Roman buildings. Even after several centuries of recovery, architecture remained simple. A late tenth-century church tower at Earls Barton, Northamptonshire, survives. There is no sign here of the Roman building tradition of *opus latericium* or indeed of fired tile or brick at all.[2] The building is of stone, but even there the more challenging techniques of Roman builders are avoided where possible. Arches, the defining Roman structural technique, are mostly not true arches here. Rather than separate stones resting on each other to achieve a wider span, the little openings that look like arches at Earls Barton are single stones carved to look like arches, or in some cases two straight stones leaning on each other as in a house of cards.

These openings are few in number and most of the tower would be blank wall but for a decoration of protruding stones. These are built well into the masonry and were probably intended to reinforce the stonework, but the pattern in which they are arranged – in so far as something so irregular could be called a pattern – is more like a timber structure than Roman masonry.[3] Motifs are copied, presumably from extant or ruined Roman buildings, but the whole tower gives the impression that the people who built it had not built many others like it before.

Although the work is cruder than that in previous chapters, it was still clearly produced with a fair amount of labour and design effort. The stones which were intended to be exposed on the outside seem to have been brought all the way from Lincolnshire, about fifty or sixty kilometres away.[4] And the tower is sufficiently well built to have survived, with relatively modest repairs, for over a millennium.[5] However, if you compare the church's level of craft with the illuminated religious manuscripts or jewellery of the period it is immediately apparent that the illuminators and jewellers of this society were operating at a much higher level of expertise in competitive artistic traditions, in which the best work was the product of not only individual years of practice, but also

centuries of accumulated skill and knowledge. The amount of monumental stone building in this period seems not to have been adequate to produce a comparably impressive body of stonemasons.

The reasons for the disappearance of Roman building technologies are complex and, like much else about the period, largely mysterious. They probably include the fact that the Germanic tribes that dominated England in this period had little tradition of monumental stone building, and did not necessarily rush to acquire one. Their wooden buildings, which do not survive beyond occasional archaeological evidence of supporting posts, may have been more refined than their stonework. Widespread political instability may also have been a disincentive to invest heavily in construction: valuables you could carry with you on horseback would survive a raid by a neighbouring king as long as you did, whereas – as Yeavering, a royal site in Northumbria, shows – buildings could be burned down and rebuilt several times in a matter of decades. Archaeological evidence suggests that the complex was destroyed by fire and rebuilt at least twice during the seventh century, before its final abandonment around 685 CE.[6]

What the church tower at Earls Barton certainly demonstrates, however, is the distinction between a region that maintained its complex energy system – like Damascus under al-Walid, where existing traditions of construction and decoration were simply turned to new uses – and a region that saw a radical drop in energy capability, accompanied by the near-disappearance of urban life and abandonment of established crafts. Why had this happened in northern Europe? No one knows for sure, but it seems likely that changing climate and reduced crop fertility played a significant part.

Rome had risen to greatness during a period of gently warming climate, with agricultural productivity rising gradually as Roman political and military power expanded. The peak of Roman power, between about 100 BCE and 200 CE, was a period of very benign and stable climate.[7] The decline of the Western Roman Empire coincided with a period of cooling and considerable climate instability, including abrupt drops in temperature and solar energy caused by volcanic eruptions. These fluctuations are thought to have contributed not only to internal crises in Rome and its provinces, but also to increasing territorial pressure from pastoralists living in the vast grasslands of the Eurasian continent.[8] These expert horse riders knew a lot about their agrarian neighbours, having traded horses and livestock for grain with them for thousands of years, and having often been employed as respected mercenary soldiers for their toughness and skill.[9] These pastoralist groups, as their own grasslands dropped in fertility, first raided and then invaded regions of the collapsing Roman Empire in the West.

However unjustly, the name of one of these steppe tribes, Vandals, is still a standard term for mindlessly destructive people. In fact, the evidence repeatedly shows that the invaders had great admiration for the achievements of Rome. King Theodoric, the Goth king who ruled much of Italy in the early sixth century CE, built his best attempt at an emperor's mausoleum for himself. It is much smaller than the mausolea of Augustus and Hadrian – as you would expect for a ruler operating only on locally grown crops rather than the surplus of almost the entire Mediterranean region – but it is not without its impressive boasts. His masons managed to transport a 350-tonne capstone from modern Croatia over the Adriatic Sea to Ravenna and raise it on to the top of the mausoleum.[10] The porphyry sarcophagus in which he was probably buried seems suitably imperial too, though in fact it was reused rather than newly commissioned for Theodoric.[11]

The level of apparent collapse in social complexity in Western Europe in this period probably has less to do with the invaders than with underlying energy changes. As the climate cooled and destabilized in the fifth and sixth centuries, and Germanic tribes raided and invaded north-western Europe, bubonic plague swept across Eurasia in a devastating pandemic, killing perhaps a third or a half of the population.[12] In north-western Europe some combination of plague and invasion led to a very substantial drop in the quantity and sophistication of agricultural activity. In what is now France, for instance, legumes seem to have fallen out of widespread cultivation, despite their value in replenishing the nitrogen levels of depleted soil; animal-drawn farm equipment became smaller and simpler, while the skeletons of draught animals from the period shrank by 50 per cent from Roman levels – the big, powerful draught animals of the Romans presumably needed more food than the post-Roman farmers could spare them.[13] With large areas falling out of cultivation entirely, forest regrew and the limited exploitation of coal that the Romans had profited from in northern Britain disappeared.[14] So did the production of architectural materials like brick, which only make sense if there is enough building activity to support the market and to keep the specialists who make and transport the bricks.[15]

Cities shrank or collapsed, no longer supported by an effectively traded agricultural surplus, and the specialists who thrive in urban populations disappeared. By the end of the tenth century, when Earls Barton tower was built, the climate had already been recovering in warmth and stability for two or three centuries, but the shadow of the slump of the fifth and sixth centuries was still evident and old specialisms were only just starting to be revisited.

There were still impressive achievements in certain arts and techniques in north-western Europe, but the architecture in stone was nowhere near the scale and quality of their higher-energy contemporaries in Byzantium or Muslim Spain.

Even the most powerful ruler of early medieval Europe, Charlemagne, who had himself crowned by the pope as a new Holy Roman Emperor, in terms of monuments could not rival the Byzantine Roman emperors. His Palatine Chapel in Aachen (c. 792–804) is a beautifully made, accomplished and ornate piece of architecture, clearly influenced by Byzantine models. Charlemagne had stone *spolia* brought from sacred Christian sites, including the slabs of Roman marble revetments or paving that he had clamped together with bronze to form his throne. Yet Justinian's Hagia Sophia, built in around half the time, was so much bigger that the entire Palatine Chapel would fit into its central void without touching the sides or the dome. Charlemagne could muster an impressive military force and could administer a complex rule over disparate people across a very large area, yet he could not scale up the complexity and size of his construction industry to the levels seen in a very large, well-established and high-energy city like Byzantium.

The High Middle Ages

As the climate of Western Europe improved again through the period 800–1200, the region's architectural ambition grew. By the eleventh century, France was enjoying a major agricultural boom and rising population levels, with new and revived farming techniques producing more calories per farmer. In a weak echo of the Song dynasty's farming reform, new areas were brought into cultivation – landlords and possibly peasants organized the conversion into arable of forested land, areas that were more marginal in their fertility and swampland that, once drained, could with investment become very productive of human and animal food.[16] Again as in Song-dynasty China, with considerably more energy available all sorts of activities flourished: trade, scholarship, art, craft and technology. French self-confidence rose so steeply that French-led armies set out to invade the powerful and technologically superior cities of the Middle East, motivated by a combination of religious conviction, acquisitiveness and bloodlust.

In France, cities grew rapidly. Landowners became rapidly richer, but none more so than those who owned the suddenly very valuable land in and around the fast-expanding cities. France's bishops and cathedral chapters (theoretically advisory bodies for the bishop, but by this period semi-autonomous governing bodies for a cathedral and its estates) were at the forefront of this boom.[17]

An early thirteenth-century visitor to the city of Bourges, in the southerly part of the region of north-central France ruled directly by the French king, would have had no doubt that something remarkable was happening. The Cathedral of St Stephen had been rebuilt at considerable expense less than two centuries earlier, on what must have seemed at the time an impressive scale – it was certainly far larger and more refined than Earls Barton. This church was then refurbished and enlarged only decades before a complete replacement was commissioned, beginning in the 1190s and possibly triggered by a fire in the existing building.[18]

The new cathedral was so large that much of the old cathedral could have fitted within it. In this case, that is probably exactly what happened: the new building is thought to have been built over the top of the old cathedral before the latter was demolished. The new nave leaped over the surviving parts of the old structure in a move that was practical (it meant there was still somewhere to worship throughout the first decades of construction) but also a spectacular piece of building-site showmanship.[19] This huge, bold new building could no longer be contained by the ancient defensive walls of Bourges. Instead it stepped over them, its east end pushing out into what was previously countryside, while the much longer, wider new west end punched its way inwards into the city.[20]

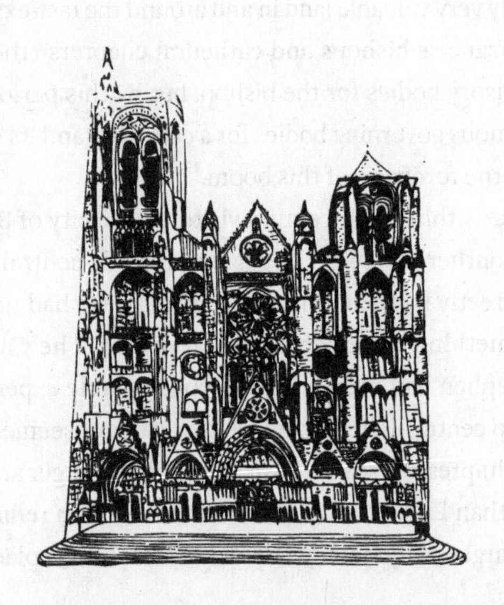

The scale of the cathedral is awe-inspiring even today – the main roof reaches the height of a modern building of ten or eleven storeys. How much more astonishing it must have appeared to locals and visitors at a time when most people lived in single-room, single-storey structures, sharing space with their entire family, their animals and goods. The medieval eye must have been used to the higgledy-piggledy mess of wood, animal dung and thatch that made up most other buildings, no two the same in each chaotic, meandering, narrow street. As they arrived at the cathedral the walls soared up in smoothly finished stone, probably painted in vivid colours when new and put together with a regularity that must have seemed almost as breathtaking as the building's size.[21]

Inside the new cathedral, order, cleanliness, light and height contrasted with the dark, sooty interior of the peasant home more markedly than any present-day building could possibly compare to normal European housing.

The construction industry of France in this period of energy boom is thought to have been the largest economic sector after agriculture itself. Construction on something like the scale of Bourges Cathedral was taking place simultaneously at tens of churches within a few hundred kilometres of each other, alongside thousands of smaller churches and monasteries, new and larger castles, expanded circuits of city walls and stone bridges being built to take the place of wooden ones as the quantity and weight of freight crossing them grew with the expanding economy.[22]

The clients for the great new cathedrals and abbeys are likely to have known a lot about the design and construction of churches, as senior figures in the Church were well networked. They travelled a lot by the standards of the day, and knew each other and each other's buildings. The presence of a culture of clients who were in a lot of contact is clear from the particular ways in which the cathedrals competed with each other. They are strikingly similar in their layout and engineering. Bourges made a rare departure in trying out a more open and spacious-feeling interior, but later cathedrals closed rank and followed established planning norms. Typically for a big building boom sponsored by clients who followed each others' progress with interest, churches built in different parts of the kingdom of France at the same time tend to be more similar than churches built in the same city but a century apart.

Each of these great cathedrals is larger than the one before: Notre-Dame de Paris, Bourges, Reims, Amiens and Beauvais each built cathedrals higher than the one before.

It is an impressive competition but a relatively subtle one. It would take a sharp eye to be able to tell from memory after a few days' travel between them which was five metres higher or two metres wider. Yet despite this subtlety, the bishops and chapters put so much effort and resources into building larger that eventually Beauvais paid the highest price of all by being built larger than the masons could manage, repeatedly collapsing and remaining unfinished. To this day it is incomplete, held together with wooden scaffolds.

At Bourges and elsewhere, the vast undertaking was placed in the hands of an established expert, a master mason who had risen through skill (and doubtless professional politics) in his trade. Perhaps as much as 80–90 per cent of the construction workforce moved from job to job depending on where there was work, so masonic expertise was as intensely networked as the client group.[23] Techniques for extraction, transport, carving and erection of stone were spread and passed down within each trade. As usual, where the energy was present to support a building boom ingenuity flourishes. The scale of simultaneous activity and the intensity of competition it fostered stimulated improvements and innovations.

Stone is the element of the Gothic cathedral that draws the loudest gasps from modern observers: how did ordinary human beings, with only wood and iron tools, wooden scaffolding and machinery, and animal- or plant-fibre ropes, raise so many stones so high, and secure them with such precision that they are still there after centuries of storms and periods of neglect and war?

Two things made the growing scale of churches particularly challenging. The first was the emphasis on light, which led masons to cut back further and further on the amount of wall in favour of larger windows.

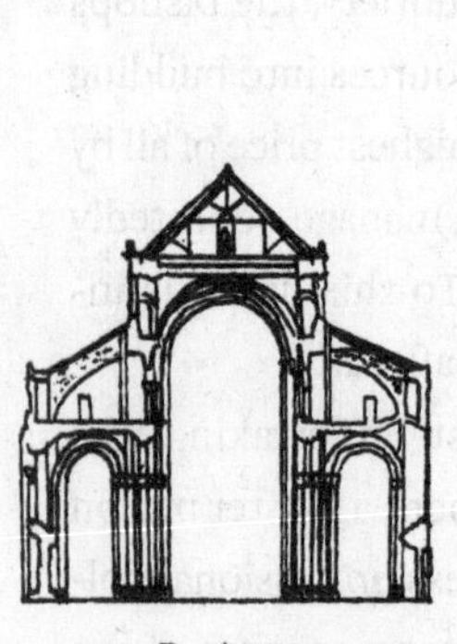

Durham

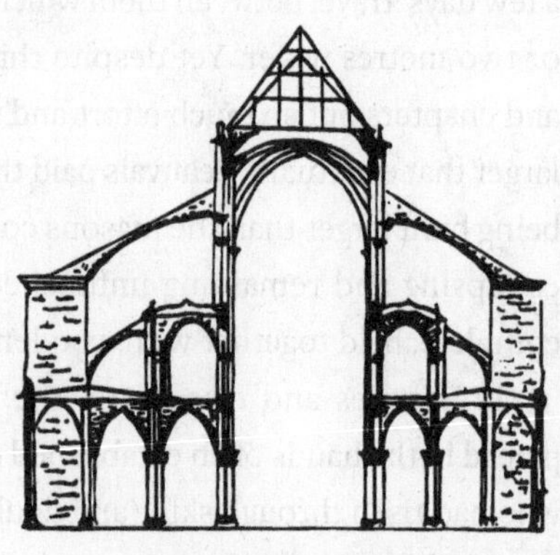

Notre-Dame de Paris

The second challenge was the stone vault – an arched stone ceiling beneath the weather-proof roof of timber and lead. Stone vaults offered some fireproofing; the disastrous fire at Notre-Dame de Paris in 2019 would have done more damage without a stone vault. Perhaps more important was a shared sense among French church grandees and their masons that proper churches had stone vaults.

The problem a high stone ceiling gave the masons was that, like any arch, vaults push out to the sides. The piers they landed on were built as high as the masons dared, and the sideways pressure of the vault tested the weakest aspect of a masonry structure – its ability to withstand twisting and pulling forces. Romanesque buildings had resisted the pushing force of the vault with very thick walls, or with extra stone buttresses thickening the walls wherever the vault pushed hardest, with further buttress arches hidden

Bourges

Beauvais

within galleries above the side aisles – look back at the section through Durham Cathedral above to see how in a building so much smaller than its French Gothic successors, the nave piers are in fact considerably thicker, not just in relative terms, but absolutely.

At the height and width of a church like Bourges these buttresses and piers would have been thick enough to shut out the light and cut the aisles into a series of mean little rooms. The solution found by Gothic French masons was to build a separate set of external buttresses, detached from the upper wall so that plenty of light could come past them to the windows, then to build an arch from each freestanding buttress to the point on the main external walls where the vault landed, pushing back against the outward pressure of the stone vault inside. These flying buttresses became, via trial and error, almost the optimal structural shape that a contemporary engineer would make them: a parabola.

The term 'trial and error' conjures up something rather terrifying, with visions of numerous abrupt collapses. While such things did happen, normally a large masonry building gives a lot of warnings that it is under strain before it collapses. As late as the nineteenth century Chichester Cathedral's tower fell, but not before a large team of workmen had spent days hammering steel cramps in around its growing cracks to try and save it. They were all able to get clear before the eventual collapse.[24] Similarly perilous was the state of Salisbury Cathedral in the late seventeenth century, when Sir Christopher Wren was engaged by the bishop to assess the spire, which, as a result of unstable foundations, had begun to lean alarmingly.[25] Wren's report survives in the cathedral's archives, and, thanks to his suggestion of strengthening the structure of the spire with iron bands, the spire has survived too.

At Bourges, it seems likely that the signs of structural insecurities began to show themselves while it was still under construction. Around the entire east end, about half-way up the innermost wall, a huge iron chain is built into the stonework, tying together the masonry and giving it the tensile strength that stone and mortar lack. The following stages incorporated iron ties built in as the building went up.[26] The accounts of Bourges reveal that a new furnace was built for the blacksmith; it must have had a busy time, not only working on reinforcing chains and bars but also making and repairing the masons' tools, iron components in the construction machinery and quantities of nails.[27]

In energy terms, the carving and erection of the stone for the cathedrals benefited from the familiar economics of agrarian societies: there was a large population and the cathedral chapter, with its extensive landholdings, had no shortage of food to feed them with. It was the same logic as the emperors' projects in Rome and the pharaohs' in Egypt. For the rich in agrarian societies, labour was cheap.

Even with cheap and plentiful workers, among the biggest energy challenges for medieval cathedrals was transporting stone from the quarry. The cost of quarrying stone could be doubled by the expense of carting it less than twenty kilometres by road, so the location of the quarry was crucial.[28] The choice of stone for a cathedral was partly based on the quality of the stone, but also on its accessibility, ideally by water. Transport by water could be in the region of five times cheaper for a given distance.[29]

A rare French case where stone was brought on a long and awkward journey from the quarry was for the most refined carving around the entrance for a great church. Here, where the viewer would be closest to the sculptures, the finest sculptors would produce their most detailed work. Any unevenness in the stone could result in pitted or lumpy surfaces, and damage to crucial elements like faces, hands or the symbolic objects that told the illiterate viewer whom they were looking at – St Peter's keys, St Andrew's cross and so on. For these sculptures, therefore, special high-quality limestone with a very smooth, even grain was quarried near Paris and transported to church building sites all over the region. Presumably to accelerate the supply chain, these Parisian limestone blocks were produced to a standard size, irrespective of the size of the doorway in which they would appear. The distinctive poses of the resulting figures are in part the sculptors' solution to the problem of how to fit limbs and symbolic objects into the standard squashed cylinder of the blocks the quarry provided.[30]

These limited steps towards prefabrication were taken much further at Amiens Cathedral. Amiens was a city which stood out for its cloth-making industry. It held a valuable monopoly on woad dyeing, which was the most successful dyeing technique in Europe before the Portuguese importation of Indian indigo took off in the sixteenth century.[31] The historian Dieter Kimpel makes a convincing case that the local expertise in efficient cloth production may have encouraged a unique search for economies of scale and systematic repetition in the cathedral project. For Amiens, a standardized kit of parts was roughed out in the quarry and finished at the site, before being assembled the same way in each successive bay.[32] It may be an indication of the limitations of the masonic network that this important idea did not spread to later cathedrals like Cologne, which returned to irregular-sized stones being shaped as one-offs each time, despite the fact that together they would make up a repeated elevation pattern.[33]

If the stone was a spectacular effort to extract, transport, carve and erect, even more demanding in some senses were the iron, lead, lime mortar and glass which made up important secondary materials in the great Gothic churches. Glass has most often been discussed by historians as a form of artistic expression or a problem of technology rather than one of energy. Of course it was all three, but the energy aspect is crucial to understanding the achievement of the glaziers.

It is typical of the low-energy period between the fall of the Western Roman Empire and the medieval warm period that sophisticated Roman glass techniques were abandoned by European craftspeople.[34] Glass requires sustained temperatures of over 850 degrees Celsius in a well-controlled setting to make in the first place, with a powerful solvent added to dissolve the sand grains which are the principal ingredient.[35] The resulting viscous liquid needs to be kept molten-hot for hours to allow air bubbles to work their way slowly up and out. It then needs to remain hot while being processed into flat, workable sheets. For glass like that at Bourges, the glass is dyed with chemical additives, some of them based on precious metals and other pricey ingredients, to produce strong jewel-like colours.[36] The resulting sheets were then broken carefully into pieces that could be composed into a mosaic. The basic shapes were then overdrawn to put the details in, using a mix of vinegar or urine, iron or copper oxide and gum arabic from acacia trees in the Middle East. Once the painting was complete, the glass pieces had to be fired again to fuse the black overpaint on to the glass surface, after which the separate pieces could be attached together with lead strips to produce the finished window.[37]

Despite the fact that much more skilled labour and special ingredients were involved, coloured, painted glass typically cost only two and a half times more than plain clear glass in medieval times.[38] This suggests that a high proportion of the cost of glass production went into the basic business of melting and settling the glass itself. For comparison, in today's fossil fuel economy, contemporary prices for hand-crafted stained glass are around forty times the price of plain glass.[39]

Medieval glass was so expensive that it seems to have been almost unknown for domestic windows below the grandest homes of the rich. The homes of the rest of the population were aired and lit in mild weather by unglazed holes in the wall. In winter these openings would be boarded up or stuffed with rags to retain precious warmth. Even in churches, the cost of real glass could be too high; there are recorded cases of waxed and painted fabric being put up in church windows in place of real stained glass.[40] Nevertheless, religious construction formed a major part of the market that in central Europe from about 1250 to 1500 is estimated to have consumed around 40,000 tonnes of expensively wood-fired glass.[41]

With military necessity and royal power behind it, construction could be even larger in scale than the cathedrals: when the English king Edward I was building four huge new castles to consolidate his conquest of North Wales in 1282, his builders brought together up to 1,600 labourers on one site.[42] For comparison, even large church projects in England

tended to call on at most around 300 workers at peak times.[43] On the Edwardian castle projects, many were also engaged in stone extraction and transport, and in mining coal in Flintshire and shipping it along the coast to fuel the kilns that provided the lime for the mortar that held the castles' stones together. In one castle 2,500 tonnes of coal were used in one summer for this purpose – its hotter temperatures were better than firewood for the manufacture of lime.[44]

To judge by the scale and complexity of the religious, royal and aristocratic building activity of the French and English High Middle Ages, it would be easy to assume that the agricultural boom that supported it was the equal of the booms that had produced the great monuments of Egypt or Rome: the scale of construction effort was comparably vast and the structural mastery in some respects even more astounding. Yet the area controlled by each cathedral chapter was far smaller than those empires, while in these smaller estates the fertility of the land was considerably lower. Historians suggest that even towards the height of the boom, in the most productive region of France, yields of grain were only around 1:4 (four times as much harvested as was planted) or 1:5, as against Roman highs of 1:10 or more.[45]

What the astonishing achievement of Gothic builders demonstrates is not the effect of the highest order of energy boom but rather the effect of the concentration of that boom in relatively few hands. The energy wealth of France in the High Middle Ages benefited landlords disproportionately, allowing spectacular competition on the pet projects of the leading landlords – the secular aristocracy and the Church elite.

The extraordinary capabilities of powerful landlords and their masons in summoning and coordinating labour are attested by the number and scale of the buildings which survive. Yet Bourges Cathedral, in common with every other great church of the high medieval period, remained incomplete. In every case, provision was made in the lower parts of the building for tall towers to rise above it, and the intention was probably for spires to soar high above even these towers. Many cathedrals have one or more towers, often added some time after the main push of building the body of the church and its grand entrance front. One cathedral, Laon in northern France (mostly built c. 1160–1230), got five of its towers, their lovely cow-shaped gargoyles boasting for miles around of the agricultural wealth of the region.[46] Even here, however, the masons appear to have planned for seven towers, and the two towers that originally got as far as the spire have lost their gargoyles since.

It is tempting to suggest that it was never likely that the will to build so high could have lasted the many decades required to complete such a project. However, on the face of it, it might seem unlikely that so many churches that were so large were completed to the extent that they were, and still saw lavish west fronts added at the end of the main building campaign. The enormous landholdings of chapters and bishops, allied to the strong monetary economy of high medieval France, and its large capital market underpinned by rising agricultural production, allowed cathedral chapters to budget with some realism for these prolonged campaigns of heavy expenditure.[47] The incompleteness of their skylines may in part have arisen from sensible caution. Around 17 per cent of major churches of the medieval period have suffered, at the time or since, from a partial collapse.[48] Those with spires and towers have a particularly dicey record, with the wind load of such height needing more supporting width than a tower could feasibly provide. And while lightning was understood as an act of God, it picked on spires more often than on lower structures.

Alongside the structural hazards of high towers for great churches, there was another shift that intervened between the economic bullishness of the Church in the 1100s and its greater caution in the 1300s. The new shift came from another change in climate.

In the later thirteenth century another cycle of gradual cooling started to have a marked effect on European agriculture. The population had expanded substantially during the preceding centuries of rising agricultural production, and by the early fourteenth century it was putting considerable pressure on the available land.[49] The price of wood rose substantially after much forest was cleared for arable and grazing, and areas of north-east England which had access to coal near the surface began to exploit it and even to ship it to the crowded and fuel-hungry capital, as firewood prices rose enough to justify the effort.[50]

Land that was intrinsically less fertile was pushed into food production, yielding only just enough food to feed its farmers. Rising food prices made draught animals expensive, further reducing the efficiency of farming, and so cutting the proportion of farming output that was available to support city life. Desperate peasants intensified the energy-poverty trap by having more children to increase the labour they could put to work and to increase their chances of some children surviving to feed them should they make it to old age.[51] High prices for land and food enriched landlords and brought the poor to a condition where they would work for very modest rewards.

The Black Death

By the early fourteenth century this self-reinforcing pattern had reached crisis point. With cooler temperatures came crop failures and famine. Among the crowded, hungry population, with whole families sleeping in the same straw as each other and their animals for warmth and security, plague was an unsurprising outcome. Bubonic plague returned in 1347 – sweeping through the populace as catastrophically as the so-called Plague of Justinian had done in the sixth century – once again wiping out between a quarter and a half of the population and hitting the poor particularly hard.[52]

This catastrophe, and further plague epidemics that followed until the seventeenth century as the climate continued to cool into what historians have dubbed 'the Little Ice Age', had major consequences for the energy economy. Suddenly the extreme pressure on land was gone: there was ample space to graze draught animals, so the smaller peasant population could farm larger areas more efficiently, even in a more difficult climate. The radical decline in the working population meant that, while peasants were still officially forbidden to move around, in reality landlords would offer incentives to attract the workers they needed. In England, pay for some low-skilled labourers tripled within years of the Black Death, despite a royal ordinance making pay rises illegal. Landowners were disgusted by the self-confidence this gave ordinary people. As the head of a Leicester monastery wrote, 'The workers are so above themselves and bloody-minded that they took no notice of the king's command. If anyone wished to hire them he had to submit to their demands, for either his fruit and standing corn would be lost or he had to pander to the arrogance and greed of the workers.'[53]

With higher energy production from peasant farming and better pay for labour, young couples settled in separate houses from their parents, had fewer children, and more often gave them at least a basic education in church schools. Increased access to horses contributed to changing the structure of town and country, too. England, for example, had been dotted with village markets, spread around close enough so that peasants could bring their goods to them on foot, sell them, and return home all on market day. After the Black Death peasants could take their goods considerably further on horse-carts, and many small markets disappeared, concentrating trade in larger towns.[54]

In the High Middle Ages, the monumental projects of the secular and church aristocracy had thrived on cheap labour from a large population, and high incomes from the land and food that they controlled. Now that labour was scarcer and more expensive, and pressure on land was reduced, the new economy of the Late Middle Ages saw energy wealth spread much more widely among a new category of wealthier peasant farmers. More efficient farming techniques and economies of scale resulted in more food being traded to cities once again, supporting a new urban boom characterized by richly productive competition and mutual learning in trades and intellectual disciplines.[55] In Italy, this period of rapid urban growth and increasing societal complexity has come to be known as the Renaissance.

The Italian Renaissance

Florence was early in showing how far upwards the rising middle of society could take their material wealth. Typically, the richer an economy is, the lower the proportion of its activity related to energy production. Across most of medieval European history, land and the energy it grew were the largest and most consistently valuable sources of wealth. In Renaissance Florence the energy supply was taken so cheerfully for granted that a new urban bourgeoisie came into existence whose substantial financial portfolios could see well under 10 per cent of their wealth invested in land, the rest being put into trading and industrial activities that developed the financial economy and promoted technical and intellectual innovation in a dazzling range of areas.[56]

Proudly a republic rather than a city-state ruled by a hereditary aristocrat, fast-growing Florence became a leading financial centre of Europe, its bankers providing a version of the international financial guarantee now furnished by credit cards, while using the impressive network of agents they had for these transactions to keep an eye on economic and political developments around Europe.

One of the leading banking families, the Medici, rose in importance, producing popes, a queen of France and other powerful European figures, but the wealth of Florence remained widely distributed. In 1427, for example, there were 1,649 Florentines with a personal fortune worth more than fifteen times the annual earnings of a well-paid skilled craftsman.[57] This new urban upper-middle class lived in ever-larger houses that showcased their success. Rich families ended up rattling round in gigantic versions of their previous house, with rooms that were absurdly large and whose purposes were ill-defined until, after decades of custom, ways were found to furnish and exploit space initially acquired out of competitiveness.[58]

Florence had several major projects under way at this time, most prominently a new cathedral, which had begun in pre-Black Death conditions when large landlords like the Church could build very ambitiously, profiting from direct incomes from landownership and cheap labour costs. After the Black Death, the cathedral remained unfinished throughout the Florentine Renaissance. What thrived instead were smaller private chapels within the city's churches, which saw a boom in competitive merchant spending as the new urban commercial rich competed with each other to hire the best artists and architects to make their chapels more magnificent or more sophisticated than those of their neighbours and rivals.[59]

It was in these circumstances of diverse and competitive patronage that the reputations of the leading artists and architects rose steeply. Previously masons and craftspeople had needed patronage from either a limited aristocracy, concentrated in a court, or from the Church. Their efforts to sell their services needed to be concentrated on this small pool of patrons in a spirit of eagerness to please. There was little point in attempting wider publicity. The architects and artists of the Renaissance city vied for a wider pool of work and began to establish what we might now call brands. Clients would pay any premium for the work of those who had come to be regarded as 'geniuses'. The leading names still resonate today: Michelangelo, Botticelli, Brunelleschi, Donatello, and so on.

In the field of construction, the new category of 'architect' emerged – designers separated from the rest of the building industry by their specialism in drawing, and adding artistic or intellectual lustre to the resulting building.[60] Up to the Renaissance, credit for buildings tended to go to the person or organization that commissioned and paid for them. Caracalla's name is associated with his baths in ancient Rome, and al-Walid's with the Great Mosque of Damascus. However heroic the achievement of their masons in getting the buildings built, and coming up with elegant solutions to the many problems of a large, complex project, their names do not survive.

In reality there are many people involved in different aspects of design and construction, and the habit of lumping together these diverse activities under one name, whether Michelangelo, Christopher Wren or Frank Gehry, is curious and often distorting. It is, however, a powerful selling tool in a large and competitive architectural marketplace. It was the broadening market for buildings in the post-Black Death centuries of more widely distributed energy wealth that gave rise to the architect as brand.

The most enduring architectural brand of any in the Western world has been that of Andrea Palladio (1508–80). He made his name in and around Venice, a lagoon city ideally placed for the easy transport of goods, which had for centuries sustained a large share in East–West trade across the Mediterranean. In the sixteenth century a series of military defeats led to declining profits from trade, plus growing risks to investment in it. The highly effective business brains of Venice began to diversify their investment portfolios into carefully chosen profitable farmland on the mainland.[61] As traders expanded their farm holdings and consolidated scattered landownership into geographically coherent estates, some commissioned magnificent houses as the focal points of their estates. Palladio exploited this niche brilliantly, designing 'villas' which, with their temple-like fronts and elegant symmetry, evoked for his clients the rural homes of wealthy ancient Romans.[62] Palladio's appeal arose from the intellectual cachet of his meticulous research into ancient Roman architecture, accompanied by an attempt to find a scientific theoretical basis for beauty: he derived the proportions of his plans and elevations from the mathematical ratios of notes that together make harmonious musical chords. Whether because of the proportions or not, the resulting buildings have exerted an extraordinary attraction ever since. What

enabled Palladio's architecture to achieve truly international reach before any other famous architect was publication of his book *I quattro libri dell'architettura* (*The Four Books of Architecture*, 1570). The book offered wealthy, literate purchasers plenty of high-quality information on technical, practical and theoretical aspects of classical architecture and city planning, and provided good archaeological reconstructions of major ancient Roman monuments. Advice was given on achieving comfortable conditions in terms of what we would now call energy: he recommended appropriate sizes of fireplace for given room sizes, and window dimensions that would offer light but not too much overheating for their orientation under the summer sun.[63] The book also included a large number of Palladio's own designs, presented, and widely accepted, as ideal models for villas and other buildings. His easy-to-follow formulae, his versatile kit of parts and the orderly perfection of his buildings all exerted a powerful influence over architecture in Europe and eventually America for several centuries.

Renaissance Rome

It was not only republics that showed exciting and rapid development, and competed to hire the best artists, in the centuries of bounce-back after the Black Death. Other Italian states ruled by princes or dukes also experienced improved energy access per head of population, and spent some of the wealth this supported on lavish artistic and architectural patronage. For most of the period since the fall of the Western Roman Empire the city of Rome had been dominated economically by its role as the centre of the Catholic Church, and politically by its head, the pope. Successive popes had fought for well over a millennium to maintain as much independence as possible from secular rulers and to maintain the high dignity that they felt was appropriate for the status they proclaimed as the successors to Jesus's closest disciple, St Peter.

From 1309 to 1417, there was no universally accepted pope in Rome, with popes either in exile or in dispute with rival claimants. When Pope Martin V returned to Rome in 1417, he found it crumbling and depopulated even compared with its shabby state when the popes had left. Its population was around 17,000, under 2 per cent of its peak ancient size, and its churches and other buildings were dangerously dilapidated.[64] The pope moved his base from the Lateran Palace, in southern Rome, to the Vatican, partly because that was where St Peter was buried, partly because it was easier to defend against attack. The great family tomb that Hadrian had built near the Vatican now served as a castle, with a secure raised walkway leading to it, along which the pope and his entourage could flee from the Vatican if invaders or major riots threatened.

The pope straddled two roles: as a Renaissance prince of a regionally important city-state and as the leader of a much larger spiritual realm, covering western Europe and indeed beyond once Portugal and Spain started to claim vast overseas empires under the flag of Catholicism.

This dual role did not always make for an attractive combination. The most infamously worldly of the Renaissance popes, Julius II (1503–13, one of several to choose as his papal name that of an ancient pagan powerbroker rather than a Christian saint), lived more like a prince than a pious bishop. Clad in full armour, he led armies into battle. When attacking a walled town, Julius II is said to have opened with a crude pun about his artillery – 'Let's see if I have bigger balls than the King of France' – and ended by scrambling up a ladder into the defeated city himself as soon as it surrendered. It is reported that he had to be dissuaded from having the defenders massacred and the city sacked.[65]

This dual status can be read on to papal buildings too.

On the return of the undisputed papal court to Rome, repair of churches, monasteries and residences was urgently needed. The popes had higher ambitions for their building programme, though. Pope Nicholas V (1447–55) gave a persuasive justification for erecting great buildings. It seems to speak for many of those we have already seen in this book. He explained his view that those whose faith was weak could be helped to believe by 'great buildings, which are perpetual monuments and eternal testimonies seemingly made by the hand of God'. He hoped that architecture could leave their 'belief continually confirmed and daily corroborated by great buildings'.[66]

Sixtus IV (1471–84) had even greater ambitions. He wanted to make Rome the 'head of the world', and conducted extensive work to improve both the functionality and the magnificence of the city, punching straight new roads through the tortuous network of old streets and re-erecting at key points in the city Egyptian obelisks originally taken from Nile temples by the ancient Romans to adorn chariot-racing stadiums and other public buildings.[67]

The high aspiration to make Rome echo in physical terms its spiritual importance to Catholics pushed the papacy to ever-grander building projects. Julius II enlarged the Vatican Palace lavishly and commissioned an imposing tomb from the famous thirty-year-old Florentine sculptor Michelangelo Buonarroti (1475–1564). To house the tomb he asked Rome's most exciting architect of the period, Donato Bramante (1444–1514), to add a new chapel to the great church of St Peter, built some 1,200 years before by the emperor Constantine and still at the time the largest church in the world.

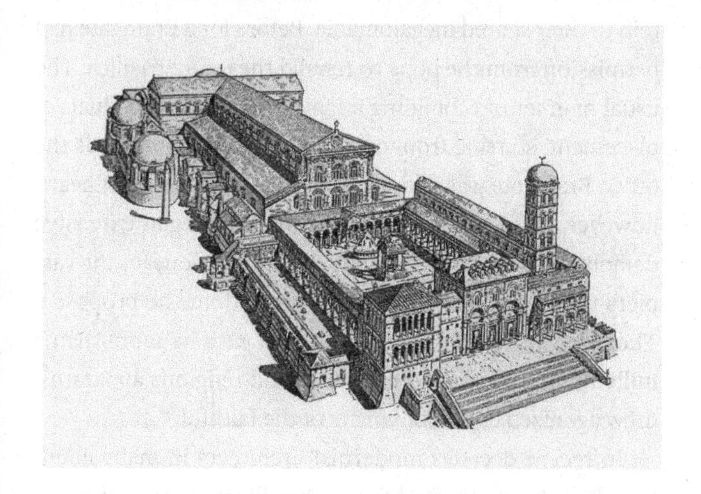

With the involvement of Bramante, everything got rapidly out of hand. He was obsessed with the architecture of ancient Rome – he and his contemporaries pored over the best-preserved ruins and buildings from the ancient city, drawing and redrawing, measuring and calculating, and reading Vitruvius with an almost religious fanaticism. The buildings Bramante designed have a mathematical perfection and an elegant orderliness that seem at first to reflect what he himself believed in: the triumph of reason and mathematics. Yet the calm perfection of his architecture conceals another story.

Once called in to work on St Peter's, Bramante pursued a personal campaign to get the ancient and venerated building entirely torn down and replaced with one to his own design. He had initially been commissioned only to extend an incomplete new chapel at the west end of the basilica, but once Bramante and Julius II came together, they gave rein to their shared megalomania. Before long Bramante had permission from the pope to rebuild the entire basilica. The usual manner of rebuilding a major church was gradual replacement, starting from one end and working towards the other. Bramante was taking no chances on a change of heart, however, and from around 1505 workers began extensive demolition, and started establishing foundations for the vast piers that would hold the huge central dome he proposed. The devastation of the great basilica, with its more than a millennium of accreted meanings and religious apparatus, drew agonized cries from many of the faithful.[68]

In recent decades modernist architects in many countries have been attacked for their willingness to tear down old shops and houses in the interests of their grand schemes, but compared with Bramante almost all twentieth-century architects look conservation-minded and moderate. So extreme was Bramante's architectural fundamentalism that, in a building whose chief purpose was to mark and glorify one of the holiest sites of Catholicism, the burial place of St Peter, he proposed digging up and moving the tomb so that it fitted in with his own architectural ideas.[69]

Bramante was also happy to sacrifice ancient patterns of worship in order to pursue his notion of mathematical beauty: symmetry, he felt, represented perfection. Thus a building that had many axes of symmetry was better than a building with only one. His design for a circular dome surrounded by a square building with symmetrical chapels around it had four axes of symmetry. Never mind that it was not the long, directional shape that had evolved with ancient patterns of liturgy. Many have subsequently expressed disappointment with the architectural effect of St Peter's as it was eventually completed – its long nave cuts off the view of its dome from nearby, where a square plan would have made a more impressive architectural impact. Yet the client was the Church and the brief was a church, so however wonderful the alternative fantasies of successive architects, if they failed to fit the liturgy and the expectations of the client they were doomed to be rejected.

The unhealthy encouragement that Julius II and Bramante gave each other, working themselves up to tear down the mother church of Western Christendom, demonstrates a damaging consequence of the rising status of the architect. It is very likely that some of the anonymous masons and organizer-designers responsible for the production of earlier buildings in this book were, like Bramante, charismatic, glittering-eyed enthusiasts whose verbal or sketched fantasies of magnificent buildings swept clients along with them into overambitious building works. With the Renaissance, though, the importance of brand within the practice of individual architects became stronger and stronger: at the upper end of the architectural world you went to Bramante for 'a Bramante' in the way you went to Botticelli for a Botticelli in the art world. There have been numerous instances ever since of this kind of mutual stirring to hysteria between architect and client, though rarely on quite the scale of St Peter's.

Successive popes inherited an impossible building site from Julius II and Bramante – Old St Peter's half-demolished with four vast new central piers starting to rise. They couldn't leave it as it was, but continuing would stretch their fundraising abilities and labour supply to breaking point.

The response of the popes embodies a tendency common in institutions: they appointed architects who were like themselves. The rational procedure for such a big, lengthy building project might seem to be to appoint a gifted, well-organized, youngish architect in excellent health, so that he could carry the project through as consistently as possible for as many decades as the fundraising and construction would take. Several popes, however, elected to the role themselves in late middle age or old age, appointed architects who were themselves elderly. Bramante and Julius II had both turned sixty when they started the project. Giuliano da Sangallo, who succeeded Bramante, was born around the same year as Bramante and died within his first year or two as architect for St Peter's. Raphael and two of the others who worked on the building in the first half of the sixteenth century were younger, but when Michelangelo was brought back to St Peter's, now as architect rather than sculptor or painter, he was seventy-two. By remarkable good fortune he lived until nearly eighty-nine, and during these last years pushed through a great part of the main body of the basilica, including enough of the magnificent dome to leave his successors no prospect of major changes.

In a pattern that again echoed what was happening in the papacy, several of these architects acted like the quintessential old man in a hurry, complaining furiously of the ineptitude of their predecessors and calling for radical changes of plan. Only a few decades after Bramante had rushed to get as much built as he could, a substantial portion of his work was demolished amid accusations of inadequate structural strength.[70] Antonio Sangallo the Younger, at the head of a team, made substantial progress, but when Michelangelo arrived on the site he took against Sangallo's work, arguing angrily that it had 'dark hiding places above and below [...] perfect lairs for crime, for forging money, raping nuns and other such roguery'.[71]

With Michelangelo the project took on more of its final form. So large were the heroic central piers

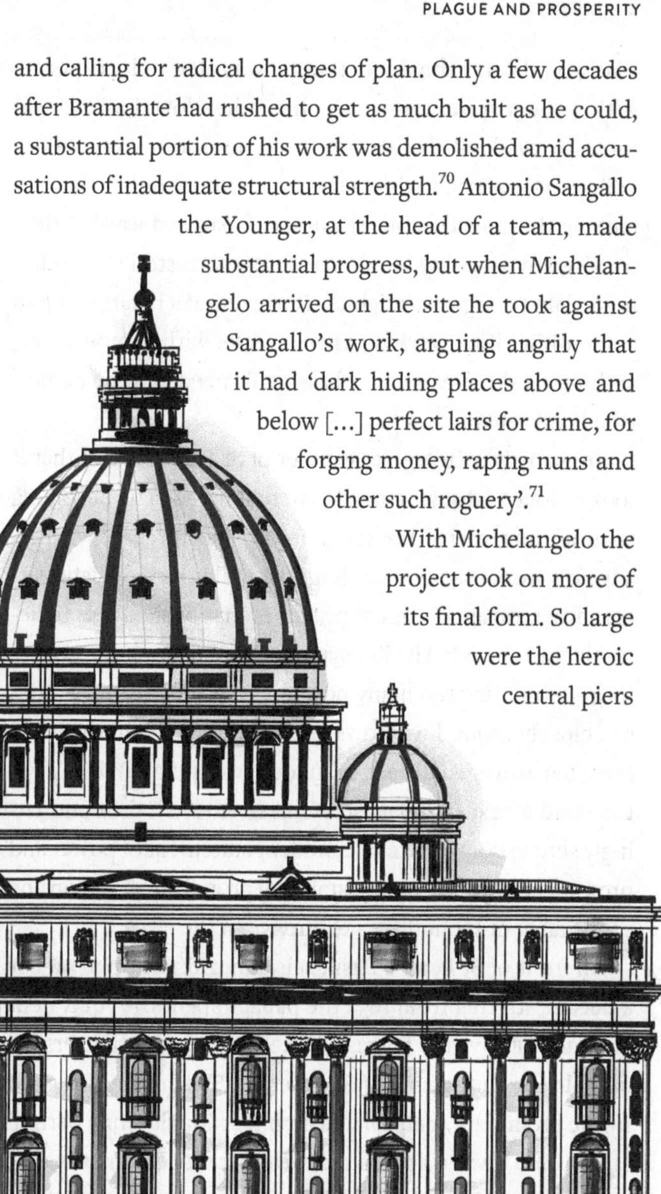

that would carry the dome that he was able to incorporate within them spiral ramps which would allow donkeys and mules to walk all the way up to the level of construction, delivering materials with smooth continuity rather than incurring the considerable machinery costs and waiting time involved in winching hundreds of tonnes of stone and other materials to a great height.[72] But even Michelangelo had some difficulties working at this scale, with a supervising architect's slip meaning a whole vault needed to be demolished and rebuilt.[73]

In many ways the great wonder of St Peter's is not that it took so long and cost so much, but that it ended up becoming such an impressive building. The visual force of Michelangelo's work is so outstanding that it almost vindicates the gerontocratic appointment practices that pushed him reluctantly into the job. His idiosyncratic and distorted architecture – cramming too many pilasters (flat columns) together too closely around windows that seem too small, the walls breaking forward and back in sudden swooping curves – gives the building exceptional forcefulness. Above it the contrastingly serene dome makes a confident statement of power and prominence for the papacy, standing proudly higher than any of the ancient monuments to survive in the city.

In front of St Peter's Basilica is a magnificent piazza. For a decade starting from 1657 the papacy spent the equivalent annually of a year's wages for 1,280 manual labourers on completing the setting of St Peter's.[74] The result, designed by the prominent sculptor and architect Gianlorenzo Bernini (1598–1680), was a huge, almost-elliptical open-air arena for vast congregations to pray together against the awe-inspiring

backdrop of the basilica itself – perhaps at the time the world's largest building. Even today, with so many alternative attractions and experiences available, it is an extraordinary feeling to stand amid tens of thousands in St Peter's Square while the pope addresses the crowds.

St Peter's is one of the most ornate and substantial buildings Rome has ever seen. By making it a high priority over twelve decades, the papacy was able to achieve Julius II's aim of surpassing the monuments of ancient Rome, despite the fact that the energy context of the ancient city was vast imports of grain supporting a huge labouring population, whereas the energy context of Renaissance Rome was far more marginal – productive local farmland over a medium-sized area of central Italy, and the potential to import a certain amount thanks to the fundraising power of the papacy and the stimulation to the market brought by religious pilgrimage.[75]

The Reformation

Yet Europe paid a phenomenal price for the popes' attempts to revive Rome's greatness without the Egyptian corn that had driven construction in the ancient city: the fundraising campaigns that made St Peter's possible fuelled existing resentments in northern Europe, especially among its growing post-Black Death middle classes. The fury at the cost of this project which offered so little to those not living near Rome, and which led to the extension of controversial money-making techniques by the Church, contributed substantially to the Reformation – a division in the Western Church that has never been healed.[76]

Many of the new church buildings of the Protestant denominations which emerged from the great split were designed in aggressively stark contrast to the lavishness and ornament of St Peter's and numerous other Catholic churches. Protestants believed that the Bible was the closest they could get to the authentic Christian message, so they designed churches that would enable the words of the Bible to be clearly heard. Ornament and architectural fuss could only distract from the all-important words, and might even encourage the worship of idols.

In response to the challenge of Protestantism, Catholic churches took the opposite tack, extending Nicholas V's strategy further than ever. They used the sensory power of architecture, music, ritual and art to make churches as close a representation as possible of heaven on earth.

Catholic churches in southern Germany and Austria, where Martin Luther's charismatic German-language Bible posed the strongest threat to the remaining Catholic German-speaking states, showed this fight-back at its most spectacular. The pilgrimage church Basilika Vierzehnheiligen, near Bamburg, Bavaria, was built 1743–72 behind a relatively austere exterior. Inside, however, the Bavarian architectural showman Balthasar Neumann (?1687–1753) created a hidden

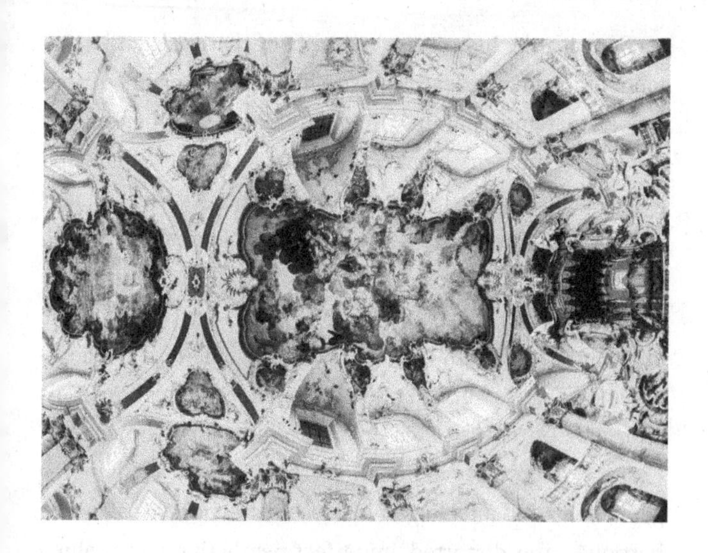

world of the pink and blue of a summer sunset. Its architecture seems to swirl and move like the clouds depicted in its painted ceiling. The walls of the church curve out and back with a mysterious organic order more like the carapace of an insect than a neatly rectangular building. Neumann took the expressiveness of Michelangelo's St Peter's, and other churches of late-sixteenth and early seventeenth-century Rome, and heightened it yet more.

Anyone coming to a service in Neumann's church was stimulated by architecture, grand robes, emotive painting and gold leaf glinting in candlelight, the hallucinatory atmosphere heightened by clouds of sweet-smelling incense, heavenly choral and organ music, and the collective experience of shared worship. It offered the best preview that Neumann was able to give them of the celestial afterlife that good Catholics could hope for.

Beneath the luxurious sensuality of materials and ornament is a hidden order for those whose faith demanded more intellectual underpinning. Bramante's circles had embodied the religious and mathematical understanding of his day: God had designed a universe where heavenly bodies were circular and moved around the earth in circles. Neumann's church was based on a series of ellipses, known by Neumann's time to be the actual shape of the orbits of most planets.[77] If the ellipse was chosen by God for the arrangement of the heavenly bodies, what geometry could be more fitting for God's churches on earth? This entire architectural movement, characterized by heightened aesthetic effect, challenging geometries and a feeling of luscious distortion, is called the baroque, after distorted, imperfect pearls that were valued for their novelty and curvaceous uniqueness.

What both Protestant and Catholic churches had in common was a new anxiety to attract adherents. The medieval church had reflected the energy-related hierarchical social structure of its period: top-down and elite. Churches were an essential part of the running of society, but the most sacred parts were closed off from the public. Medieval churches had a screen to protect the religious ritual from the smelly throng of unimportant peasants who crowded into the nave, and also to keep celibate priests and monks from catching distracting glimpses of lay people. It was the quality of the prayer that counted, not the subjective experience of the powerless farm labourer.

Improving levels of education in Europe in the centuries after the Black Death could be a threat to a church that would not adapt. The Protestant churches made direct appeals to these individuals through offering them direct access to religious learning; Catholic Christianity's offer was both theological and sensory.

By the time Neumann's church was going up, a new force was on the rise in the north-west corner of Europe. It was to change the world far more radically than even the greatest transformations of the agrarian millennia. Over the thousands of years covered in the first half of this book humanity had struggled and adapted to increase the amount of useful energy they could extract from fields and forests, streams and breezes. Chance and intelligent manipulation of circumstances could periodically allow one group to thrive conspicuously, but every boom was followed by decline or bust, as the remorseless cycles of crop fertility and changing climate imposed scarcity and instability. In the latter half of the eighteenth century, however, the greatest innovation since the invention of farming was well under way in England: the exploitation of coal.

PART TWO

CHAPTER 7

The March of Bricks and Mortar

Coal and the City

Whistle While You Suffer

Walt Disney's seven dwarfs live in the archetypal popular dream of the medieval house: a gabled thatched roof over walls made of timbers and something whitewashed. There are simple stone benches out the front and a chimney at one end, its old earthenware chimneypots poised to smoke cosily. Over the door is a simple porch of tree trunks and thatch. It is a relaxing spot representing a simpler, older way of life. The windows are small, made of little diamond-shaped pieces of glass (quarries) held together by a lattice of lead.

Many northern and western Europeans might imagine their ancestors living simple but good peasant lives in this sort of house before the complexities of modern life swept away such earthy and natural housing in favour of the intimidating modern city and its suburbs full of boxes made of ticky-tacky that all look just the same.

Disney's depiction of pre-modern times makes for a tempting dream but has very limited historical reality. The dwarfs' house could have existed for a brief period in the centuries after the Black Death for the most successful farmers in a few of the better-off regions of Europe, but for almost all of agrarian history peasants' houses have been infinitely less charming and comfortable. The last of these typologies still built in Britain is what is known as the blackhouse of the western Highlands and Islands of Scotland – areas remote enough, and with marginal enough farming fertility, to be late to share in the rising energy wealth that most of mainland Britain has enjoyed since the arrival of fossil fuels. Blackhouses were being built as late as the second half of the nineteenth century, and inhabited until the 1970s.[1]

There is some debate over the origin of the name 'blackhouse', but it seems likely that the reason it stuck is that inside they were very black indeed. Low, single-room dwellings with stocky stone walls, a small door, few if any windows (boarded up or stuffed with rags for draught-proofing) and a thick turf roof to keep the freezing winter out. In the middle of the room, on the floor, was an open fire for cooking and

heating. Chimneys like the one Disney gave the dwarfs are very good at getting rid of smoke, but in the process they get rid of around seven eighths of the heat produced by the fire.[2] For most of agrarian history, most ordinary people in cold weather could not afford to gather or buy eight times more fuel, or to procure the fireproof materials for a fireplace and chimney: smoke-free interiors were an unaffordable luxury. Instead, the smoke from the fire clouded the air, coated their meagre possessions in soot and contributed to deposits of toxic material in their lungs. Lung disease was one of the many afflictions which ensured that few grew old.[3] Sleeping together in the same bundle of straw, close to animals, helped diseases and parasites to spread.[4]

In winter, a diet only of things that could be preserved made scurvy an annual epidemic.[5] Food was even more grindingly tedious and repetitive than in summer – it was typical for peasants to get the overwhelming majority of their calories year-round from the local staple crop, consumed as bread, beer, porridge or gruel.[6]

Labouring jobs were few in winter, with many roads deteriorating beyond use, and icy conditions making some crafts impossible (mortar must not freeze while setting, for instance) and others too slow and hard to be worth paying someone to struggle with. Cathedral workforces dropped to a few skilled masons carving in their shed and even they were paid less in winter to reflect the limited daylight hours they could work.[7] Clothes were not waterproof and roads were not properly paved, so slogging through mud or snow to an unheated church, though a welcome social moment, risked dangerous levels of cold. No wonder Christmas and its pagan predecessor festival were organized around twelve days of drunkenness and feasting on any extras the family could muster, to help them build up their morale before facing the worst months of cold.

Of course, across so much human history there were numerous times and places where proportions of the population did substantially better – plenty of examples of multi-roomed houses, of people with leisure time and the means to enjoy it, and of urban trading populations living in houses we might still find comfortable and attractive. The best-off had houses with sophisticated adaptations to weather and climate that allowed them to enjoy something more like modern standards of comfort.[8] Even in better times, however, vermin were ubiquitous, hunger was rarely more than one bad harvest away, plagues far more lethal than Covid-19 appeared recurrently, no one knew why, and economic booms tended to result in crowded, unsanitary conditions for the substantial urban poor.

These tough conditions for most were a direct product of what energy historians call 'the photosynthetic constraint': any society that is dependent on new-grown plant matter for almost all its energy is constantly having to balance the areas of firewood and food crops on its finite land, with more of one meaning less of the other.[9] While the Black Death adjusted the maths in favour of ordinary people by reducing the number of mouths to feed, it was the rise of coal from the later sixteenth century that was to break the photosynthetic constraint and initiate centuries of continuously rising energy wealth, with the material changes that it has brought to every corner of human life.

Georgian London

Someone transported from London before the Black Death to eighteenth-century London might well have found the Georgian city more astonishing than an eighteenth-century visitor would find contemporary London, despite all our modern technological miracles. The medieval eye would have been amazed by Georgian London's miles of paved streets flanked by houses of a height, regularity and solidity achieved in earlier centuries only by religious buildings and castles.

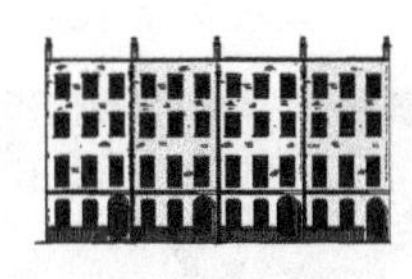

Who were the thousands of (by medieval standards) princely people who could afford such large and well-built houses? What mighty organizing force had enabled entire areas to be built to harmonious plans, with broad straight streets stretching for hundreds of metres before opening into welcome green squares or crescents? What technical wizardry allowed such large panes of glass, and so many of them, identical in house after house? It was not just buildings that had changed. Everywhere there were horses pulling numbers of goods carts and iron-sprung passenger carriages that would have boggled the mind of our medieval visitor, their iron-bound wheels and the iron-shod hooves of the animals sounding loudly on the sturdy cobbles.

The medieval eye might also have found deficiencies in Georgian London. How come most buildings were so expensively big and solid, yet so austerely under-decorated? In a medieval context of much cheaper labour and much pricier materials, opulence had tended to be shown in abundant carving and painting, whereas Georgian houses were predominantly composed of plain brick walls. Churches too had clearly lost much of their relative importance. They had not kept anything like their proportional dominance in the cityscape, in terms of either quantity or size relative to the many big new houses. If thousands of Londoners were living in private houses twenty or thirty times the volume of the typical urban house of three centuries earlier, churches had stayed much the same size, and were more widely spaced in the new suburbs than they had been in the tight confines of the medieval walled city.

Such new churches as there were were based not on earlier English churches but on ideas derived from Renaissance Italy and France, with Roman-style columns and fronts evoking the architecture of Roman temples.

The grander houses too picked up this language of classical pomp, for porches, window details and, more rarely, whole façades. In some places, from the later eighteenth century entire terraces of houses were unified behind a façade that makes them look like a royal palace until you notice the number of front doors dotted along the bottom.

In Georgian times, one pattern of medieval hierarchy was still entrenched: magnificent housing for an emerging wealthy class was underpinned by poor or appalling conditions for the many. Very large numbers of lower-class people worked inconspicuously in domestic service, paid and in theory free to leave, but with a very dramatic difference in power and wealth from their employers. The mean annual pay of unskilled female workers was less than £9 in the 1860s, whilst the rich had incomes measured in tens of thousands of pounds.[10] In towns, many lived in slums, crowded into shabby older houses or purpose-built tenements, one family or more to a room, sometimes without even access to a window. Lavatories were communal, as was a shared water pump. In many tight, dark city courtyards rainwater and leakage would carry sewage into the drinking water, causing bouts of cholera and dysentery to sweep lethally through the urban poor. It is these inhuman slums that would most strike and offend the modern rich-world visitor to Georgian British cities, but for the medieval eye these might have seemed the most familiar and unchanged aspect of the city.

It was not only the British poor who suffered. Slavery occupied a significant role in societal structures of the time, as the politically advantaged attempted to silence the disenfranchised at home by endowing them with a sense of superiority to slaves abroad. The slave trade and colonial expansion, in many ways perhaps at one remove from the day-to-day life of many in Georgian England, were instrumental in providing the energy and establishing the wealth of Georgian England. In Britain's colonial expansion, the coal-fuelled industries of the aggressor supplied the economic strength, and the weapons, tools and tradable goods that allowed such asymmetrical power and such appalling abuses.

Feeding and Fuelling the Metropolis

How had this radically new type of city arisen so quickly? The answer starts with food.

The great energy explosion of Britain in the wake of the Black Death took place against the surprising background of a cooling climate. Despite the reduced crop yields caused by colder seasons, the drop in population following the Black Death proved sufficiently steep to reshape the energy economics of the country. Resilient new crops introduced to Europe from the Americas, in the context of European colonial exploitation, revolutionized the productivity of previously unpromising land. In countries like Britain and Ireland potatoes are said to have accounted for as much as 47 per cent of urban population expansion in the eighteenth and nineteenth centuries, and elsewhere in Europe too populations rose, with maize bringing bountiful new calories to areas that were too dry for potatoes.[11]

The result was a self-reinforcing circle of more efficient farming, more surplus to support city dwellers and more income for farmers. Rural changes could be cruel and unjust at an individual level. From the sixteenth to eighteenth centuries English landlords trampled with increasing confidence over complex ancient patterns of land rights in order to enclose larger and larger areas of common land. Yet, especially from the mid-seventeenth century, many landlords also made real improvements in the productivity of their land, turning to ever more effective evidence-based farming methods like crop rotation and new fertilizers that increased yields substantially. Profiting from a range of agricultural improvements, medieval wheat yields per hectare had doubled by 1800.[12]

While improved agriculture was providing more food for the rising population, many peasants were driven from their fields to seek a new life in cities. By 1801 around a third of the population of England and Wales was living in towns and cities rather than in the countryside.[13] Food could be bought by the new urban population if they could find paid work; the urban economy thrived through abundant cheap labour and adequate food supplies to fuel it.

One of the biggest challenges for cities in temperate climates had always been how to provide their population with enough fuel for cooking and winter heating. It has been estimated that in medieval times a city of 10,000 people as far north as London needed 10,000 hectares of land given over to firewood to supply its domestic and industrial needs without depleting its woodlands.[14] A large, cumbersome industry was required for the tree-felling, transport and distribution involved, with between thirty and fifty cartloads of firewood per day clunking through the city gates, pulled by animals whose appetite for fodder represented a further land commitment.[15] London's population was much bigger – probably over 100,000 by the later sixteenth century. The firewood required to fuel such a city might occupy a circular area of land more than thirty-five kilometres across.[16] The coldest cities in northern Scandinavia required ten times more fuel per person than the mildest cities in southern Italy.[17] Given the superior fertility of warmer climates, it is hardly surprising that northern Europe had been so much less rich in useful energy and in urban size and complexity than southern Europe until the rise of coal.

London had, since at least the thirteenth century, supplanted some of this firewood requirement with coal.[18] In the area around Newcastle coal was so abundant that seams of it stuck up above the ground and could be easily extracted, then shipped cheaply down the coast to the Thames. Coal, particularly coal from near the surface, is a dirty and foul-smelling fuel. As early as 1288 people were bothered enough by the smell of coal fires for campaigns to arise against their use.[19] Stink or no stink, however, coal had its adherents: energy-hungry industries like brewing and dyeing were early adopters, and its greater intensity of heat was better for blacksmithing and for making lime.[20] It took far less land to produce coal than firewood and coal was more economic to transport than wood, since the same weight held two or three times more energy. Thus coal became a relatively cheap form of heat when land was under pressure before the Black Death.[21]

Demand for coal dropped as land became available for growing firewood during the century or so after the Black Death, but then, as industrial activity ramped up and the climate cooled further in the sixteenth century, coal made a comeback. Even with transport costs, it tended to be cheaper per unit of heat than wood, and with a growing emphasis on shipbuilding for trade, naval warfare and colonial expansion, by the early seventeenth century monarchs were explicitly legislating to encourage industry to move from wood to coal. So, from 1580, iron-making was not permitted to use wood from within thirty-five kilometres of outer London, where demand for building timber and firewood was at its most intense; from 1615 it became illegal to use anything but coal as the heat source in glass-making.[22]

Windows

The English glass industry struggled at first to move across from firewood to coal. The byproducts of burning coal can stain glass unattractively and its heat is more localized than the tall flames of wood, so the design of glasshouses (the premises used for glass-making and glass-blowing) needed considerable adaptation.[23] Yet once these difficulties were negotiated, the transformation wrought on the English glass industry by cheap intense energy was dramatic. Through much of the seventeenth century the best-quality window glass had been imported from continental Europe, even as the English glass industry took off.[24] The very best glass was from Venice.[25] By the 1740s, the coal-fired English glass industry was so sophisticated and substantial that it was producing, for export, mirror glass – perhaps the most technically challenging of all glass given the large panes required and the glaring obviousness of an uneven surface. English glass makers claimed as early as 1706 that their large plates of glass were selling in Venice itself.[26]

Through the seventeenth century, coal made glazing affordable for many more of the increasingly comfortably off English population. By the late 1670s it was worthy of comment when a rural area was so impoverished that the poorest houses still did not have glass windows.[27] While more and more people had some glass windows, however, they remained sufficiently expensive for the number of windows in a house to be a good proxy for the owner's wealth, such that it became for a time an ill-conceived method of assessing taxes in Britain and France.[28] The grim but unsurprising result was landlords and owners bricking up their windows to dodge the tax.

As the production of window glass rose sharply with coal, so did the size and quality of panes and frames. English wood-fired glass had generally been bad and its framing inadequate; the great scholar Erasmus of Rotterdam, who had made repeated visits to England in the early 1500s, concluded that terrible draughts from England's ill-fitting windows were a major cause of health problems there.[29] Before widespread use of coal furnaces, English window glass was made by taking a substantial blob of molten glass, as viscous as thick honey, on the end of an iron pipe and blowing through the pipe while spinning it to form a large glass cylinder. Working in a hurry, before the glass became too solid, the ends were then cut off, the side slit and the resulting fast-cooling curved sheet was flattened as best the workers could manage on a metal table. The results were prone to bubbles, distortions and uneven thickness, and thanks to impurities in the sand from which the glass was made, a pronounced green or yellow colour.[30] This poor-quality sheet glass would be carefully broken into small quarries, which were then leaded together to make windows in rattling metal frames. Views through them were limited, and they leaked cold air.

Through the seventeenth century this poor tradition was overtaken by huge improvements. The increased production of coal-fired glass fostered new ways of blowing, flattening and polishing window glass. The size and quality of individual panes increased, making ever-larger rectangular panes easier to achieve in place of the quarries of the previous century. Framing improved too. The rusting and distorting iron frames that Erasmus had experienced were replaced by increasingly expert carpentry producing close-fitting wooden window frames. A leading architect of the late seventeenth century, Sir John Vanbrugh (1664–1726), claimed that his windows were so airtight that even in a storm the candles did not flicker from any draughts in the huge central hall he had designed for the Earl of Carlisle at Castle Howard. So well-fitting were Vanbrugh's windows that 'all his Rooms, with moderate fires Are Ovens'.[31]

The figure at the heart of the English window revolution was Robert Hooke (1635–1703). Hooke was part of an impressive group in London who were at the forefront of scientific inquiry. Hooke's contributions were many and diverse, but among the most important was his work designing and helping to make the scientific instruments which gave the likes of Isaac Newton the data to develop and test their mathematical models of nature. Hooke worked closely with London's coal-fired metalworkers and glass manufacturers to produce better telescopes and microscopes, laboratory apparatus and measuring instruments.[32] After the disastrous Great Fire in 1666, he was appointed Surveyor to the City of London and collaborated with Sir Christopher Wren on the reconstruction of the city.[33]

Hooke's buildings included most of the earliest examples of a type of window that was to become ubiquitous in England and Scotland: the wooden sash window with a counterweight, such that the two panels can be raised or lowered to any point and remain there securely. This allowed precise control of ventilation and avoided the near-universal tendency of casement windows (the ones which open like a door) to come free during windy weather and slam open and shut.[34]

Sash windows – with heavy lead weights on systems of pulleys, hidden within a slender frame – are intricate pieces of machinery, the more so since a distortion in the timber could jam them permanently, and it seems that Hooke contributed to perfecting them. Initially this major advance in glazing technology was available from only a few of the best woodworkers in London, transported carefully round the country to the sites where the richest landowners were building their palatial country houses. In 1701, a wealthy merchant even shipped sashes from London to be used in his house in Boston, Massachusetts.[35]

Good-quality air in quantities adjustable by sash windows that could open anywhere from a crack to wide, joined many other new luxuries for the affluent. Today, opening windows are useful for keeping the temperature comfortable in hot weather or for bringing in fresh air and dispelling odours. In seventeenth-century housing, window ventilation was much more critical. The growing use of coal fires for warmth and the burning of candles or lamps for artificial light could rapidly make the air indoors unpleasant or even dangerous due to waste gases. However, too much air was a problem also. On top of the discomforts and dangers of cold, the air of city streets could be noxious. Sewage and other domestic waste continued to be dumped into the street in most areas until the nineteenth century. To this long-standing threat was now added that of an urban atmosphere poisoned by the sulphur and soot of thousands of coal-burning fires, domestic and industrial. With little understanding of germs and the spread of disease, dirty air or air at the wrong temperature was widely believed to explain outbreaks of illness. Good windows appeared to be a matter of life and death.

So when Hooke designed a new headquarters for a leading London medical institution, the Royal College of Physicians, after the Fire of London had destroyed their home, he was concerned not only with demonstrating stylistic sophistication – using the latest classical motifs from the Netherlands and France – but also with the technical demands of a scientific meeting place on a hemmed-in site in a dirty city.

He based his college, built in the mid-1670s, round a courtyard to secure a little space and air that could be kept relatively clean, and put the main room one floor up at the end of this court so that it would get as much skylight as possible through its tall sash windows. As a rough rule of thumb, vertical windows provide useful light to a depth of about one and a half times their height, so the emphasis on height in much seventeenth- and eighteenth-century room design was not purely for the grandeur of high ceilings.[36] Tall rooms also became stuffy less quickly from the candle and fire gases.

The new college had a dissection theatre (one of its members, William Harvey, had recently been the first to work out that blood circulated through the body) and Hooke treated the room almost as a large-scale optical instrument, working out good sightlines from every seat and lighting it carefully through large windows up above the roof height of the surrounding city to capture as much open sky as possible, and a big, heavily glazed lantern roof directly over the dissection table. The quality of glazing and framing required for a large area of glass roof like this was only newly available. Previously rain would have poured in with every shower.

Hooke's assiduous efforts to secure daylight indicate that it was still at that time the best form of lighting. A print of 1808 shows an elderly member of the college almost pressing

THE COLLEGE OF PHYSICIANS.

himself up against the window to get every bit of light for reading a letter. Beeswax candles were the gold standard for artificial light, burning cleanly and with a pleasant smell, unlike rendered animal-fat (tallow) candles or smoky oil lamps. They cost between three and three and a half times more than the next-best candle, tallow.[37] Yet even the hugely expensive beeswax candles were dim; an incandescent 100W light bulb of the late twentieth century produced over 125 times more light.[38] A great many activities were confined to daytime by these conditions, and those which took place after dark needed to take lighting costs into account, as suggested by the old gambling expression 'The game's not worth the candle' – when the stakes are too low for the host to make back the cost of keeping the lights burning.[39]

Bricks

If artificial lighting remained poor until the nineteenth century, the structural materials of London became very much higher-performing from the seventeenth century, thanks to coal. London, built in an area singularly lacking in durable building stone, had throughout the Middle Ages been a city of wood, straw, thatch and animal dung. Stone was imported by water for churches and the grandest houses and public buildings, but ordinary houses, as so often in medieval European cities, were wobbly-looking affairs of unseasoned, quick-to-distort timber and other easily procured organic matter. Rats and other vermin thrived in a cityscape that held no obstacles for them. The bare earth floors also presented a health hazard: even in the twenty-first century a health programme providing concrete floors for poor people in Mexico has resulted in children having 20 per cent fewer parasites than those living on earth floors.[40] Before today's widespread medicines and vaccinations, the death rate among children in insanitary urban conditions was appalling, for instance, around 15 per cent of infants living in the Cheapside area of London in the first quarter of the 1600s died before their first birthday.[41]

Disease was only one lethal threat. Cities like London resembled a continuous patchwork of bonfire materials, hardly separated by the alley-like streets, especially on upper floors, where many buildings stepped out and almost touched their neighbours. Industrial premises from bakeries to forges to breweries maintained large fires under or near thatched roofs. Every house had at least one domestic fire, with naked

flames for supplementary light on winter evenings. The poorest still cooked their food on open fires in rooms where the combustible walls and roof were just feet from the flames. Drunkenness was widespread, exacerbating the usual quotient of human clumsiness and stupidity. Urban fires of the early seventeenth century included one started by clothes drying too near a fire, one by an attempt to warm a cold bed and another by someone searching for something under a bed using a candle to see by.[42]

Everyone knew that fire would come suddenly and frequently, and they knew what to do when it did – tear down a set of houses round the edges of the area affected in the hope that the fire would burn itself out within that quarter rather than spreading further. Fire was sufficiently inevitable for some property contracts to agree medium-term boundaries 'until the first fire', after which they could be redrawn for more rational rebuilding on the flattened ash field, through mutual agreement between neighbours.[43]

Although Roman Britain had seen the use of fired brick, the material had disappeared with the end of Roman rule and remained rare throughout the Middle Ages. From the early sixteenth century it made a reappearance as a form of conspicuous consumption for the rich, but it really took off in the seventeenth-century city, initially as a fireproof membrane to be built round the outside of wood-structured houses in order to prevent any fire inside the house from spreading to neighbours.[44] This fireproofing became compulsory in London after the Great Fire, which destroyed around two thirds of the city. At first, brick was not much respected as a material, with an angry polemicist writing as late as 1703 that the standard of most brick houses going up in London was so low that the majority would not outlast their lease of fifty or sixty years.[45] This mistrust seems comical to a modern English reader, indoctrinated as we are with the idea that Georgian houses represent a gold standard for elegant design and timeless quality. Yet the early mistrust was not entirely misplaced. The best bricks tended to be reserved for show, used for the visible work on the front, while the party walls that increasingly came to take the structural weight of floors and roof would be made with shoddy 'place bricks' that cost around half as much and were considerably less even and robust.[46]

Clay is very widespread in London and, thanks to cheap coal, large numbers of bricks could be made quickly and locally for building projects of any size. After the Great Fire, astute brick-makers noticed that clay which had been adulterated by ash from the burned city could be fired faster and needed a lot less heat, roughly halving the amount of coal needed.[47] From then on all manner of urban rubbish – known, presumably through Protestant xenophobia, as

'Spanish' – was mixed into the clay to speed up and reduce the cost of production.

Manufacture of lime for the mortar that bonded the bricks into secure walls was also cheaper thanks to coal. In an aggressively competitive marketplace, builders would save pennies by buying it dry and mixing it themselves, despite its tendency to spit life-ruining gobs of highly caustic sludge at those who were stirring it.[48]

In five years of reconstruction after the Great Fire, around 8,000 houses were built, requiring somewhere in the region of half a billion bricks, each handmade in moulds and then coal-fired. The total demand for this scale of brick production must have been around 14,000 tonnes of coal per year.[49] Lime for the required quantities of mortar cost around half the energy of the bricks themselves, giving a total estimate for coal consumption for bricks and mortar for the 8,000 rebuilt houses as over 100,000 tonnes of coal, not including that used to produce window glazing, at around six tonnes of coal to every tonne of glass produced.[50] The 300,000 tonnes of firewood that, in an earlier era, would have been required to generate the same amount of heat would have been absurdly uneconomical – better to live in flammable wooden buildings.

The post-Fire reconstruction within the historic bounds of the city was just the start. London grew rapidly beyond its old walls. A much later satirical cartoon shows 'The March of Bricks and Mortar' (1829) as a great battle, the beleaguered trees of the countryside suffering defeat and death at the merciless iron hands of beings made of chimneypots and mortar-filled hods, while a brick kiln shells the surrounding

LONDON going out of Town. — or — The March of Bricks & Mortar.' —

hayfields with a barrage of hot bricks. The construction processes of the new terraces, such a familiar sight in London from the late seventeenth century, are each transformed into aspects of the military offensive, with demonic workers riddling the lime for mortar by throwing it through propped-up sieves, the noxious powder resembling the smoke of musketry and 'mortally wounding' (the pun on 'mortar' being one of many in the captions) a tree beyond. Even the rows of new houses in the background appear to be marching in step, their walls badly made and already cracking, their TO LET signs forming the battle standards of the triumphant army that is crushing the countryside.

The sight of massed ranks of brickworks, concentrated on the edge of the suburbs in proximity to building sites, and pouring black smoke into the skies, gave them a prominence that led critics to argue that they and other industries were responsible for the blackening soot and filthy air of the city. In reality, however, it appears that more than three quarters of all coal burned in seventeenth-century London was for domestic fires and cooking rather than for industrial purposes.[51] In the stone paving in front of each house a cast iron manhole cover was frequently lifted to pour substantial coal deliveries directly into the basement coal store, and each house contributed multiple chimneypots to a city skyline composed overwhelmingly of chimneys and church towers. The age of architecture in which very high operational energy costs are met by fossil fuels had started. It has not yet ended.

The Rise of the Property Developer

The Georgian London that emerged from this coal-fuelled transformation was unlike any earlier city. Already by the 1720s Daniel Defoe could write that nothing like it had been seen since ancient Rome.[52] It is almost an article of faith for many English people to admire the Georgian areas of cities like Bath, London and Bristol, and throughout the later twentieth century they were held up as ideals of public-spirited civic pride, in contrast with the work of heartlessly profit-motivated developers, who many felt were ruining British towns in the 1960s and thereafter. In fact, however, the developers of the late seventeenth century, who set the patterns of the Georgian city, could have taught even the toughest modern developers a fair amount about ruthless business practices. These early speculative developers operated in a context of general suspicion about the expansion of the city. London had risen from 5 per cent of the national population in 1600 to 10 per cent by 1700.[53] Some compared this disproportionate growth of the capital with a child suffering from rickets (an increasingly common disease caused by vitamin D deficiency under the smoky skies of London), its head too big for the body to support.[54] They proposed that building lots of new housing in London would suck agricultural labourers into the city, causing problems for farming and producing new slums with tendencies to criminality and disease that would harm existing residents and cause property prices to collapse. Monarchs had actively resisted the expansion of London since at least 1580, with the support of major agricultural landowners, and perhaps also motivated by fear of the unpredictability and radical thinking that cities tend to foster.[55]

Medieval London, its population constrained by limited access to food and firewood, had remained largely within the walls of the original Roman city. By the 1680s and 1690s it was importing an average of around 425,000 tonnes of coal every year.[56] This amounts to 3.4 million kWh of supplementary energy for London, with next to no farmland lost to its production. The commercial growth and industrial and technological advances that this flood of new energy provided produced a boom that could no longer be contained. The incentives to defy the law and build over the countryside became too powerful to hold back. As usual, when the energy is available the ingenuity will follow, whether technical or organizational.

The leading innovator of London's spectacular explosion was Nicholas Barbon (c. 1640–c. 1698). He was the son of a prominent Puritan preacher of the Cromwellian period, and according to some sources the younger Barbon had the pious but unwieldy middle name 'If-Jesus-Christ-Had-Not-Died-For-Thee-Thou-Hadst-Been-Damned'.[57] Barbon had picked up his father's charismatic crowd-stirring skills and was exceptionally intelligent. When he wrote a defence of the expansion of London he drew on an extraordinary range of clever arguments which tend to come to economic and historical conclusions decades or centuries in advance of those of his contemporaries. He worked out population estimates for ancient peoples that showed how Britain was now more populous by far and then explored the economic effects of meeting the housing demand this caused.[58] Although he did not ascribe its emergence to coal, Barbon's understanding of the new fossil-fuelled economy was dazzling.

In previous centuries, industries had been limited by the photosynthetic constraint. If the glass industry grew larger, so did its firewood demands. With rising demand came rising prices for a finite local supply and expensive transport costs for firewood from further away. The industry's very success was prone to strangle its future growth. With coal, this logic was reversed. The more you used, the more investment the coalmine owners could afford to make in mining infrastructure, so the price of coal could remain similar or even drop. To put it in the simplest terms, agrarian growth was prone to contain the seeds of its own destruction, whereas fossil fuel growth could (it seemed until we became aware of the greenhouse effect) continue until the world's reserves of coal, and later oil and natural gas, ran out.

Right at the birth of the fossil fuel economy, Barbon articulated the notion of growth that has dominated economics from then until now. By generating extra activity, trade makes richer the parties in control of production, transport and exchange. With fossil fuels removing the long-standing limit on the amount of energy that could be put into industry without driving up energy prices and reducing profits, Barbon's insight was fundamentally important: London's growth, economic and physical, did not need to mean the productivity of the countryside or any other city shrinking. Assuming that farming continued to improve, fossil fuels could potentially mean everywhere getting more populous and more prosperous simultaneously. The entire economy responded to the new conditions, from industry to commerce to transport, in the increasingly networked island. Between the 1650s and 1837, road improvements and the better logistics that came with increased demand meant that travel times fell considerably. The coach journey from London to Chester, for example, dropped from six days to twenty-two hours in that period.[59]

Barbon also applied the same intelligence that allowed him to see to the heart of the new coal economy to the many smaller, practical challenges of large-scale housing development. He knew it was 'not worth his while to deal little'.[60] Buying up, or sometimes stealing, plots of land around the fringes of the City, Barbon trampled over property rights and restrictions on development to throw up his new terraces of brick houses. When the young lawyers of Lincolns Inn Fields drove off his workforce by throwing stones, hoping to protect their quiet, green surroundings, Barbon came back next day with a larger force and fought them in a pitched street battle.[61]

Barbon was completely without scruple: anything that worked he would do. He exploited the inability of London's

creaking legal system to keep up with the crimes and disputes of a much-expanded population. Rather than borrow at 10 per cent interest he would leave bills unpaid and delay in every way possible. When at last he was obliged to repay the original sum and the legal costs of extracting it from him, this typically amounted to the equivalent of only 4–5 per cent interest.[62]

Similarly, Barbon occupied a brickworks to which he had no right, calculating that by the time the courts got round to making him give it back he would already have managed to extract all the worthwhile clay from it.[63] He served as a Member of Parliament for a spell, because it offered him immunity from prosecution.[64] Alongside these outrageous malpractices, Barbon's unending search for new ways of making development more profitable included some which are by now familiar from earlier energy booms in this book: he standardized the dimensions of windows and doors in his houses to cheapen and speed up production.[65]

Like many later developers, Barbon lived like a rich man, though the true extent of his fortune is not in fact known and probably rose and fell dramatically.[66] Again like later developers, he was hated by many, but he and others of his kind played a major part in allowing the City to break dramatically beyond its ancient limits. And again, even though some of his contemporaries professed to dislike him and his activities, they were in fact quietly hand-in-glove with him and others like him when they could benefit from it: aristocratic landlords partnered with experienced developers to maximize the profits they could make by building houses on the land around their London estates.[67]

London changed beyond recognition in a matter of decades, and kept up a comparable rate of change for centuries. The long-standing relationship between city and countryside was disrupted, with those living nearer the city centre within their lifetime ending up an hour or more's walk from the fields that had once been just minutes away. For the richest city-dwellers, squares with central gardens (safely gated against non-residents) and wide streets mitigated the effects of urban growth. For those in the slums, life became increasingly nightmarish. And as the city grew, the diseases nurtured among the poor killed even the children of the well-appointed rich. Due to the pressures of urbanization and terrible housing conditions, almost two thirds of children born in mid-eighteenth century London were dead before their fifth birthday.[68]

Perhaps the most conspicuous visual effect of coal in the city was on air quality. Coal smoke was deeply unpleasant to breathe, blackened everything, and corroded building materials. The diarist John Evelyn wrote as early as 1661 of how London was more like 'the Suburbs of Hell, then [meaning 'than'] an assembly of Rational Creatures'.[69]

Already by the time that Christopher Wren, Robert Hooke and others were rebuilding the City churches and St Paul's Cathedral after the 1666 fire, they were aware that coal would take its toll on the stone of their new buildings. The cathedral and its contemporaries seem to have been designed with soot in mind, their strongly modelled surfaces meaning that, as the smoke turned the exterior black, rain kept parts of the stone lighter than others and some sense of the decoration was retained. Even after decades of coal-free air, and many cleanings, St Paul's Cathedral is still beige from the soot leeching back out from deep in the grain of its porous limestone. It provides a poetic reminder of the way that its construction was originally funded: by a coal tax.[70]

CHAPTER 8

'That which all the world desires'

Victorian Liverpool

Coal into Motion

'I sell here, Sir, that which all the world desires to have – power.' This was the 1776 marketing pitch for the first economically viable steam engines, invented by James Watt (1736–1819), after whom the unit of energy is named.[1] It was no exaggeration. Through the chance of its abundance in Britain, cheap coal had already largely replaced firewood there, but with the advent of steam engines it could stand in for muscle power as well. The reach of the fossil fuel revolution increased enormously, liberating manufacturing and transport from the limitations of feeble humans and animals, and their space-hungry requirement for farmed food.

The first step towards the steam engine revolution was an increase in the availability of iron. Humans had been using iron for perhaps 5,000 years, but had tended to employ it sparingly because of its hunger for intense heat, which was difficult and expensive to achieve when provided by slow-growing wood that had been inefficiently processed into charcoal.

Even with today's highly efficient industrial processes,

making cast iron from iron ore takes around 5kWh of energy for each kilogram of iron.[2] In the conditions of the early eighteenth century this equated to at least 40 kilograms of wood, which had to be burned into charcoal, and then consumed, for each kilogram of iron.[3] Unless there was a very particular reason why iron was the only material that would do the job, why not just use the wood in the first place? As late as the sixteenth century, one of the grandest houses in the Liverpool area was built using a wooden structure, the beams held together not by iron nails but by wooden pegs. Over the course of a hundred years from the mid-eighteenth century, once the challenges of coal's contaminants had been resolved, pig iron production in Britain rose around sixty-five-fold, from around 35,000 tonnes to 2.25 million tonnes.[4] To do so without coal would have required an area of woodland big enough to cover almost the entirety of England.[5]

Iron produced with coal heat became cheap enough to use for whole building structures. The church of St George's, Everton, was built in 1813–14 on a beautiful hilltop just north of Liverpool, to serve a wealthy suburb of villas in big gardens. Behind traditional masonry outer walls, the familiar arch shapes of Gothic architecture appear transformed as if in a fairy tale. It is a church spun by an enchanted spider rather than, like its medieval precursors, lumped into place by grudging donkeys and sweaty workmen. The French Gothic aspiration to lacy, apparently lightweight structures is brought to a perfection medieval masons could only have dreamed of by the strength of iron, allowing arches just centimetres thick to support the roof and land delicately on the slenderest of columns.

The ambition behind this church had never been merely to make one beautiful building. It was the first of a group of churches using variants on the same system – the second moved away even from the solid masonry walls, using slate to infill much of the free-standing iron frame. The hope of the iron founder John Cragg (1767–1854) was that these kits could be mass-produced and shipped anywhere in the large and growing British Empire, to supply the colonial demand for Anglican churches.[6]

During the agrarian millennia, even buildings which showed off materials that had travelled a long way, like the Baths of Caracalla's display of exotic stones, had been largely composed of very local materials with minimum heat inputs. By the early nineteenth century, coal had changed the rules. It was worth making iron buildings in Liverpool for export to anywhere on earth. It is not clear whether any of Cragg's churches were ever exported, but as the century wore on iron church systems did become a feature of religious life in

British colonies. The age-old practical pressure to use local materials was starting to give way to the might of coal-fuelled industry, establishing a pattern that is still dominant world-wide, where cheap, bulky materials routinely travel thousands of kilometres in fossil-fuelled ships and lorries, to be delivered to construction sites.

By the time Liverpool's iron church components were in the foundry, the rotary steam engine was becoming well established. Only 250 years ago, before the steam engine, almost everything that needed to be moved for human life was moved by human or animal muscle, or not at all. Leaving the specialist niche of sail power aside, wind and water power amounted to 1–2 per cent of human power consumption in medieval Europe, 50–60 per cent was firewood and the rest was split more or less equally between food for people and fodder for animals.[7] The more work people and animals did, the more food they needed, and the more land was needed per head – land which became unavailable therefore for other functions.

Throughout the period covered by Part One of this book 'energy' tended to mean crops and the human and animal strength that they fed, or the heat of new-grown fire fuels. Rising populations throughout most of the period covered in Part Two required ongoing increases in the world's food supply, and the use of horses increased throughout the nineteenth century in Britain. But even as they continued to rise, these organic energy sources were overtaken in scale during the nineteenth century by fossil fuel power.

Already by 1800 the stationary steam engines installed in Britain had a combined force of around 22,000kW – the equivalent of almost 300,000 extra labourers working steadily. By 1840 that had risen to 150,000kW, and by 1900 they offered 6,700,000kW. By 1900, in other words, steam engines in Britain could do the equivalent amount of work of around 90 million extra labourers working steadily, or an average of more than two extra labourers for every resident of England, Wales and Scotland.[8] And this was just stationary steam engines. Coal was also powering trains and steamships, and millions of domestic and industrial fires up and down the country. By 1900 human and animal muscle together provided only 4 per cent of energy used in England and Wales; coal furnished 95.5 per cent. Food remained crucial, but the country was now coal-powered.[9]

Every aspect of human life was changed by it, and at the forefront of the changes were architecture and city development. Factories, powered first by water and then by steam engines, increased the scale of manufacturing from cottage industry to twenty-four-hour mass production. As the quantity of manufactured goods rose, new demand for bulk transport rewarded substantial investment in building and upgrading roads, canals, docks and warehouses.

By the 1830s a network of canals was well developed, offering transport ten or twenty times cheaper than taking the same weight and bulk of material by road.[10] There tends to be a kind of simple, sturdily built beauty to canal infrastructure, the aesthetic emerging confidently and unapologetically from robust materials, solid construction and patterns of use. The engineering of canals was clever, but it had all been worked out almost a millennium earlier in Song-dynasty China, just as the improved road network

was a familiar feature of ancient energy booms like that in Achaemenid Persia.[11] Why did these widespread infrastructural upgrades come so late in Britain? Because before the eighteenth century there was not enough being produced and transported to repay the initial cost of construction. By the late eighteenth century the pace of change was accelerating fast. At the heart of the changes were towns that, through accidents of location and resource wealth, grew rapidly into major industrial cities: Newcastle, Glasgow, Birmingham, Manchester and, the new super-port and northern commercial centre of the Industrial Revolution, Liverpool.

From Backwater to World Port

Though founded in 1207, Liverpool remained little more than a village by 1650 – just three or four streets, a castle and a natural inlet.

In 1715 a first dock was built into the natural inlet, allowing easier and safer unloading of goods, and by 1725 the city

Liverpool, 1650 — 1725 — 1785 — 1836 — 1896 — 1947

was growing fast, profiting from the colonial expansionism of a country growing in wealth and ambition with the coal boom. The appalling trafficking of enslaved people from Africa to the Americas proved a lucrative trade for ships registered in Liverpool.

By 1785 the city had gained more docks and continued to grow steadily, but the real expansion of the next fifty years, as steam engines began to convert coal energy into motion, was much faster.

The railway and tram expansion of the city by 1896 served a long string of new docks, built to accommodate ever-larger steamships. By 1947 car suburbs were joining the existing bus and train expansion of the city.

The most remarkable of Liverpool's surviving docks is Albert Dock, built all of a piece, the dock itself constructed in 1841–5 and the warehouses, which represented around half the cost of the project, finished in the following years.[12] The basin is over 31,000 square metres and the warehouses provided almost 120,000 square metres of space. It cost over £782,000 – more than 18,000 times the annual earnings of a building labourer.[13] Its grand scale shows the confidence Liverpool felt that trade would keep on booming. Its chief designer was Jesse Hartley, Liverpool's dock engineer from 1824 to 1860. Hartley was clever, hard-working and practical, seeking out solutions to a huge range of problems that might confront his docks, from legislative changes to silting caused by the Mersey estuary's constantly shifting sand and mud.

The new Albert Dock was secure, both to protect valuable goods and to reassure tax officials that they need not implement obstructive measures to prevent smuggling. Alongside theft and bureaucracy, another major brake on efficient trading was the cost of insurance. After major recent fires in Liverpool's wooden-floored warehouses, one of which had caused £323,000 worth of damage, insuring goods against fire was a substantial expense for traders.[14] Hartley made sure that no part of the new dock could burn. He learned every trick of fireproof construction from the cotton mills that had pioneered it, replacing wooden beams and columns with cast-iron ones and building each floor on shallow brick vaults.[15] Hartley's warehouses have since survived fires, wartime bombing and other incidents causing serious structural damage.[16]

Hartley secured a Scottish granite quarry to provide very strong stone for the dock walls. He used cheaper, smaller stones for the less vulnerable parts, with very big, robust stones where ships and carts would thump into them.[17]

He stood the warehouses on the very edge of the dock wall – convenient for raising goods straight from ships into the upper floors, but a considerable structural challenge.[18]

The architecture of Albert Dock is tough. It is immediately clear that these are serious working buildings. Yet their continuous cliffs of reddish-brown brick, their remorseless rhythm of windows and goods doors, and the orderly spacing of arches and columns together convey the power and confidence of one of the world's greatest ports. A kind of sombre, simplified classical architecture underlies the column shapes and the rhythms. This is the warehouse learning the lessons of the Georgian-style houses still being built just up the hill at the time. It impressed many, with Herman Melville, the author of *Moby-Dick*, comparing the Liverpool docks with the pyramids of Egypt, and even a normally matter-of-fact cotton merchant letting slip the poetic reflection that the Albert Dock warehouses were 'for Eternity, not time'.[19]

There are further pleasures to be found in the workmanlike details of Albert Dock. The grey Scottish granite of the dock front continues up the corners of the warehouses as quoins – emphasized stones at the corner of a masonry (brick or stone) building. The hardness of the granite was to protect the structurally crucial corners from careless carters bashing into them. The higher storeys use local pink sandstone instead, as there was less risk of damage there, and both transport and carving was easier and cheaper with sandstone than with granite. The corners of the warehouses that would have ships passing close to them are gently rounded to avoid catching a yardarm, which would have caused delays and potential damage to both ship and warehouse. A swing bridge has a neat slot in the stone dockside into which it can rotate to allow ships to pass, and when it did so its railings folded down to avoid catching loose ropes or rigging.

Hartley's search for technical improvements was unceasing. In 1847 he went to Newcastle to view a pioneering new crane technology, and promptly ordered two for Albert Dock at a cost of £1,000.[20] These new cranes solved a major problem with steam engines. Steam power is good for sustained, continuous effort, but very wasteful when used for stop-start tasks – the fire keeps burning but the power is not needed. By using a relatively modest steam engine to continuously pump water up into a tank with a weight on top, lifting the weight, the power was then available as water pressure, powerful and instant, whenever it was needed for operating the cranes. Albert Dock was the first of many docks along the Liverpool waterfront to adopt hydraulic power transmission, which had the additional major advantage of not increasing fire risk – the potentially explosive steam engine and flammable coal store were in a separate building to one side.

Its builders' confidence in Liverpool's future prosperity shines through in the Albert Dock. It turns out, however, that they underestimated the speed and scale of future change. By the 1880s 36 per cent of the shipping that docked in Liverpool was made up of the steam-powered, iron-hulled ships that were the new gold standard of international goods transportation, and by 1900 this had risen to 93 per cent, with only 7 per cent still sail-powered.[21] Albert Dock's water entrances were too narrow and awkward for bigger steamships, and from the 1880s its usefulness declined. In its short period of real success, however, Albert Dock's well-lit, fire-proof secure warehouses handled an impressive proportion of luxury imports.[22]

With the docks, as well as the region's industry, scaling up fast, the inland transport network needed upgrades too. The canal from Liverpool to Leeds was too busy, suffered theft, stopped dead when cold weather froze it and dropped capacity when drought lowered water levels and boats had to travel half empty. Meanwhile the monopoly held by the canal on bulk transport on this critical route was being gouged sufficiently aggressively that shares in the canal had risen from £70 to £1,250.[23] The first steam railway between cities, from Liverpool to Manchester, opened in 1830. Characteristic of the cities' spirit of innovation and experimentation, a race was held between possible locomotives to see which design would serve best. This event marked the final displacement of the horse by the steam train after six millennia of horses as the fastest and most powerful form of land transport. The competitor who used two horses on a treadmill as the power source encountered total failure when, with bleak symbolism, one of the horses fell through its floor.[24]

Within twenty years, railways had connected all the country's major cities and were extending into smaller towns and villages, linking the entire nation, with unprecedented power to move freight and passengers. Proximity to navigable waterways had long determined where it was worthwhile producing things for sale – coal in Newcastle had easy coastal access to the London market and so was worth mining, but in remoter coalfields expensive road transport had made mining uneconomic until a canal or railway was built. The sprawling rail network revolutionized the productiveness of almost every corner of the country.[25]

Rapid passenger transport shrank Britain, allowing architects to practise nationally, with clients from Liverpool becoming familiar with London buildings and vice versa. The immense terminus stations at which dozens of passenger and mail trains arrived every day were a thrilling new building type for the architects and engineers fortunate to be entrusted with their design. As one commentator put it in 1850: 'Nothing in the history of the past affords any parallel to such a spectacle.'[26]

Liverpool's main station was covered in 1867 by the largest column-free span of any building ever built, a vast arch of iron and glass leaping sixty-one metres across the platforms (this is eighteen metres further than the span of the Pantheon's dome).[27] But the hectic competition between rival railway companies did not let Liverpool enjoy the world record for long. The last of the major railway terminals to be built in London, St Pancras Station, beat it by more than twelve metres only the following year.

'Possibly *too good*'

St Pancras was the London base of the Midland Railway Company, one of the most profitable of all the nation's railways, linking a complex grid of thriving industrial towns across the Midlands.[28] Yet up to the 1860s the Midland Railway Company had suffered the indignity and expense of its London-bound trains having to arrive in a rival company's terminal at King's Cross.

The Midland Railway Company spent thirteen years and

over £5 million – in the region of 100,000 years of annual pay for an ordinary construction worker – threading and smashing their new railway from Bedford to central London, through or past a string of challenging obstacles including a canal, a gasworks, an old church with a crowded graveyard, a river and a mess of crowded and insanitary slum housing.[29] Around 32,000 people were displaced from their housing, uncompensated because poorer renters had no right of tenure. Even the judge who confirmed that the evictions could proceed protested at the injustice of the law he was bound to implement.[30] In a period when most believed in the physical resurrection of the body at the Last Judgement, the digging up of a long-serving and overcrowded graveyard caused widespread disgust and scandal. The poet Thomas Hardy later recalled in horror the fate of the bodies slopped around by the workers like 'human jam'.[31]

The ruthless displacement of living and dead was counterpointed by extraordinary achievements. Coal-produced iron and glass made the

vast train shed bright and lightweight, leaping in a single arch over all the platforms and tracks. The outward thrust produced by such a wide span would normally, as at Liverpool, have been resisted by lots of expensive wrought-iron ties criss-crossing in the air beneath the roof. At St Pancras's train shed, however, the outstanding engineer William Henry Barlow (1812–1902) rethought the problem and held the two sides of the arch together by running continuous girders underneath the platforms and railway lines – an iron bowstring keeping the slender iron trusses of the curving roof above from slumping. The cleverness of the structure made it possible to use ordinary riveted iron plates rather than the expensive wrought-iron pieces needed for less ingeniously engineered ironwork. The station itself (not including the grandiose hotel that fronted it) cost only £436,000 – under a tenth of the price of the line that connected it to Bedford and considerably less than Albert Dock.[32]

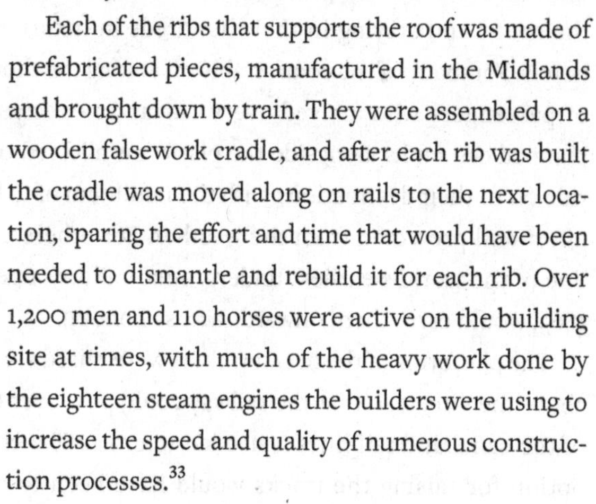

Each of the ribs that supports the roof was made of prefabricated pieces, manufactured in the Midlands and brought down by train. They were assembled on a wooden falsework cradle, and after each rib was built the cradle was moved along on rails to the next location, sparing the effort and time that would have been needed to dismantle and rebuild it for each rib. Over 1,200 men and 110 horses were active on the building site at times, with much of the heavy work done by the eighteen steam engines the builders were using to increase the speed and quality of numerous construction processes.[33]

The contrast between the overwhelming scale and

bustle of London railway termini and the coaching inns that were state of the art in the previous century could hardly have been greater: the new infrastructure served thousands of times the number of passengers, travelling at several times the speed and staying in their hundreds in luxurious new hotels. For many of the comfortably off among the thousands of passengers who arrived every day at St Pancras, well-planned cab ranks alongside the platforms allowed them to ride on to their destinations with the minimum of delay and effort and protected from the weather. For others the first line of the world's earliest underground railway, opened in 1863, would carry them onwards across London.

The platforms of the new terminus were raised six metres above the street, so that the railway could pass over the top of the canal and rail lines it needed to cross. The cheapest option for raising the tracks would have been to use the rubble from tunnel excavation elsewhere on the line to build up the ground level for the new station. In the event, however, the platforms and lines were raised on slender iron columns and supported on girders, in order to free the ground level for a vast beer warehouse. At the time, beer was the staple drink of many working Londoners, a habit left over from before the new sewerage system and clean water supply made drinking water safe. In the 1881 census London had more than 4 million residents. The Midland brewing town of Burton upon Trent supplied a full train per day of beer barrels to help slake their thirst. Arriving in St Pancras, the barrels rolled down a ramp from platform level into the ground-level warehouse under the station (now converted into a shopping centre and the main concourse for Eurostar trains), whence they were distributed to the pubs of London by horse-drawn carts.[34]

Milk also arrived in trainloads in the capital, brought speedily from farms tens or hundreds of kilometres away to supply London's ballooning population. Fish from the coast and meat, fruit and vegetables from a wide radius were now able to arrive fresh in London's great central markets. By the time St Pancras opened, the sprawling transport networks of the British capital were feeding what was by some way the largest population ever gathered in a single city anywhere in the world. With ever more food and drink brought in by train and ship, much of it canned in production centres around the world, the population would double again over the following thirty years.

This population explosion opened up a seemingly insatiable market to the businesses of the Midlands. The aspirations of the Midland Railway Company did not stop at passengers, food and drink. The station hotel they commissioned at the front of St Pancras acted as a giant billboard, advertising Midland building materials to London's buzzing construction industry. The flaming red Nottingham and Loughborough brick of the Midland Grand Hotel now seems merely Victorian – it was to become a common material in London – but when it first went up it was shocking amid a city still predominantly built of the beige and yellow bricks, swiftly smoke-blackened, of London and Bedfordshire.

The architect for this built advertisement was selected by competition in 1865, with different architects submitting their ideas for the design based on a published brief. With the increasing professionalization of the role of architect, the newly founded Royal Institute of British Architects had established rules for such competitions in 1839. The winner of the competition for the Midland Grand Hotel, George Gilbert Scott, made the ambitious decision to propose a design two storeys taller, and therefore much more capacious, than was specified in the brief, at an estimated cost of £316,000. The lowest estimate for any of the competition entries had been well under half the price, but the Midland Railway Company was attracted to the quality and size of Scott's scheme and chose it despite the expense.[35]

An imposing wall of bright red brick resulted, punctuated by towers and sweeping round at one end in a broad arc, with a large sloping carriage drive allowing vehicles to reach the station's platform level. It was given rhythm by abundant windows whose lavish decoration, like the fine fireplaces and columns inside the hotel, showcased a range of stones from the Midlands that could now easily be machine

cut, and brought to London by train to adorn buildings and monuments. The fantastical decoration mostly referred back to medieval European architecture, with Gothic arches and tracery, and a turreted and dormer-windowed roofline that would have been the envy of most French châteaux. As we will see below, the massive architectural production of the Victorian period was to result in a lively spread of styles and architectural ideas. St Pancras confidently set out its stand for Gothic, and for the beautiful, high-quality brick and stone that its trains could now furnish for other London projects.

Not everyone rejoiced at the building's conspicuous magnificence. One critic attacked it for being too grand – and above all too much like a sacred building – for its purpose: 'a complete travesty of noble associations, and not the slightest care to save these from sordid contact; an elaboration that might be suitable for a Chapter-house, or a Cathedral choir, is used as an "advertising medium" for bagmen's bedrooms and the costly discomforts of a terminus hotel; and the architect is thus a mere expensive rival of the company's head cook in catering for the low enjoyments of the travelling crowd'.[36] Even the architect himself mused that his architecture was 'possibly *too good* for its purpose'.[37]

To this way of thinking, the great Gothic churches of the medieval period had been and remained tributes to the most important consideration in human life: the relationship with God. To evoke their architecture with comparable ornateness in a mere hotel and station was tantamount to sacrilege. In energy terms, however, the elaboration and dominance of St Pancras made sense: medieval cathedrals had been so spectacular because of the patterns of agrarian landownership; Victorian stations reflected the new power of coal, and of the industries that turned fossil energy into material improvements in human life. However uncomfortable it might make the pious observer, the bagmen (travelling salesmen) of the 1870s were shaping the world far more forcefully than the bishops.

Churches and Their New Rivals

Even if religion was declining in economic importance relative to other institutions, religions profited too from the energy boom, with a rush of new sacred buildings springing up to serve the expanding population. Back in Liverpool, within a 300-metre stretch of one road in the new southern suburbs, five places of worship were built: a Greek Orthodox church, unmistakably Byzantine in its stylistic influences and shape, paid for by a small but affluent population of Greek merchants brought by trade to the thriving port city; an Anglican church which revived the feeling of mystery and ornamentation that its architect believed medieval Gothic churches had had; an octagonal Gothic-Revival chapel for an institute for deaf people; a Welsh Protestant chapel in the style of a grand medieval church and, designed by the same architects, a fine synagogue showing a number of loosely 'Eastern' architectural influences. A minute's walk further on, a Methodist church completed an impressive concentration of religious buildings for such a small area.[38] Elsewhere in the city were Britain's first mosque and a growing number of churches for the substantial Catholic population. Through the twentieth century two gigantic cathedrals were to be built, one Anglican and one Catholic, competing to dominate the city skyline.

The spectacular boom in church building was supported extensively by industrial and commercial private fortunes floated by Britain's energy wealth. The family of Liverpool merchants and stockbrokers who had commissioned the first-ever iron church were to go on to pay for eight more churches, most of them architecturally distinguished, in and around the city.[39]

Yet the boom in church building could easily disguise the fact that the role of religion in English life was increasingly subject to competition. Not only was coal-fuelled industry dominating the economy, in place of the landownership that had always given the Church its core revenue and worldly power, but the role of religion in social welfare and in education was also seeing powerful new competition, reflected in new buildings all over the city centre and the fast-growing suburbs. Board schools, independent of the Church, provided compulsory elementary education from 1880, a university college was founded a year later and numerous public libraries were built to serve suburban populations from the 1890s.

Everton Library

While the social welfare role of the Church remained important throughout the Victorian period, here too there was increasing secular competition from an expanding local government. A city with greater energy resources was able to support a wider range of institutions and to furnish them with a considerable diversity of buildings, grand or ordinary.

Workhouses systematized parochial provision for the desperately poor. They fed and housed their inmates, but in intentionally grim conditions to avoid their becoming more attractive than very bad industrial working conditions. Less punitive institutions were founded to help blind people and deaf people to support themselves through useful work.[40] A substantially enlarged hospital was built next to the university in the late 1880s to combat the threat of public ill health – diseases spread fast through badly housed working populations and there was the ever-present terror of a foreign plague arriving by ship in Liverpool from any corner of the globe.

Far more effective than the early medical efforts of the city were sanitary improvements. The country's first borough engineer, medical officer of health and inspector of nuisances were all appointed in Liverpool by 1847.[41] They tackled the worst housing and forced through the installation of proper sewers across the city. A new public water supply provided healthy water from a reservoir built in the 1880s, over 100 kilometres away in clean rural Wales. Arriving at the city, the water was steam-pumped up into water towers to feed the city's water main. Epidemic infectious disease and resulting death rates dropped substantially.

Living in Liverpool

Until the intervention of the city's energetic officers, the state of housing for many in Liverpool had been appalling. Dr Duncan, the first medical officer of health, had described it as 'the most unhealthy town in England' in 1843, having studied the spread of cholera through poor populations there since 1832.[42]

In the first half of the nineteenth century, the worst slums in Liverpool offered conditions even worse for many residents than pre-Black Death peasant houses. These new working classes had considerably more spending money than their peasant ancestors, but this could not compensate for lethal levels of overcrowding, disease and pollution. In a classic example of market failure, working-class housing was built by speculators wanting to secure profits and by employers who used it to claw back a proportion of their labour costs as rent.[43] The result was, even in economic terms, disastrous, with much labour lost to ill health. The human cost is incalculable.

Housing for the poor in early nineteenth-century Liverpool was generally built using superficially decent materials – brick, with glazed windows. Yet tenements typically had only one windowed front, either as back-to-backs or walled in tightly with warehouses or industrial premises next door. Narrow courts were accessed through narrower 'entries', or at best open only at one end. Ventilation was therefore very poor, with the lack of sewers, the many coal fires and the absence of washing facilities for clothes or bodies making for a fetid and noxious atmosphere, worsened by ubiquitous damp.

More than 20,000 of the poorest lived in cellars, which suffered from the worst conditions of all: to avoid paying for the removal of sewage from communal privies it was often spread over the court, and from this or rainwater making privies overflow, the cellars were frequently awash with human waste.[44] Cholera, typhus and tuberculosis flared up without warning as bacteria spread through the air, or trickled down from the cesspool into the drinking water supply to the local shared pump.[45]

Crowding and conditions got dramatically worse in the 1840s, as cruel mismanagement plunged Ireland into famine after the spread of potato blight through what had become the island's staple crop. Irish people fleeing starvation sailed to nearby Liverpool, in the hope of either finding work there or boarding a long-distance voyage to some part of the New World where they might find a viable life. Those who could not afford the onward voyage were stuck in Liverpool's worst slums, with up to sixty people cramming themselves into a four-room dwelling, frantic for any source of food or income.[46]

Religious and humane interventions by the better off and from within working-class communities attempted to improve the physical and spiritual well-being of the poor and varied from church building and temperance movements to Kitty Wilkinson's heroic work to help her neighbours avoid an 1832 cholera epidemic by opening up her kitchen as a washing room to everyone who needed it, allowing them to sterilize clothes and bedding.[47]

From the 1840s, systematic intervention by the new city officials led initially to the evacuation of the worst housing, followed by regulation to set minimum standards for new houses. The modest scale of these tens of thousands of new houses built from the 1850s onwards, and the social problems they have given rise to in some instances, have led to them having a mixed reputation in subsequent decades. Yet compared with ordinary people's housing from any other place or earlier period, the terraces of later nineteenth-century industrial towns in England, and the tenement flats that were their counterpart in Scotland, were astonishingly good. Liverpool's rules insisted on a roadway 7.3 metres wide, with pavements each side protected by kerbs and drainage channels, bringing the street as a whole to nearly eleven metres in width – easily enough to let air and light reach every house's front windows. The houses had, by the laws active in 1912, to be 5.5 metres wide – which makes for modest but not unpleasantly small front and back rooms, heated by coal fires, and on the upper floor two adequate bedrooms and a tiny box room. The light and air to the back rooms are assured by a minimum distance to the terrace behind and by a compulsory pathway for sanitation and fire access.

The results for each individual house were decent. The combination of a minimum standard with private development for profit, however, effectively determined the design of every working-class street: the developer would build to the highest density within the by-laws. By building tight-packed parallel streets devoid of green areas, developers could achieve a density of nearly seventy-seven houses per hectare. So that is what they built, and fast, over tens of kilometres of new suburbs around Liverpool.[48]

In 1869 Liverpool City Council was the first in the world to take the next logical step from regulating the market provision of housing: the council itself commissioned a block of modest flats for poor residents. It was the start of a movement that was to last over a century, with social housing coming to be seen by many as a necessity of industrial societies. Public housing projects were built in huge quantities

through the twentieth century, not only by socialist political regimes but also by centrist regimes, and even right-wing dictator states like Nazi Germany and Franco's Spain. This tendency for governments to improve ordinary people's housing arose in part simply from the increase in the amount that industrial societies could do. With so much more energy available for making and transporting building materials, it was at last possible to build systematically for everyone.

There may have been another factor at work too. Energy historians point to a tendency for coal-fuelled industrial societies to experience rising power and status for the working classes. Industrial patterns increased the demand for skilled and semi-skilled labour, and concentrated workers together, generating a sense of common cause and collective power.[49] Expensive equipment could feasibly be blockaded by striking workers, meaning that for investors and managers compromise was frequently better business practice than confrontation. Clashes between workers and employers could become bitter and entrenched, but overall the conditions of most industrial jobs and the life that went with them improved substantially over the period when coal was the dominant industrial fuel.

If the quality of housing for ordinary people rose sharply through the nineteenth century, so did the quantity of grander housing for the better off. As the city got bigger and dirtier with growing industrial and domestic coal burning, rich merchants began to abandon the age-old custom of living by their place of trade. A typical eighteenth-century merchant might live in a house near the docks with his own warehouse built on to or next to it. His own goods would be brought to his warehouse and sold on from it.[50] As the nineteenth century got under way warehousing became increasingly flexible: merchants would hire storage capacity from specialist warehouse owners only when they needed it, making the use of warehouse space considerably more economically efficient and allowing faster scaling up of provision.[51]

In the first three decades of the nineteenth century, these merchants and other affluent citizens moved to imposing new terraces on broad streets on airy hills outside the city centre. These handsome houses took their cue from London's Georgian housing: urbanity and a sense of privilege coming from orderly repetition and wide streets. Each house was built with fireproof brick walls, large windows and

elements of classical trim around doors and fireplaces. These fine streets were clean and weatherproof, with stone pavements for pedestrians and the roads themselves armoured by granite setts – shaped paving blocks able to resist the ferocious grinding of iron-bound carriage- and cartwheels. They were also safer, with street lighting provided by coal gas piped beneath the pavement to each lamp standard. Until then the nocturnal darkness in urban streets had been so extreme that even the Prince Regent and his brother had been successfully robbed at gunpoint just a few hundred metres from his palace in the early nineteenth century, and a learned society in Birmingham, attended by Boulton and Watt (of steam engine fame), had been called the Lunar Society as they met during the full moon to make their journeys home safer.[52] Through the nineteenth century, as gas lighting spread to less affluent streets and into building interiors, the scale of gas production rose. The skylines of Britain's cities gained the distinctive forms of gas holders rising and falling as they pushed the gas out into the network of distributary pipes. Demand rose continuously through the second half of the century. South London's main gasworks, near Vauxhall Bridge, built a telescopic gas holder of almost 17,000 cubic metres in 1847, the largest in the world at the time, but replaced it only thirty years later with one five times larger. Just ten years later that was adapted to double its capacity again to almost 170,000 cubic metres.[53]

By the 1840s Liverpool's most fashionable rich people were on the move again, this time to villas further outside the town, set in idyllic large gardens.[54] An ambitious development, Prince's Park, to the south of the city, erected large villas and terraces around a landscaped park, their light-painted stucco shockingly white after the sooty brick of the city centre. Prince's Park is unusual in surviving substantially intact. More often than not, these luxurious enclaves lasted only a few decades. By the early decades of the twentieth century they found themselves absorbed by poorer suburbs as tram and train networks grew and the large gardens of the villas began to secure high prices from speculative builders seeking to cram more housing stock on to the land. Once one villa was sold to speculative terrace builders, the neighbours had little incentive to stay. The illusion of a rural idyll that had drawn them there was shattered by the hundreds of working-class residents now arriving to live in the new houses nearby.[55]

It shows how cheap construction had become relative to working-class renters' wages in the late Victorian period that developers so often demolished without trace fairly new, well-maintained, large houses in order to secure a few more small new houses on the site. In most earlier periods, existing buildings had been converted or cannibalized for materials wherever possible.

As the number of Liverpool's rich continued to grow, it became profitable to lay out entire large estates for them. The grandest of these, in southern Liverpool, was centred on the new Sefton Park – 109 hectares of Arcadian beauty, the design of which was awarded by competition in 1867, ringed all around with fairly large plots of land for individual houses. The scale of the homes which the rich built on these plots from 1872 was astonishing. Each house was comparable in scale with the houses of the leading aristocracy of pre-industrial times, but whereas the great houses of earlier periods had been widely spaced, each presiding over the miles of countryside that provided the owner's energy wealth, in Victorian Liverpool mansions were packed in tightly along the front of the park, all but touching in some cases. The gardens of these cotton brokers, shipowners, international merchants and so on were only gardens, not fields for crops; the wealth and power of these new plutocrats came not through land but through coal-fired industry and the vast money economy that it supported.[56]

The bizarrely large size of the houses was clearly related to competitive display of wealth rather than to function – even with numerous children and a household of servants, the size and number of rooms in the largest Victorian homes is hard to explain in terms of need. Enjoyable but eminently dispensable separate functions abounded, to fill the space – billiard rooms, breakfast rooms,

grandiose stair-halls and even ballrooms.[57] Most of the individual houses in Sefton Park have now been converted into eight or more big flats.

The same considerations of self-definition and competitive display encouraged the Victorian rich to choose architectural styles which flattered their sense of who they were. These industrial wealthy were a new phenomenon in world history – thousands of people of common birth finding themselves richer than ancient aristocrats within a lifetime – and their architecture can be read as part of an attempt to explain to themselves their position in the world. Their architects drew on the great houses of the English past for staircases, dining rooms, salons and ballrooms, stained glass and external architectural styles, and on the exciting new technologies of the day for orchid houses and comfortable heating arrangements.[58] The results are often lumpish in massing but impressive in size, with some first-rate building craft in the areas where the owners chose to show off.

The park itself offered all the delights of the countryside, but heightened and exoticized: deer rather than sheep, a huge range of rare and beautiful trees and plants, an aviary, full-blooded artificial waterfalls and fountains, a grotto, a large boating lake, a heated iron-and-glass palm house and spaces for newly fashionable middle-class sports including archery, cricket, croquet and tennis. The whole was laid out ingeniously in a series of paths which curve so gently that they deceive the strolling visitor into believing that the park is even larger than it is. The planting of different trees was calculated to maximize this feeling of distance by avoiding monochrome expanses of identical foliage.[59] Typically of self-confident, international Victorian Liverpool, one of the two designers of the park was a leading Parisian landscape designer.

For the richest of all, the palatial country houses of the earlier agrarian aristocracy remained the ones to beat. Thomas Henry Ismay (1837–99), the founder and president of Liverpool's White Star shipping line, commissioned from the prominent London architect Richard Norman Shaw (1831–1912) an enormous Tudor-style palace in pink sandstone. Dawpool (1882–6) had interiors richly stuffed with wooden panelling, decorative plasterwork and gargantuan fireplaces, all lit by the brand-new technology: electricity. Beneath it was a railway that moved the large amount of coal required to fuel such a big establishment.[60] Shaw's historical styling did not indicate any nostalgic hostility to new technology. Elsewhere, for an armament manufacturer (Lord Armstrong, the inventor of the hydraulic cranes that Jesse Hartley had ordered from Newcastle), Norman Shaw was working on another country palace, built in several campaigns from 1869

to 1895, which combined an exaggerated Tudor appearance with the wonders of hydroelectric gizmos to light the house, run a lift and even power a rotisserie.[61]

Dawpool was built with a solidity that exemplifies Victorian Liverpool's preoccupation with durability. The city had grown suddenly from humble origins, but it would not, they hoped, be a flash in the pan.[62] At Dawpool, Ismay was eager to achieve the permanence of a great aristocratic country seat. 'Only the best would do.'[63] Solidity and permanence were the major form of ostentation in this impressive construction, right down to the rejection of commonplace iron nails in favour of brass screws, even in the stables, that would hold for ever.

The great irony of this extra emphasis on durability was that only eight years after Ismay's death the family abandoned what they saw as a white elephant of a house. Shaw wrote: 'Poor old Dawpool! I am sorry. Perhaps it can be turned into a sanatorium or a small pox hospital! I remember Mrs Ismay saying to me more than 10 years ago, that even then it had more than answered its purpose, for it had interested and amused Mr Ismay every day of his life for 15 years!'[64]

Mrs Ismay's joke is thought-provoking, suggesting as it does that thousands of tonnes of materials were extracted, transported, worked and assembled by dozens of labourers to satisfy a rich man's dreams of posterity. In the event, the lasting fame of the Ismays lies not with Dawpool but with their son, who became a celebrated hate figure in the censorious moral climate of the early twentieth century for surviving the sinking of his company's ship the *Titanic*. As for the house, some outbuildings survive, and two of its enormous fireplaces were repurposed as building façades elsewhere. Otherwise, even Shaw's wry hopes for its afterlife were to be unfulfilled. After a brief spell as a hospital during the Great War it was demolished in the 1920s, with great difficulty and abundant dynamite.

Steam and Styles

Coal-fuelled economic growth caused an astonishing boom in architectural activity. It was not only the very rich captains of industry and commerce, like Ismay, who were able to commission new buildings. Very large populations of industrial workers with a little money in their pockets, and substantial middle-class communities with considerable wealth, made available vastly increased resources for new buildings to meet their diverse needs: new places of worship, new facilities for entertainment, new civic buildings to scale up the governance and legal system of the growing city, new and larger workplaces and new kinds, sizes and quantities of commercial premises. The number and diversity of clients and building designers grew accordingly. Where architectural patronage in earlier periods – even in Renaissance cities – had rested in the hands of a restricted elite, the nineteenth-century explosion in quantity and ambition of construction activity supported greater diversity of approach and technique.

Much of the richness of Victorian architecture came from the ways in which architects and clients adapted a wide variety of older architectural styles to the new functions, new materials and new technologies made available by the energy bounty of coal.

Classical architecture remained a prominent influence throughout the nineteenth century in Liverpool, especially for civic buildings which sought to bring to mind the widely venerated artistic, literary and political traditions of ancient Greece and Rome. Liverpool's ruling class was determined that the wealth of the city should convert into appropriate prestige. Thanks to their ambitions, arriving in Liverpool by train was, and remains, an experience of architecture at its most elegantly ostentatious. The wide train shed which shelters the disembarking passenger was loosely classicized, standing on cast-iron Doric columns, but it does not feel like a Graeco-Roman building. No earlier period had been able to build anything like its span and transparency in iron and glass, and its message of engineering prowess drowns out its echoes of the ancient world. As you emerge from the front of the station, the classicism really takes over. In front of you, standing in open space like a sculpture, is St George's Hall (1841–54), a stone-faced building that looks like a

much-enlarged Greek temple. It was built as a grand concert hall and law court, put up by a mixture of city council money and private subscription. Its function is conspicuously civic – not commercial or industrial, but there to ensure the highest standards of urban civilization. A Latin inscription above the entrance boasts prominently that 'The citizens built this for the arts, for law and for counsel'.[65] Behind St George's Hall runs a line of wonderful civic buildings dressed in the pomp of stone and classical orders: another court (1882–4), a distinguished art gallery (1874–7), a large public museum and library (1857–60), with a beautiful circular reading room (1875–9), and a technical school (1901). The whole set piece feels like a proudly heightened contrast to the Victorian experience of getting out at the other end of the train line in Liverpool's neighbour and rival, Manchester, where grimy, profit-hunting warehouses and factories assailed the visitor from all sides.

Classical architecture was the obvious choice for the young architect of St George's Hall, Harvey Lonsdale Elmes (1814–47): this was a celebration of reason, culture, the rule of law and civilized city life. Built while the British Empire was expanding voraciously, it also evoked the Roman Empire's military and commercial success. Its handsome central hall is ringed internally by substantial pink granite columns that reflect ancient Roman monuments (though even steam-powered Liverpool did not use monolithic shafts here, like those the Romans transported by human muscle and sail power from Egypt). The granite may have been polished by steam-powered machines in Aberdeen, where it was quarried.[66] The main space of St George's Hall is roofed by a magnificent plastered brick vault influenced by the Baths of Caracalla. In case anyone thought that the echoes of Rome were casual mimicry, the fine wrought-iron gates to the room make it clear that Liverpool was laying claim to being a new Rome. Where Roman public projects often bore the initials SPQR (an acronym of the Latin for 'the Senate and People of Rome'), St George's Hall is emblazoned with the letters SPQL.

Liverpool's ambition to rival the glories of the ancient world went beyond style alone. During the construction of Exchange Buildings (1803–8), a new trading facility, Liverpool's leading architect, John Foster Senior (1758–1827), used monolithic stone columns more than 6.5 metres high. Weighing nearly ten tonnes, they had to be raised almost eight metres to their place on the façade. A regional newspaper boasted: 'We believe no pillar of that weight, composed of one entire stone, has ever, in any former instance, been raised to such an altitude, in any part of England.' Liverpool was much smaller than London, but clearly eager to show that it was capable of even greater wonders in the field of construction.[67]

St George's Hall saw 299 men working on the site at its busiest – slightly more than the maximum workforce on the Pantheon.[68] They harnessed the power of coal and steam to speed up the work; a tramway built parallel to the wall under construction carried a steam-powered hoist that worked 'with wonderful rapidity and precision'.[69]

Steam engines were also used in the finished building to achieve things Roman builders could not. Where the public baths of the Roman Empire had achieved underfloor heating through huge fires, St George's Hall was the best-implemented early exploration of a new possibility: buildings around which air was pumped mechanically. A steam engine in the basement drove fans that carried air from outside through a purification apparatus that sprayed water into it to remove the soot and smell of city-centre air. Next the air passed through a battery of water pipes carrying hot water in winter and cool water from the city mains in summer. The temperature-controlled, purified air was then pumped up through elegantly integrated holes at ground level in the rooms above. Once the crowds and the gas lighting had contaminated the air, it would be smoothly evacuated through further ventilation holes concealed in the ceiling decoration.[70]

It was the very latest technology, implemented by a Scottish doctor who had become a ventilation engineer out of conviction that dirty city air, especially at the wrong temperatures, was one of the biggest dangers to human health. Dr Reid's schemes ventilated the new Houses of Parliament in London and the inner areas of the increasingly large passenger ships that were now plying routes between Liverpool and other continents.

The dominance of classical architecture, almost total in the eighteenth century, was subject to significant competition by the time St George's Hall went up. Religiously motivated admirers of Gothic architecture not only criticized St Pancras for debasing the Christian architecture they loved, but also attacked the use of classical architecture. For these critics it was pagan, formulaic, and superficial. In Gothic architecture, wrote its most influential revivalist architect, A. W. N. Pugin, in 1836, 'alone we find *the faith of Christianity embodied*'.[71] To the Gothic Revival's leading theorist and critic, John Ruskin, classical architecture was all style and no substance: 'It mattered [...] nothing what was said, or what was done, so only that it was said with scholarship, and done with system. [...] A Roman phrase was thought worth any number of Gothic facts.'[72] Early Gothic Revivalists saw the dispute as moral rather than merely aesthetic, with Pugin contrasting the dignified provision for the poor that he believed had been provided in medieval almshouses with the intentionally degrading austerity of contemporary workhouses. He also contrasted an idealized view of a late medieval townscape, dominated by fine Gothic churches, with the abundant chimneys and the gasworks of an English town in the 1840s.[73]

If classical architecture was an obvious way for St George's Hall to lay its claim to ancient reason and civic greatness, the new university that Liverpool built from the 1880s chose Gothic architecture. In an English context Gothic architecture represented not only Christian piety but also the country's oldest universities, Oxford and Cambridge, whose medieval origins were part of their claim to a long and distinguished tradition of learning. Sure enough, Alfred Waterhouse's design for the university's Victoria Building had at its centre a version of one defining feature of an Oxford or Cambridge college, the dining hall, dominated by a grand fireplace and lit by tall windows decorated with stained-glass panels.

Once again, though, the Victorian tribute to earlier styles was proudly transformed through the glorious new powers of coal-fired industry. Waterhouse's materials are glaringly

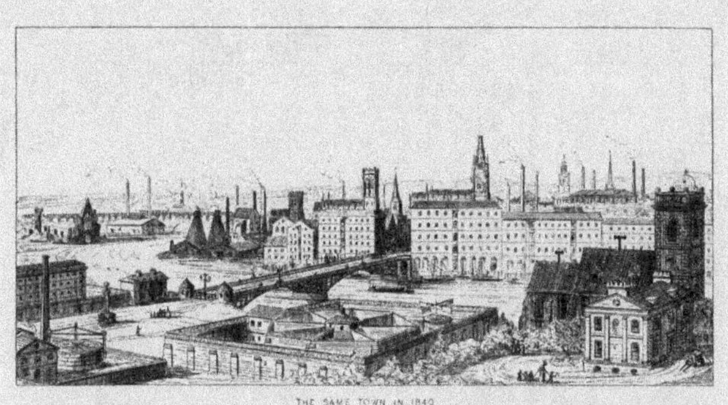

THE SAME TOWN IN 1840

1. St Michaels Tower rebuilt in 1750. 2. New Parsonage House & Pleasure Grounds. 3. The New Jail. 4. Gas Works. 5. Lunatic Asylum. 6. Iron Works & Ruins of St Maries Abbey. 7. Mr Evans Chapel. 8. Baptist Chapel. 9. Unitarian Chapel. 10. New Church. 11. New Town Hall & Concert Room. 12. Westleyan Centenary Chapel. 13. New Christian Society. 14. Quakers Meeting. 15. Socialist Hall of Science.

Catholic town in 1440.

1. St Michaels on the Hill. 2. Queens Croft. 3. St Thomas's Chapel. 4. St Maries Abbey. 5. All Saints. 6. St Johns. 7. St Peters. 8. St Alkmunds. 9. St Maries. 10. St Edmunds. 11. Grey Friars. 12. St Cuthberts. 13. Guild hall. 14. Trinity. 15. St Olaves. 16. St Botolphs.

– almost polemically – contemporary. Externally, the familiar local bricks used for plain walls are shouted down by a shocking red brick and terracotta (larger components of good, smooth brick clay, shaped before firing in a kiln) made in Wales and brought sixty kilometres to the site by train.

These hard, almost shiny red bricks exploited steam power not merely for transport, but for their manufacture. Georgian bricks had been produced on modest local brickfields, handmade at each stage, from the preparation of the clay, through its formation into bricks by pressing into individual wooden moulds, to its drying and firing. Victorian brick producers had scaled up their enterprises through the power of steam engines. Now the clay was worked mechanically, then forced through a brick-shaped orifice, producing a continuous rectangular sausage of clay which could then be cut by wires further along the production line into individual bricks. These were then fired by coal heat in large batches, producing bricks of far higher technical quality and regularity than Georgian bricks. They were ideal for the arches of railway viaducts that would need to resist the forces of hundreds of tonnes of speeding train, but also loved by many architects for their crisp edges and (at least until the air-borne coal soot darkened them) their vivid, consistent colours.[74]

The energy cost of these high-quality bricks was very substantial. One source suggests that even with today's more efficient machinery the energy cost of engineering bricks is almost 2,300kWh per tonne – the equivalent energy of more than 30,000 hours of steady human labour.[75] No wonder most agrarian projects, where heat and food were in direct competition for limited land, avoided materials like engineering brick in favour of mud brick, stone and wood, which could be procured and worked with labour alone, not requiring large amounts of expensive firewood.

Inside the Victoria Building too the materials are resoundingly new: tiles and glazed bricks whose patterns, colours and hard shine could not be mistaken for any genuinely medieval building. These tough, impervious surfaces were easier to keep clean and sanitary than the porous organic materials of older buildings. More modern technologies and materials are hidden in the structure. The building stands on the edge of the deep railway cutting into Lime Street Station, so an iron structure held up the walls and the fireproof concrete floors – an obvious improvement on the flammable wood or bulky stone vaults of medieval flooring. There is even steel used in the clock tower.[76]

The Victoria Building was facing a new challenge characteristic of its century in Liverpool and elsewhere: the types of facilities clients now needed had few precedents among the historic architectural styles that were so venerated. Erudite copying would not provide all the answers. In particular, the number and variety of rooms required in the new buildings were typically considerably higher, and the complexity of planning the rooms and movement round the buildings was proportionately greater. Gothic cathedrals and Roman monuments like the Pantheon had typically been planned around fewer, larger internal spaces, with less ancillary accommodation to fit in. Often, as in the cloisters of medieval great churches, additional functions had been housed in separate smaller buildings in the surrounding area rather than being integrated in the main building, as they had to be in the dense-packed institutional buildings of the industrial city. There were some precedents – grand houses and palaces of the Renaissance, for example – but some of the most interesting work resulted from Victorian architects trying to establish principles based on the architecture of the past and then applying them to the new shapes and functions of buildings. St George's Hall seems to evoke a classically derived ideal of making a clear, magnificent external statement. The complex brief for law courts, concert halls, prison cells and offices required a messy maze of different sizes and shapes of rooms, staircases and corridors. Yet the statement externally is clear, simple and powerful: vast sheer walls, largely without windows, given rhythm and interest by the tall Roman columns that stand before them. According to this way of thinking, why should a building, any more than a person, display its innards when it could be dressed in an elegant suit of formal clothes?

The university's Gothic building is, by comparison, almost bitty. Lots of different-looking, different-sized windows compete for attention. The rowdy red brick and terracotta nearly clash with the purplish brown of the ordinary local brick between. There is little consistent rhythm and no grandly over-scaled repeated element to stand comparison with St George's Hall's columns. In fact, where St George's Hall really looks quite like the ancient temples which gave it its inspiration, Waterhouse's building doesn't look like any actual medieval building. What it does well, however, is to abstract a principle from medieval Gothic architecture which it then applies to an entirely new building type. Gothic architecture, its Victorian admirers felt, was very good at allowing the fundamental shapes of the interiors to make the exteriors candid, interesting and beautiful.

Waterhouse's building incorporated a library, a grand staircase, a hall, a large semicircular lecture theatre and some smaller offices. Each is legible from outside – not hidden for the sake of architectural order, but giving rise themselves to a complex new architectural order that pursued the beauties of honesty and incident. For them the bombastic exteriors of buildings like St George's Hall were dissimulating what went on within. Arguments based on morality heightened the conflict: Gothic architecture was honest and classical architecture dishonest. These concepts of honesty and dishonesty have retained a great deal of power to flatter and wound architects across the many decades since, as they have fought to tame ever-more-complex briefs with the continual rise in size and sophistication of buildings driven by further energy revolutions.

As the century wore on, ever more styles and theories were proposed by growing crowds of architects and critics, as better ways of reviving the cultural wealth of earlier golden ages within the context of the steam-powered city – 'exotic' styles from beyond Western Europe, eclectic mixes of styles, Romanesque and every variety of Italian Renaissance. One of the most radical of these new sets of ideas was oddly under-represented in Britain, the birthplace of industrialism and its architecture. Art nouveau – new art – was an explicit attempt to design novel forms that reflected the materials of the industrial age, in particular the architectural use of metals. Charles Rennie Mackintosh's Glasgow buildings reflect some of these ideas, but it was in fast-industrializing Belgium and elsewhere in continental Europe that the style was to achieve its most luxurious and hallucinogenic peak of swirling opulence.

Architecture, with its high cost and public prominence, has always aroused strong feelings. The numerous factions within Victorian architecture fought angrily with each other in pamphlet and drawing, and the controversy extended into wider intellectual and political discussions. The argument about whether the new Palace of Westminster should be classical or Gothic was extraordinarily vehement, but ended in a surprisingly successful compromise – planning by a well-established classical architect, Charles Barry, and details by the exciting young Goth, Pugin, all presided over by the impressive ventilation stacks required by Dr Reid.

The extravagant Gothic architecture of Scott's Grand Midland Hotel arose at least partially from another political clash over styles. He based his 'too good' station hotel on the elevations he had proposed earlier for the Foreign Office building in Whitehall, rejected by the committed classicist Lord Palmerston, the prime minister, in the most celebrated clash of the 'battle of the styles'.[77]

Each successive movement had, behind its angry, ideologically committed avant-garde, a much larger body of architects who were less fanatically devoted to the ideas but found the style convenient or attractive. Most architects in Victorian England worked in every style or none, depending on the job, the surroundings or the client's preference. Even for the ideologues, architecture was a business and any architect who could not get enough work to cover wage bills and expenses was in trouble.

On the side of the many competing architects was the sheer amount of work thrown up by the rapid expansion of so many cities. As well as churches, housing and educational and civic buildings, new buildings intended for leisure pursuits were underwritten by the enormous cash economy of cities with many industrial workers, each poorly paid but collectively capable of spending power unlike anything seen in agrarian cities. One such industry was brewing, an activity that had been conducted in medieval times either in and for the households of landowners, or by private individuals at a more or less domestic scale, to supplement a meagre income by serving drinks to neighbours. In the Victorian period the scale of working-class spending, and the cheapness of the heat energy required for brewing, permitted larger scales of production that were considerably more efficient than older, smaller breweries. To a greater extent than ever before there were fortunes to be made from providing cheap things to many ordinary people rather than expensive things to the rich.

A leading brewer in nearby Warrington, Sir Andrew Barclay Walker, had amassed enough wealth by 1877 to be the leading donor to Liverpool's new art gallery behind St George's Hall. In Liverpool itself, the Cains Brewery was competing hard for market share. The brewery building itself was large, and encouraged drinkers to feel that the beer must be especially good because it came from such a fine and ornate building. As so often in advertising, the Cains Brewery promised the best of both worlds: lots of pseudo-medieval coats of arms and Latin mottoes hinted at a long-standing tradition, while hard, vivid red brick and terracotta promised the highest new standards of hygiene and technology.

The brewery building was only part of Cains's architectural marketing. All over the city they constructed lavishly decorated pubs. Perhaps the most spectacular, the Philharmonic Dining Rooms (1898–1900), is a Victorian fantasy of a grand Jacobean feasting hall. Everywhere is ornate plasterwork touched with gilding, rich-coloured, carved wooden panelling, mosaic floors and stained glass. Gaslights gave the interior a bright glow and glittered through windows frosted and decorated using a new industrial process to resemble the older handcraft of cut glass

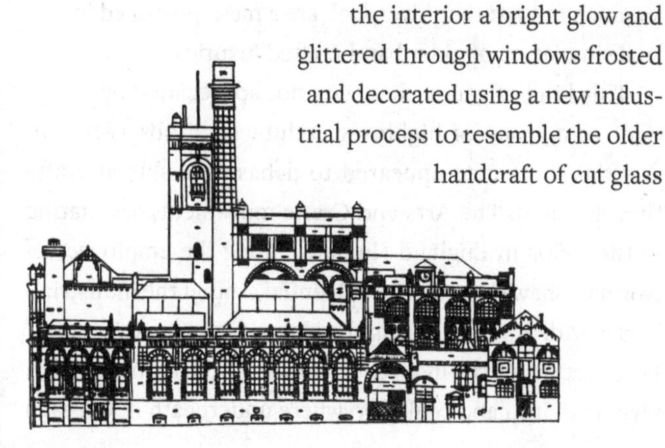

(pub windows were always obscured to preserve the privacy of men drinking away their pay packet).

Even the lavatories had a new opulence: the urinals are today widely believed to be marble and are certainly bulky and solid-looking. In fact, though, you can see the fine print pattern in the marble swirls on the glazed ceramic: these, like so much in Victorian Liverpool, are a mass-produced industrial simulation of older hand-crafted luxuries.

This kind of gaudy show was not appreciated by everyone. To a group of highbrow architectural observers, industrial short cuts appeared to debase the skilled crafts they imitated. The Arts and Crafts movement, originating in the 1880s in England (in part among the employees of Norman Shaw) but rapidly influential around the industrialized world, developed from the ideas of Pugin and Ruskin.[78] Ruskin attacked the industrial rich who lived in a bogus rural idyll 'with iron and coal everywhere underneath it. On each pleasant bank of this world is to be a beautiful mansion, with two wings; and stables, and coach-houses; a moderately-sized park; a large garden and hot-houses [...] At the bottom of the bank, is to be the mill [or factory]; not less than a quarter of a mile long, with one steam engine at each end, and two in the middle, and a chimney three hundred feet high.' In it, a workforce of up to a thousand 'who never drink, never strike, always go to church on Sunday, and always express themselves in respectful language'.[79] Ruskin and his follower William Morris saw the poor living and working conditions of the industrial working classes as a disgrace to society. They believed, with rose-tinted idealism, that before the catastrophe of coal-fuelled industry craftspeople had derived dignity and satisfaction from handcrafting their wares. Handmade crafts offered the labourer variety and opportunities for personal expression, as seen in the unrepetitive decoration in medieval churches. Factory production, they felt, had reduced humans to mere servants of machines, repeating the exact same action hundreds or thousands of times per day in a mind-numbing, slavish fashion unworthy of human dignity.[80]

In 1901, amid the idyllic countryside of the English Lake District, a believer in the artistic ideals of the Arts and Crafts movement built his country seat, Blackwell. Its architecture rejects the frenetic, morally corrupting modern city. It was designed by Mackay Hugh Baillie Scott (1865–1945), an architect who believed deeply in the superiority of medieval crafts and the historic relationship between the craftsperson and architecture.[81]

Blackwell is characterized by rich wood interiors and

simple-looking, solid stonework, all of it showing the variety and human touch of an architecture that aspired to collaborating with craftspeople rather than dictating to them. The interiors are dominated by fireplace openings big enough to house intimate family moments. Baillie Scott attached a poetic significance to these fires and the coal they burned:

> In the house the fire is practically a substitute for the sun, and it bears the same relation to the household as the sun does to the landscape. It is one of the fairy-tale facts of science that the heat and brightness from the burning coal is the same that was emitted from the sun on the primeval forests; and so the open fire enables us to enjoy to-day the brightness and warmth of yesterday's sunshine.[82]

Externally, Blackwell has the simple architectural expression of many medieval manor houses: good-quality stone dressings round the windows, the less prestigious materials of the walls concealed and waterproofed by white-painted render. The irregular windows, dotted around where the interiors required them, in the medieval fashion, are far smaller than those of Georgian houses. The relative gloom now seemed cosy to a generation whose familiarity with big windows by day and bright gas lighting by night made natural light into a tool for shaping mood and architectural expression, rather than the longed-for necessity it had seemed to the Georgians.

Like Shaw's houses (see pages 259–61), Blackwell suffers from gigantism. In particular, there are more big rooms than there would be in any real medieval house below the highest nobility. Large spaces were draughty and greedy for firewood, in medieval conditions, whereas Blackwell is centrally heated, so the fireplaces are for atmosphere and flickering cosy light as much as for heat.

Blackwell's Arts and Crafts evocation of a romanticized past is delightful, but in reality it was no more independent of coal-powered industry than were the houses round Sefton Park. The client was a Manchester industrialist and he could only build a house more than 100 kilometres from his business because it was conveniently close to the new railway that had opened up the Lake District as a location for weekend breaks. The basis of his fortune could hardly have been less in sympathy with Arts and Crafts ideals: coal-brewed beer, sold cheaply in gaudy pubs to the large working class brought together by Manchester's coal-fuelled factories.[83]

This underlying irony plagued the Arts and Crafts movement. Its adherents' resistance to the coal-fuelled industrial reality was only possible at all because of an affluent industrial middle class. William Morris's beautiful fabrics and wallpapers were produced using handcraft techniques that preserved older traditions and allowed independence and satisfaction to their makers, and as a result were only affordable to those who had access to substantial income from the factory-driven economy. Morris's argument that quality and durability would make his expensive products good value in the long run cannot have convinced many factory workers.[84] Even Morris's own ability to devote his life to art was funded by industrial wealth from his family's part-ownership of a company which mined lead and arsenic.[85]

The coal-energy boom of Britain in the nineteenth century was the biggest humanity had yet seen. As architectural historian Alex Bremner has put it: 'the machine had triumphed: the architect had little choice but to respond.'[86] Architects did. Even at the time, however, Victorian architecture had vociferous critics. Not only did the Arts and Crafts movement object to the underlying mechanisms of the boom, but leading architects saw the entire project of adapting historical styles to contemporary purposes as a sign of poverty of invention and imagination. George Gilbert Scott, whose magnificent Midland Grand Hotel put the materials and techniques of its moment to such good use, still felt reservations about having to hark back to Gothic architecture so much: 'I do not advocate the styles of the middle ages as such. If we had a distinctive architecture of our own day worthy of the greatness of our age, I should be content to follow it; but we have not.'[87]

Even as he wrote, however, a new building type was taking shape in the industrial cities of Britain, and increasingly America, that would eventually force architects out of their long-established habit of copying and adapting precedent. It was the office building.

Liverpool Offices

Already by 1844 it was clear that Liverpool was one of the world's most trade-focused cities. One observer felt that 'every house in Liverpool is either a counting-house, a warehouse, a shop, or a house that in one way or other is either an instrument or the result of trade ... and the inhabitants are nearly to a man traders or the servants of traders.'[88]

As we have seen, until the nineteenth century, most traders and artisans worldwide lived and worked in the same building or group of buildings.[89] With the expansion in Liverpool's trade, functions began to separate into different parts of town: warehouses by the docks, rented as needed, a grand house in the new suburbs and, increasingly, office space in the city centre.

As early as 1786 the Corporation had taken the boldly interventionist step of clearing city-centre land to build offices for brokers.[90] These unusually early purpose-built offices turned out to be the first swallows of a world-changing summer. Up to the mid-nineteenth century the new office buildings had a somewhat unglamorous functionalism, resembling some of the plainer Georgian houses. As one cotton broker later recalled of this period, 'anything in the shape of ornamentation, or in the least degree artistic, raised grave doubts as to the business qualifications of those who indulged in such fripperies.'[91] This hostility to show, combined with high land prices, meant that most offices were 'dark, damp, dismal, inconvenient and badly-ventilated places, situated in all sort of out-of-the-way and incommodious localities'.[92]

For the most important commodity, raw cotton, a special new building broke with this culture of bare-bones functionality.[93] With its monolithic columns, Exchange Buildings brought a new level of architectural and urban ambition to Liverpool's commercial architecture. It formed a grand public square just behind the town hall, the front of the new building having an architectural elaboration worthy of a royal palace.[94]

Its grandeur was particularly remarkable as it was an entirely commercial venture, built at the expense of shareholders who expected to profit from their investment.[95] The new Exchange Buildings contained a newsroom allowing merchants to keep abreast of international events that might affect their dealings. The rest of the building was largely composed of ground-level offices, with warehouse space for cotton above.[96] Warehousing might seem a wasteful use of prestigious space, but in fact the cotton trade relied on buyers personally inspecting the raw cotton – quality and density of bales varied considerably at this date, so the amount and value of the fabric that could be made from it differed too.[97]

This magnificent building was demolished and replaced after only fifty-five years. The cotton trade had become more industrially consistent and purchasers could buy it from reputable dealers after inspecting just a sample. The warehousing in Exchange Buildings was now an anachronism and maximizing the amount of well-lit office space repaid the waste and the cost of reconstruction.[98]

The new building (1863–70) reflects the changed tastes of Liverpool's businessmen over the intervening decades. As a satirical article put it with a sneer in 1868: 'offices are [now] designed by high-art architects at high-art prices, and are furnished by high-art upholsterers and cabinet-makers in a style of "princely magnificence".'[99] Many of the new offices and

banks had stone revetments that might have graced a Roman temple. Coal-powered extraction and transportation of Britain's own decorative building stones were accelerating, and some of the ancient quarries of the Mediterranean were being rediscovered and reopened.[100] Others were industrialized. From the 1850s, white marble was being extracted from Carrara in Italy by the firm of an American entrepreneur, using powerful new cranes and twenty-eight sawing frames, each able to hold 100 saws, producing building revetment at a speed the Romans would have found astounding. He shipped the resulting stone from a new pier he had built.[101] The glories of the ancient world could now be revived without having to strip *spolia* from revered monuments, and at a price that was affordable for upscale but by no means exceptional commercial buildings.

By this time the needs of office workers were firmly at odds with established thinking about what 'good' architecture looked like. Britain's leading architectural journal felt, reviewing the new Exchange Buildings, that 'the problem how to make an architecturally successful building, and yet give the amount of light which cotton salesmen seem to expect, is a task almost beyond the ingenuity of any architect'.[102]

The fundamental clash was between the need for daylight to work by and the desire of architects to produce façades which worked within rules of proportion and detail that architectural theorists derived from their favourite older buildings.[103] Historical-style buildings with very much larger windows than their precedents can indeed look odd, and architects toiled to reconcile function and form. Behind some of the stolid-looking stone-clad façades, with windows of a size sanctioned by architectural tradition, more substantially glazed back walls fought to bring in as much daylight as possible.[104]

In a handful of buildings, the architects dared to bring similar technologies to the street façade too – most famously in Oriel Chambers, by Peter Ellis, built in 1864. Ellis was viciously condemned for this infraction against good taste: Oriel Chambers was a 'vast abortion – which would be depressing were it not ludicrous', 'a sight to make the angels *weep*'.[105]

In order to capture light from front and back, the fireproof floors of Oriel Chambers stood not on solid brick walls pierced by windows, but on a frame of slender iron columns and beams. This idea of separating structure from enclosure was to prove revolutionary. Thanks to cheaper, industrially produced iron and then steel, a building could be very strong indeed, but with the structure taking up much less space than was needed by a brick wall. Not only could more of the wall now be given over to window, but higher buildings no longer needed thicker walls on the lower floors – floors which could be particularly valuable for shop premises with large display windows made possible by metal frames.

Another crucial innovation of the period, which made higher buildings suddenly attractive, was the passenger lift. The typical mid-nineteenth-century office in Liverpool had three main storeys, an attic and a semi-basement. Above three storeys the stairs were too much of a slog to attract good rents, so the attic was reserved for lavatories and a caretaker's flat. The bottom floors could be shops or more offices, and beneath the semi-basement was often situated a bonded warehouse. The density of offices was limited by this height restriction.[106]

With passenger lifts (once business decision-makers trusted their safety), considerably higher office floors became not only possible but prestigious, thanks to their views, their distance from street noise and their cleaner air. Early lifts appeared from the 1860s in the new Exchange Buildings and elsewhere.[107] Some were powered by water pressure from the new high-pressure hydraulic power main installed in 1888; a 570 kW pumping station north of the city centre forced filtered canal water through twenty-nine kilometres of pipes, providing an instantly responsive power source, free of fire risk, for cranes, goods lifts, passenger lifts and some industrial machinery.

Other new office blocks were denied high-pressure hydraulic power by the council, who refused permission for the mains extensions that would have connected them. These buildings were already running lifts raised by the power of the council's own drinking-water main. The council's main could manage less than half the speed for passenger lifts and cost twice as much, but the council was reluctant to lose the revenue.[108]

Liverpool's most striking example of the office architecture which developed from lifts and frame structures is the Liver Building, built between 1908 and 1911 on a waterfront site right in the city centre, made available by infilling a dock that was now too small to be useful. With cautious local authorities limiting building heights for fire safety reasons, at ninety-eight metres the Liver Building was to remain the tallest office building in Britain until the 1960s. The Royal Liver Friendly Society, for whom the building is named, started in 1850 as a cooperative mutual insurance scheme whereby numerous working-class people could pay a modest subscription and the society would cover their funeral and burial costs. With a large industrial population on modest cash incomes, such schemes became wildly successful and the leading insurers grew fast.

The Royal Liver Building was spectacular in its size and architectural display, with two large towers sporting four 7.6-metre clock faces, and surmounted by statues 5.5 metres tall and covered in gold leaf when new, depicting mythical birds. It was also a spectacular construction site, pulling together new technologies, including concrete reinforced with steel used as the structural frame (it was the first major building in the UK to use a reinforced concrete structure) and an electric construction railway which was moved from floor to floor as the building rose. The ambitious building site brought the average time for producing the structure of each floor down to just nineteen working days.[109]

This spectacular demonstration of the construction qualities of concrete overlooked the passenger embarkation point for Europe's leading transatlantic port: the business elite of two continents passed by in their thousands during construction, and if their trip took them a few weeks they would find when they returned through Liverpool that the Liver Building had grown substantially taller. The granite cladding was left off two of the lower floors until late in the construction process, enabling passers-by to see that the spindly-looking concrete frame was genuinely carrying all the weight of the building.

By the time the Liver Building was completed, however, the cutting edge of office design had left Britain and Europe for the United States. Indeed, the Liver Building itself looks like a crossbreed of the two great traditions of US office

design, Chicago and New York City. The influence is hardly surprising: many of the numerous coal-powered ocean liners that steamed away from the dock-lined waterfront of Liverpool in the late nineteenth century went straight to New York. Their powerful engines had seen crossings on the route drop from around seventeen days in the 1830s to five days in the early years of the twentieth century.[110] British architecture in the nineteenth century had been the most energy-hungry and fast-changing. That baton was passed across the Atlantic around the turn of the 1900s.

CHAPTER 9

Form Follows Fuel

Industry and Construction in the USA, 1850–1920

The Rise of US Coal and Farming

Since the arrival of Europeans in North America, their architecture had typically had clear roots in their countries of origin, with houses and churches reminiscent of those of Georgian England or other European traditions of architecture and building craft.

Through the nineteenth century, very substantial reserves of wood, coal and potentially rich arable land, allied to technologies derived from the British and European Industrial Revolution, transformed the USA with explosive speed into the greatest industrial and economic power the world had ever seen. In the 1890s the USA overtook the UK as the world's leading coal producer. Over the following two decades the population of New York rocketed from 1.5 million to 4.8 million. Many of the newcomers were recent immigrants, including a proportion of the 9 million Europeans who sailed from Liverpool to start again in the New World between 1830 and 1930.[1] Some died of disease, some were killed by the criminal violence and unsafe working conditions of a city that was changing far too fast for legislation and policing to keep up. A great majority worked their lives away with more or less success in the frantic boom-and-bust labour market. A tiny but highly conspicuous handful became rich beyond the fantasies of any earlier generation.

Land was cheap in America. In the mid-nineteenth century population density in Great Britain was around thirty times higher than in the USA.[2] Before fossil-fuelled industry this very large area and modest population would have offered only limited scope for bringing about an energy boom; making land into productive farmland had been acutely labour-intensive and generally a gradual process – generations of modest enlargements of the cultivated area. When it was done on a more systematic scale, as at Isfahan or Parsa, for example, it required organized migrations of populations of knowledgeable agricultural workers. The manual labour involved in making new roads and other infrastructure had historically been another significant brake on economic gowth. By the mid-nineteenth century in the USA, industrially produced metal tools were conquering difficult soils and factory-made barbed wire was turning the erection of huge lengths of livestock fencing into the work of hours rather than weeks, while reducing its maintenance to almost nothing.[3] The labour-intensive cash crops of the Southern USA, farmed by enslaved people, remained an important source of wealth, but, as the Civil War showed, by the 1860s the coal-driven industrial might of the North was even greater. The pre-Columbian peoples of North America, who had adapted with impressive effectiveness to the earlier energy technologies they had acquired from the European invaders (notably horses and guns), were increasingly harshly oppressed by the scale of the industrial economy and industrializing agriculture. The incomers were able to forcibly expropriate their land for the remorseless spread of European-style agriculture.

Railways were extending rapid bulk transport across thousands of miles of inaccessible hinterland, linking remote farmland with major population centres like New York and Chicago. In 1836 William S. Otis patented a steam-powered earth shovel which accelerated the construction of railways and reduced the requirement for manual labour: steam-powered diggers speeded up the construction of coal-smelted steel railways for coal-powered trains to carry people and goods huge distances in short times.[4] Even the agricultural buildings of the many new farms springing up across the land were in many cases starting to profit from cheap energy and large, competitive industries: corrugated-iron sheets, galvanized with zinc to slow down rusting (a technique requiring sustained heat over 420 degrees Celsius to melt the zinc), provided plentiful quick-to-build, weather-resistant roofing and walling for barns, sheds, housing and even churches.[5] Meanwhile, in the railway and lake terminals of the USA and Canada, ever-larger grain silos were being built of experimental industrial materials like steel-reinforced concrete to store bulk cereal crops in transit.[6]

Crafts and industries had always tended to foster rivalry between producers, even under the limitations of the agrarian energy economy. The amount of space and cheap heat available to US industries meant that thousands of times more activity could be competed over by thousands of producers at all scales. A clever new idea might make a company hugely profitable at a stroke, and the process of innovation and the improvement of systems at all scales became increasingly specialized. The amount of money changing hands meant that there was abundant investment available for promising businesses wanting to start or expand, and very wealthy individuals and groups would systematically buy out groups of industries to establish economies of scale from unified industrial production systems bigger than anything ever seen before.

Millions of new houses were turned out at unprecedented speeds for the industrial populations of city and suburb. America's vast areas of forest often meant that wood was the cheapest and readiest structural material, but even here industrialization was revolutionizing the speed and ease of production. A single steam engine in Peter McKone's workshop near downtown Hartford, Connecticut, in the late nineteenth century powered several saws, two planers, a moulding machine and a mortising machine.[7] For centuries the dominant European tradition of timber-framed architecture had employed thick pieces of wood, since thin ones split too easily when joined together with complex nail-free joints. The price of nails dropped 88 per cent in the half-century after the invention of a viable nail-making machine in 1795, and in 1832 a builder in Chicago, responding to the fast growth of the small town, the cheapness of nails, and a shortage of skilled carpenters and large local timbers, came up with a thin-membered construction system so lightweight that it became known as the 'balloon frame'.[8] Machine-cutting could get many of these slim pieces of wood out of a single tree trunk.

This cheaper housing construction, aided by the accessibility of suburban trams and railways, allowed US conurbations to house in capacious suburbs the very large populations of industrial and administrative workers attracted by growing urban wealth and opportunity.

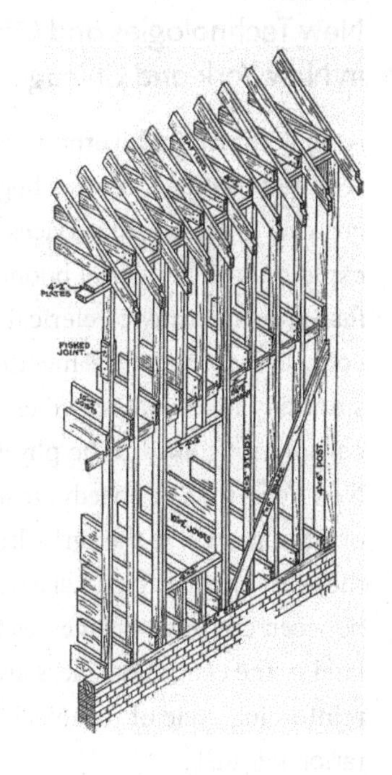

The cheapness and technical superiority of the new timber methods were not the only benefits perceived by some architects. By the early 1900s a young Frank Lloyd Wright (1867–1959), who was to become America's most admired architect of the twentieth century, was making a virtue of the country's distinctive emerging tradition of mechanically cut timber, praising it even above the great agrarian European traditions of woodcarving: 'All wood-carving is [...] likely to destroy the finer possibilities of wood [...]. The machines used in woodwork will show that by unlimited power in cutting, shaping, smoothing, and by the tireless repeat, they have emancipated beauties of wood-nature.' Moreover, they do it 'with such extraordinary economy as to put this beauty of wood in use within the reach of everyone'.[9]

New Technologies and Office Building in New York and Chicago

As in commercial Liverpool, the amount of administration required to run America's huge enterprises, plus the ever-growing professional services industry that supported and exploited the industrial boom, demanded desk space for a fast-growing army of clerical staff and managers. For the countless face-to-face meetings that made business run smoothly in an era before emails and video conferencing, corporations liked to be physically close to each other. In New York, with its speedy steamship links to the Old World, or Chicago, where the agricultural and industrial railways of the West met the Great Lake transport network, competition between cash-rich businesses for limited office space made land in the central business area extremely valuable. A self-reinforcing cycle of technical innovation and market adaptation began.

Initially, denser office buildings, not unlike those of Liverpool, began to overtake housing and other land uses that occupied prized city-centre space. Soon this first generation of commercial premises, often put up quickly and opportunistically, was replaced by larger and better-built office buildings, but these too were to be short-lived. By 1898 an insider could write of New York's office district that, even when large, fireproof and attractive, 'buildings of more than twenty-five years of age are all doomed to destruction'.[10] Indeed, buildings over thirty years old were bought and sold at the value of their empty site: 'They are utterly unsuited to the requirements of our modern day, yet they cost nothing to remove.'[11]

The new buildings that went up in their place were, in the famous words of one of America's great architects, Cass Gilbert (1859–1934), 'the machine that makes the land pay'.[12] If a new technology allowed more lettable space to be crowded on to a site, or permitted the completed offices to be let out a few months sooner, a lot of money would be generated for the client. Accordingly, there were big premiums available for construction firms and architects who could meet these targets, and for the patent-protected industrialized systems that helped designers and contractors to make the largest fireproof buildings possible in the shortest possible time. New energy technologies were rapidly harnessed to increase productivity, and to make each new block more tempting, capacious and profitable than its older rivals in the real-estate market.

With ever higher buildings, fossil-fuel-powered lifting technology revolutionized the construction site. One observer wrote of the construction of a new building in Manhattan's financial district:

> the first application of steam power to derricks used for the raising of iron, was on the Morse Building, in Nassau street, by Post & McCord, in 1878. One day Mr. Post, of that firm, sat impatiently watching the incalculably slow motion of the derrickmen in hoisting a girder. He began to think about plans for getting up some show of speed in that branch of the work, and, after figuring a little on the problem, very promptly concluded that steam power was what he wanted. That same day he purchased a boiler and engine, and the next day had it in operation on the job. In

speaking of it, Mr. Post said: 'I saved the cost of the boiler and engine on that job.'[13]

From the late 1870s, steam-powered chain-ladder elevators (a faster way of raising materials to the height they were needed) were reckoned to save between 50 and 80 per cent of the labour cost of taller buildings, and steam derricks were capable of twenty times the work rate of the old muscle-powered ones.[14] The labour and time savings were so immense that some claimed that the cost of constructing each cubic foot of good-quality office space dropped from $3 to 35c in the period 1873–98.[15]

If temporary construction lifts were changing the building site, permanent passenger lifts were equally vital in reshaping the skyline. Steam-powered and hydraulic lifts like those of mid-Victorian Liverpool were safe but slow. Initially, hydraulic lifts required a hole beneath the lift shaft just as deep as the lift would go high, so that the piston which carried the lift could sink into it. The advent of telescopic shafts, and from the 1880s electrical lift machinery, rapidly increased the speed and flexibility of lifts.[16] The limit to the number of storeys became the point at which the space taken up by the number of lifts then necessary started to eat uneconomically into the space that could be let.

As buildings got taller, their foundations needed to go deeper. Old techniques of bashing long supporting piles down into the ground from the surface level were no longer adequate and a new technique, caisson piling, was invented in which workers would dig the hole for the pile, shielding themselves and the hole from the pressure of the surrounding soil and groundwater by erecting metal tubular sheeting around the edge. For really deep piles the pressure could be enough to cave in the sheeting, so another energy-hungry solution was found: the top of the hole was sealed and then air was pumped into it to increase the pressure pushing back against the tube from inside. A mysterious lethal condition called caisson disease afflicted workers emerging from these pressurized holes, until doctors discovered that the rapid drop in air pressure was causing bubbles to form in their blood – a phenomenon known to divers today as 'the bends'.[17]

The structure above ground was equally challenging. The metal framing pioneered at Oriel Chambers was scaled up to tens of storeys in US office buildings. By the later 1880s, America's railway-building boom was slowing and the large steel industry that had grown up to provide the railway lines was looking for new markets. Structural beams and columns proved a lasting new sector for the industry, which continued to expand and to improve techniques and products, allowing building heights to rise further even than with the more brittle iron structures that went before.[18]

Construction sites were remarkable places. Steel erectors

and others working high on the buildings operated in conditions that combined an ethos of risk-taking personal heroism with competition between teams, later made famous by the photograph of the Rockefeller Center's steel teams lunching on a girder at breathtaking height. The dark side of these buccaneering years was poor site safety and little economic security for labourers. Other prominent US industries, including steelmaking, saw violent clashes between workers striking for improved safety, pay or conditions and paramilitary strike-breakers hired by management. With a premium on construction speed, New York's leading office construction companies went to a lot of effort to secure their labour supply, unionized and not, against strikes.[19] The rewards for effective management were rapid, high-quality construction projects and enormous profits.

In New York the race for height combined with restrictive building codes to produce some perverse results. George B. Post's 114-metre-high Pulitzer Building (1889–90) was lumbered with walls 2.74 metres thick on the lowest storeys, shutting out light and wasting lettable space. In Chicago the building regulations were more enlightened, allowing

steel-framed buildings to use only enough masonry to heat-proof their skeletons in order to prevent buildings from collapsing in the event of a fire. This allowed even the tallest offices to have thin façades at street level, avoiding the extra weight and expense of surplus façade masonry. Post's clunking lower walls could have been less than fifty-one centimetres thick under Chicago's building code.[20]

Building high made all sorts of things less straightforward. The history of skyscrapers has often been told as a structural challenge, but the extra height introduced major new difficulties, for instance in providing fire hydrants and water supplies at appropriate pressures at such different heights. Air flow in high buildings can become a howling vertical wind, forcing open doors and scattering papers. The revolving door solved the problem at ground level and clever ventilation systems furnished the right amount of air at a comfortable temperature to all floors. Huge numbers of ingenious patented solutions competed to supply better pumps, regulators, sealed components and pipes, which could be installed rapidly, but which would deal with the pressure differences between basement level and tens of storeys above the pavement, while not breaking down or occupying too much potentially profitable space.

A huge, highly competitive industrial sector, powered by coal, revolutionized every aspect of mechanical and electrical servicing (the provision of power, ventilation, heating, lighting, fire sprinklers and hydrants, and so on). Architectural decoration too could be produced by subcontracting businesses like the Architectural Sheet Metal Works, operating on New York City's East 144th Street in the 1890s. Its proprietor, Mr Westergren, who had himself started out as a workman, ran, among other machines of his own invention, what was thought to be the largest sheet-metal press then existing. At almost 5.5 metres long and weighing twenty-five tonnes, it could press decoration into a metal sheet almost four metres long using a pressure of up to 300 tonnes. With speedy construction increasing profits, these large decorative components could be fixed on to new buildings far faster than traditional forms of craft decoration and were widely adopted, ornamenting Carnegie Hall and many other New York buildings.[21]

The pattern we have seen throughout the book, of increasing energy levels promoting greater specialization, remained the case in skyscraper production. With so many subspecializations involved in these big, complicated buildings, the all-rounder building contractor, who directly employed all the specialists required to build, was a thing of the past. Now subcontractors would bid to provide the aspects of the building in which they specialized and would manufacture and install their components under the overall direction of the 'general contractor' – a new coordinating role that emerged in these decades.[22]

So successful was the ability of coal-fuelled industry to produce the technologies needed to build fast and high on valuable sites that office buildings grew rapidly in elevation and bulk. The economically perfect skyscraper of 1915, Equitable Building, returned 5 per cent to its investors annually by managing to pack 110,000 square metres of floor space into only 4,000 square metres of site. The largest office building in the world at the time of its completion, the bulky, looming

presence of its forty storeys on Broadway caused dismay, prompting fears that, as the market drove the replacement of lower existing offices with thumping beasts like this, the streets of New York would become dark and airless.[23] In response, a new zoning ordinance was brought in in 1916 to ensure that new high offices had to step back as they went up, allowing more daylight and air to penetrate to street level far below.

On building sites and in finished buildings, electricity was starting to make its presence felt from the 1880s. A new way to carry energy rather than a new source of energy, electricity had been the subject of scientific experimentation since the sixteenth century. By the later 1800s it was well understood in theory, and it was possible to turn steam engine power into electrical power on a fairly large scale. However, to use electricity for lifts, lighting and other functions that could already be served by existing technologies required a substantial investment in infrastructure, and in the early stages the advantages offered by electricity were relatively marginal (bulbs burned out fast, expensive equipment needed expert maintenance and the price of the energy itself was very high).

New York and London both saw their first area electrification schemes in 1882, and despite initially slow uptake, the new technology gained ground over the following years.[24] Many felt nervous of this mysterious new phenomenon, but it was safer than familiar gas for lighting and far better than steam for mechanical power: electric motors could flip on and off almost instantly as needed, whereas steam engines came on slowly after much wasteful warming up and required a skilled operator's constant attention. Electricity, once it became more than a conspicuous toy for the richest individuals and organizations, was considerably cheaper to install than a pressurized hydraulic power main. It was ever more clearly the technology of the future.[25] Electrical power was also essential for the telephones that were to change the course of communications from the 1880s on.

Electric lighting would radically reshape the office building, but not for several decades. Until the Second World War, daylight remained the gold standard. The dominant type of office work was paperwork, requiring clear light to read, write and, increasingly from the 1880s, type by. On winter evenings the gas lighting that had run off a main in New York since 1825 remained a good option until the early twentieth century. Gaslight was fairly powerful, relatively low-maintenance and not prohibitively expensive. It did, however, produce heat, together with waste gases that could lower concentration or even damage health. In other words it was expedient, but daylight from big windows was far better.

Electric lights were an improvement – free of waste gases

– but fairly dim. A typical desk lamp of the early twentieth century produced around three to four times the light of a candle thirty centimetres away. For contrast, outdoor light on a cloudy day is around 100 times brighter and even through windows daylight was twenty to thirty times brighter than the typical desk lamp.[26] To get up to the light levels recommended for reading in a 1916 report would require the equivalent light levels of two or three desk lamps. Incandescent bulbs got very hot and, used in large quantities, could make an office intolerably warm.

To maximize natural light, office buildings were designed with tall, broad windows. As light levels diminish further away from the window, deeper offices produced lower-value space, leading to an overwhelming tendency for office buildings to be designed with 6 to 8.5 metres from window to back wall.[27] The maths of daylighting, land value, plot size and shape, setbacks mandated by the 1916 Zoning Resolution, and space for lifts and services tended to determine the most profitable size and shape of New York skyscrapers. That in turn established the value of the site. The distinctive shapes of New York's 1920s–30s skyscrapers, including the tallest of them all, the Empire State Building, arose from this series of formative financial, technical and regulatory pressures.

The Cathedral of Commerce

In the early 1910s, decades of technical, industrial and organizational innovations came together to make Cass Gilbert's Woolworth Building break record after record. It was the tallest building in the world at its completion in 1913 and a prodigious achievement in construction. Its steel was

placed by teams of 180, working eight-hour days. Without spare land on which to store and prepare materials, steel needed to arrive on site in the correct order and at the time it was needed, whereupon between thirty and forty workers unloaded, sorted and secured it in its approximate final location in the structure. Two storeys beneath them as they went up were ten erectors, adjusting the beams and columns to a tolerance of less than ten millimetres. They were accompanied by four groups of four riveters, who averaged 300 rivets in a shift, their speed considerably increased by an electric compressor in the basement, which pumped air into long hoses that fed their powerful rivet guns. The building process had changed so much since the careful joiners of the agrarian age, skilfully handcarving their complex nail-less joints to connect wooden structural beams. Even the much shorter time taken to bolt metalwork together was too much of a delay for the Woolworth Building. Rivets were very quick indeed, but even they were superseded within years. By 1930 an even more energy-hungry connector was starting to take over in steel construction: electric arc welding, where a powerful current creates a short arc of plasma between 3,000 and 20,000 degrees Celsius, which can then be applied to the area of metal to be melted, joining elements robustly.[28] A typical modern arc welding machine has a power of around 3.84kW – the equivalent of over 50 labourers working steadily.[29]

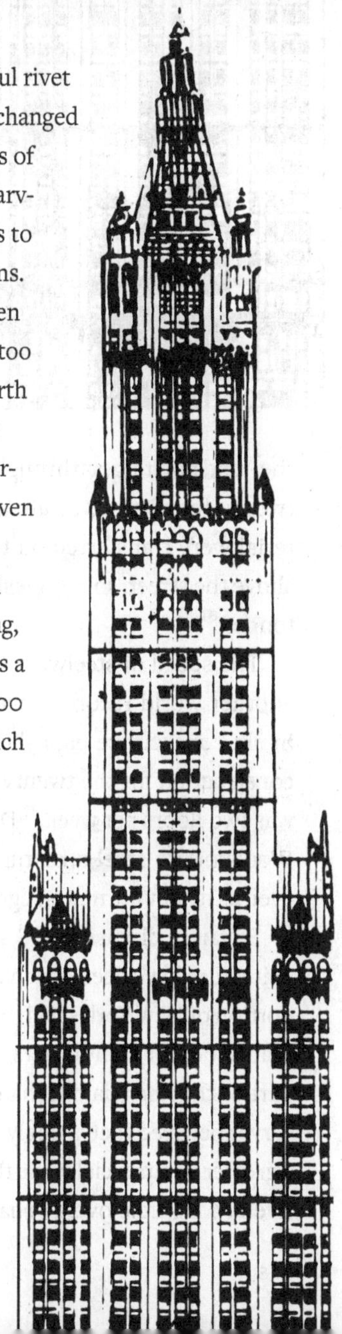

Painters, floor erectors and a range of other highly specialized teams followed the steel workers up the

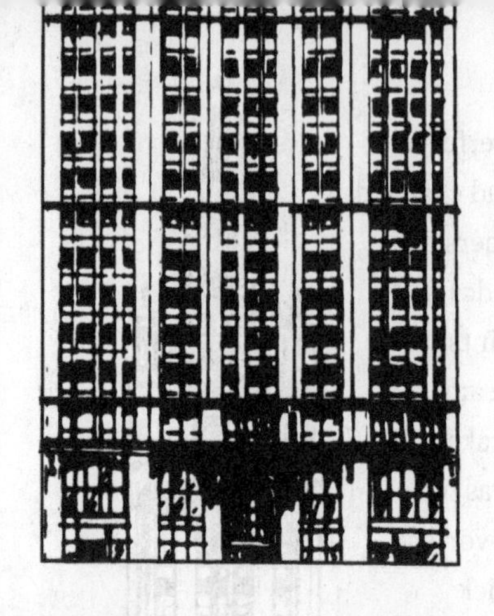

tower, each in their turn, as it rose.[30]

Electrical-powered rising scaffolding was first used on this job, as were new, faster, more powerful hoists, capable of raising twelve-tonne members at over 300 metres per minute. The boilers powering the fast hoists got through twenty tonnes of coal every twenty-four hours. In all, around 200,000 tonnes of material were transferred on to the site's storage area, raised above the street, sometimes in consignments as large as fifty tonnes.[31]

The speed of steelwork on the Woolworth Building was another world record broken on the project: 1,046 tonnes in six consecutive eight-hour days. The average speed of construction for the twenty-eight storeys of the main block was two floors per week.[32] Despite this staggering speed, the Woolworth's steel structure was becoming old-fashioned even as it shot up: newer, lighter steel frames were coming in.

So far, in discussing US skyscrapers, the aesthetics of architecture have been conspicuous by their absence. In a proportion of New York's high buildings this is a reflection of the reality, but for many others the architect's artistic skills were part of the financial mechanism of the scheme. An 1896 observer believed that 'a finely designed office building has now a greater commercial value than one that is badly designed.'[33] Even though the overall shape of buildings could largely be determined by financial and technical calculations before the architect was appointed, architects had an important role in the fast-developing new realm of marketing.[34]

Marketing was a new weapon in the fierce competition within the largest industrial boom the world had yet seen. In a marketplace crowded with the products of coal-driven factories, getting noticed was a major challenge. Frank Woolworth, the man who commissioned the Woolworth Building, was at the forefront of innovative marketing in the shops that had made his wealth. His business model informed the attitudes that come through in his great tower.

Woolworth was an archetypal new powerful person of the coal economy. His shops sold mass-produced factory goods, with all products priced at either 5c or 10c, making it easy for the vast industrial working class to budget their limited funds with confidence, knowing they wouldn't be embarrassed into an overspend by smooth-talking sales staff. Simple pricing and self-service displays deskilled staff so that Woolworth could pay them less, train them less and not worry about poor staff retention. He exploited the labour market's lower demand for women workers to pay his 'sales girls' only $3 per week – well below the 'poverty line indicator'.[35] His business model ran on tiny savings and profits, each multiplied by the enormous scale of the company. With each cent on an individual transaction mattering so much, Woolworth micromanaged tiny sums. The habit of watching for cents was so ingrained that he was rumoured to have kept his private secretary back late to search behind his grandiose furniture for a dropped quarter.[36]

The vast wealth he amassed through these tiny savings allowed him to live at a level of opulence that many agrarian monarchs would have envied. His thirty-room house – nearly palace – on Fifth Avenue and 80th Street, New York City, was fabulous in its magnificence, complete with an automatic organ that played tunes from punched cards.[37] He toured Europe, visiting great buildings and attending leading operas.

When he decided to build a new company headquarters in New York, his approach embodied both sides of his coal-fuelled commercial wealth: his Europhile sense of grandeur and his aggressive financial trimming. He tried to chisel prices down wherever he could and at all scales, from his complaint that a telephone messenger boy's pay of $2.50 per day was excessive to his attempt to persuade the contractors not to charge a profit margin (he argued that the prestige of the job was all the profit they needed). The head of the contracting firm, Thompson-Starrett, later recalled that he 'had the feeling that Mr Woolworth was turning on me, as if it were a fire hose, his customary way of buying goods for his five and ten cent stores'.[38]

Yet, at the same time, Woolworth wanted his building to be a 'cathedral of commerce' that would show off to the world his personal success and that of his company, and which would attract a good class of tenants paying high prices – as with the Liver Building, his company only used one floor of it, the rest being rented out.[39] Thus the Woolworth Building was designed to achieve a string of superlatives: highest in the world, best fire precautions of any (the power of the hydrants was shown by soaking streets around for three hours, with water sprayed from the building's upper storeys in a wide arc more than adequate to put out any office fire) and

GENERATORS AND MAIN SWITCHBOARD IN THE ENGINE ROOM

BOILER ROOM

ONE OF THE WORKSHOPS

a vertical sewerage system 'as large as that of a small town'.[40] In the basement of the Woolworth Building a coal-fired power station, viewable by curious visitors, burned anthracite coal brought by train regularly from Pennsylvania and stored in a 2,000-tonne coal bunker in the building. It could generate up to 1,400kW, lighting thousands of bulbs, running ventilating fans and 850 exactly synchronized clocks, and rushing twenty-seven high-speed elevators up and down the sixty storeys.[41] The exhaust steam from electricity generation was used to heat the building.

The public relations campaign for the new tower cost $100,000. Over 200 newspapers worldwide had already run articles on the Woolworth Building more than two years before its opening.[42] When it did open, 80,000 electric lights on the exterior were turned on not by Woolworth or any other dignitary on the spot, but by a remote switch set up in the White House, over 300 kilometres away, pressed by the new president of the United States, Woodrow Wilson.[43]

Perhaps Woolworth's bullish sense that his business experience equipped him to compete in real estate led him into overdoing the publicity drive. The letting agent for the office space had warned that 'if you get too spectacular, you don't get the tenants you want.'[44] Sure enough, the Woolworth Building was disappointing in purely financial terms, returning on investment at only just over half the rate that the clunky Equitable Building was soon to do.[45]

The glitzy showmanship surrounding the Woolworth Building was uncomfortable too for its architect, Cass Gilbert. Gilbert was New York upper class and having his work publicized so flashily was not to his taste.[46] When he met Paul Starrett, the head of the giant Fuller Company, another major construction firm, Gilbert had the contractor come to see him at his club, the Union, and arranged that he would be shown in through the tradesman's entrance. The gesture indicated that a contractor – even the head of perhaps the world's most impressive and successful building enterprise – was still not the social equal of an upper-crust architect.[47] The absurdity of this little snobbery is greater for its contrast with the reality: the contractor wielded vastly more day-to-day power than the architect, was considerably richer and ran a business that was many times bigger and more profitable.

In fact elitism, social and cultural, was part of the market position of Gilbert and many of his architectural competitors. Architectural design was yet another of America's crowded marketplaces, with large firms willing and able to expand enough to take on any amount of work, while a constant succession of new young practitioners fought to establish themselves. The prestige of many architects came through their family and social connections, as with Gilbert and his club membership.[48]

Another major source of prestige for American architects was to establish a claim to Old World artistic credentials. The gold standard was to have trained at Paris's famous École des Beaux-Arts, which taught students classical architectural details and grandiose formal planning. Fairly extensive travel round the famous old buildings of France, Italy and Britain would also give lustre to an architect's position as a purveyor of good taste. Rapid, comfortable and safe travel over the Atlantic by steamship and around Europe by train meant that not only architects but many of their clients had personal acquaintance with the great architecture of the European past.

The Woolworth Building is a glorious compendium of the

styles that Woolworth himself admired, and hoped would attract the sorts of tenants he wanted. The grand Gothic massing and external decoration gave way internally to a variety of different styles: a spacious entrance hall bringing together Roman coffering and Byzantine gilded mosaics, a Germanic *bierkeller*, and still other parts that seem to jump to the Romanesque or the Renaissance, all rendered with brio and real beauty by Cass Gilbert and his team. At the heart of it all was Woolworth's own office, with its seventeenth-century Flemish tapestry and Italian Renaissance fireplace.[49]

Precedents for the Unprecedented

In many ways, Gilbert's work is clearly closely related to the pattern of much in Victorian Liverpool, as outlined in the last chapter: historic precedents chosen for their relevance, their associations or simply personal aesthetic preference, reconfigured to meet new needs and to exploit new technologies. Yet in fact Gilbert's interpretation of Gothic architecture in his skyscraper is very extensively different from the medieval buildings from which it drew inspiration. Gilbert's Woolworth Building is, in fact, the culmination of a long period of experimentation by New York architects on how to design the façades of much taller buildings.

Victorian pubs in Liverpool were a comparable size and shape to some older houses and other city buildings; even when models were scaled up, the proportions tended to remain close enough to the historical precedent that they required little distortion. The very tall American office block was a new phenomenon in architecture, brought about by the rampant new fossil fuel technologies and economy. Most of history's tallest buildings had been monuments or had housed very big internal spaces. It was not an easy matter to bend these precedents elegantly to the new grids of similar-sized windows that office façades required.

As New York's offices grew above eight or so storeys they achieved a height that had very few useful precedents. The earliest tended to try and stretch the normal architectural language of offices up the façade, squeezing in extra floors as if by stealth. The results often looked odd and uncomfortable, but not as comical as those of the following generation, when tall towers started to appear, looming over the tops of these already unattractive blocks. Several successive holders of the record for the world's highest building based themselves on European bell towers, including the campanile of St Mark's in Venice. There is something oddly cartoonish about them, and even the architect of one of the most

celebrated, the Singer Building (1908), shown here, was later to say in awkward self-defence, 'We have a lurking inward consciousness that [tall buildings] do not belong to the highest type of art.'[50] His certainly did not.

The aesthetic training of architects sat oddly with the commercial pressures of high-value office development. Architects had to adjust to severe restrictions on their power to design their own buildings: economic and planning calculations substantially dictated the shape of the building. Within the remaining design work, they had to collaborate with structural engineers and contractors to ensure the stability, functionality and speed of construction of the building. Building designers had always worked within restrictions imposed by budget, site, custom and technology, but these new limitations felt to some like a more significant loss of status. Architects and their detractors could find themselves worrying that all the architect now did was to decorate buildings which were actually designed by real-estate accountants, engineers and contractors. The ungainliness of some of their efforts seems to broadcast their discomfort about the pace of change. Continued stylistic borrowing from older European architecture can seem, with unjust hindsight after the radical architectural changes of the twentieth century, like a confession of lack of imagination or cultural inferiority complex.

These impressions are strikingly at odds with the reality of technical advancement that the buildings demonstrated so spectacularly in their height and functionality.

If some historicist skyscrapers were comical, others could still be beautiful and powerful. Gilbert showed how much could be done with historically influenced design in these difficult conditions. The Woolworth Building's massing is not slavishly like that of a Gothic cathedral, but does successfully evoke some of the soaring verticality of cathedrals. Its exterior has far too many identical windows for medieval Gothic architecture (they are terracotta-clad; repetition was essential for reusing moulds), but the mix of extensive decoration seen from nearby and remorseless orderly repetition seen from a distance produces both a feeling of specialness and a forceful presence when seen on the city skyline. Its impact when new was even more striking, towering alone over much lower neighbours. For those who actually used the building, the pomp of the interiors was tempered by charming humorous details that borrow from the tradition of Gothic gargoyles: caricature portraits of the architect and others involved in the production of the building, including Woolworth himself, fittingly counting nickels.

The Chicago School

Although many creative and original architects continued to work in New York, the place that has become famous as the centre of the most exciting American architecture

of the decades around 1900 is Chicago. There, liberated by the superior building code that enabled lighter steel-framed façades, a small group of architects defied the widespread obsession of American elites with the European past and started to design high buildings whose form could be claimed, in the famous words of the leader of the movement, Louis Sullivan, to follow their function.[51] That is not to say that these

towers were undecorated – Sullivan used large amounts of surface decoration. The excitement of his approach lay in the fact that, rather than looking at the shape of the building and trying to work out how to fit a European-style façade on to it, he worked to dramatize, decorate and refine the functional and engineering form of the building into something new and beautiful.

Sullivan and his pupil Frank Lloyd Wright produced a series of buildings that looked unlike anything ever built before. They themselves presented this as a pure artistic endeavour, but of course the marketability of what they did was just as important to their careers as the European pretensions of their New York rivals. Their market position was one that appealed strongly to Chicago's industrial and commercial rich: Chicago was considerably further away from Europe than New York was and patently a success in its own right. Why *should* it defer to styles based on outdated technologies and built in very different economic, political and social contexts? What was wrong with being American and industrialized? Wright flattered his Chicago clients by pandering to their rivalry with New York: 'New York is a tribute to the [classicism of the École des] Beaux-Arts so far as surface decoration goes, and underneath a tribute to the American Engineer. [...] The real American spirit, capable of judging an issue upon its merits, lies in the West and Middle West.'[52] He called the houses he built for Midwesterners 'prairie houses', evoking white American foundation stories of brave, self-sufficient farming families striking out into the Midwest in the 1800s.

With their clever geometries, attractive materials, pretty decoration and striking exteriors, the sequence of suburban and rural houses that Wright built remains very influential and much admired among architects. More unusually for a leading twentieth-century architect, it has also proved unendingly popular with a wider public, who pay inconvenient pilgrimages to houses spread all around the USA and buy the numerous lavish photographic books of Wright's delightful interiors and striking exteriors, even as many of the houses themselves come under threat from America's weak protection of architectural heritage.

The Future, Delivered by Horse Cart

The story this chapter has told, of a professional, competitive construction and architecture industry serving a lucrative real-estate market, feels very contemporary. It is now the normal urban pattern across most of the globe. The 1920s skyline of New York that emerged from America's boom of coal energy remains one of the most stirring images of dense urban modernity even most of a century later.

Yet the world on to which the Woolworth Building first opened its doors was very different from today in many important ways.

The position of the early twentieth century in architectural history is perhaps nowhere clearer than in the transportation of the 23,000 tonnes of Pittsburgh and Philadelphia structural steel to the Woolworth Building site. Late success in acquiring the whole of Woolworth's desired site had meant last-minute alterations to the structural design. The biggest load-bearing columns of the tower now came to earth not on top of a foundation pile but between piles, and the messy structure was corrected simply by upping the amount of steel that bridged between piles to support the weight. The biggest of the prefabricated steel pieces was designed to carry a weight of 4,700 tonnes. The girder itself was sixty-five tonnes – perhaps at the time the heaviest single piece of steel ever used in a building – and had to be brought to the site along a route planned carefully to avoid caving in the road and falling into the sewers or the subway system. Many of the streets through which the steel travelled already looked somewhat like Fritz Lang's film fantasy *Metropolis* of over ten years later. Yet this modern challenge – moving a piece of steel which could not possibly have been manufactured anywhere on earth 150 years earlier – was handled in a way that might have seemed familiar to most of the builders in this book so far: the truck was hauled by a team of forty-two horses.[53]

In the First World War, soon to break out in Europe, the past met the future in a way that can seem shocking to modern observers: tanks and aircraft shared battlefields with horses and donkeys.[54] The twentieth century was ready for the scaling up of its next major energy revolution – the internal combustion engine.

CHAPTER 10

'The beauty of speed'

The Rise of Oil and Electricity, 1914–39

Oil

Already by the time the Woolworth Building was going up, it was clear that fossil-fuelled transport was starting to break free of its rails. It was clear too that a new fossil fuel, petroleum, was going to play a huge role in twentieth-century technology and geopolitics. The British and German navies were switching their fleets from coal to oil in these years and the international competition for control of major oilfields was one of the forces pushing the leading industrial countries towards war – both Germany and the UK had significant coal deposits within their borders, but neither had access to significant domestic oil supplies in this period.[1]

Petroleum, like coal, was formed by the compression and heating of dead organisms over millions of years. Again like coal, it had been known since ancient times, but considered largely as a foul-smelling pollutant that emerged unaccountably from the earth in certain places. Hard to control and terrifyingly flammable, oil was a much less easy gateway fuel than coal for making the transition from firewood to fossil fuel energy. It had seen occasional applications, both as a fuel and as a medicine, but its use remained marginal. By the mid-nineteenth century, the sophisticated condition of the coal-fuelled industries in much of Europe and the USA made it possible to find solutions to petroleum's technical challenges and people began to investigate expanding its use as a lamp fuel. Once it began to be systematically extracted in the 1850s, its energy intensity and ability to flow smoothly from tank to combustion chamber made it a more versatile and potent fuel than coal could ever have become. Oil not only supplemented existing coal and coal gas, but offered specific advantages peculiar to its high energy-to-weight ratio and liquid form: it could permit powered human flight and make motor vehicles much more economically viable.

The later nineteenth century had seen extensive experimentation with new modes of road transport. Electric cars and coal-powered steam cars competed in races with those running the earliest petrol and diesel motors. By the outbreak of the First World War the fastest petrol cars were capable of sustained average speeds of over 100mph in race conditions (faster even than the trains of the time), and huge competition existed between the leading manufacturers to beat speed records and win lethally dangerous races.[2]

These thrilling new vehicles could excite the most fanatical addiction in those who were exposed to them. One Italian car addict wrote in 1909 about a night-time adventure with his friends. His article is shot through with exhilaration at the new energy technologies just then getting a grip in Italy. The author and his friends had stayed up all night under new-fangled electric lighting, proud to be awake when almost everyone else was sleeping. They were 'alone with stokers feeding the hellish fires of great ships, alone with the black spectres who grope in the red-hot bellies of locomotives launched on their crazy courses'. The author marvels at the noise and lights of a double-decker tram that passes, then the young people head to their cars, frothing with excitement at the sexiness, but also the lethalness, of their powerful new machines: 'We went up to the three snorting beasts, to lay amorous hands on their torrid breasts. I stretched out on my car like a corpse on its bier, but revived at once under the steering wheel, a guillotine blade that threatened my stomach.'

This unsettling account, written by a young poet called Filippo Tommaso Marinetti (1879–1944), prefaces the eleven-point 'Futurist Manifesto', which advocates danger, courage, speed, aggression, destruction and machinery. Perhaps the most famous proposal of the lastingly influential document was its fourth, comparing new petrol-powered machinery with classical sculpture: 'We affirm that the world's magnificence has been enriched by a new beauty: the beauty of speed. A racing car whose hood is adorned with great pipes, like serpents of explosive breath – a roaring car that seems to ride on grapeshot is more beautiful than the Victory of Samothrace.'[3]

The underlying misanthropy and nihilism that are a feature of this text are poisonous ('art, in fact, can be nothing but violence, cruelty, and injustice'); but the headlong excitement at the new world that the new energy sources were bringing into being is palpable.

In May 1914 Antonio Sant'Elia, one of Marinetti's fellow

Futurists, was to exhibit a series of drawings for a 'New City' that remain thrillingly modern-looking even now, after more than a century of headlong change.[4] The shock must have been all the greater at a time when so much architecture remained firmly historicist. Sant'Elia's images make powerful abstract monuments out of large-scale infrastructure for trains, planes and motor vehicles, and out of the elevators, vents and chimneys that the future fossil-fuelled city would require. It is a fanatical and prophetic architectural celebration of the new energy sources.

Main Post Office, Berkeley, California, completed 1914

Academy Gardens, Kensington, completed 1914

Sant'Elia did not live to see the influence of his ideas. The Futurists repeatedly called for an annihilating mechanized war to wipe out the old order and its buildings and technologies. When the First World War started Sant'Elia seized his chance to pursue that appalling dream and was duly killed in combat.[5]

At the point when Marinetti and Sant'Elia began dreaming about the future of the oil-powered world, cars were a little like private aeroplanes now – the ultra-luxurious toys of the very rich. Already, though, various producers were

Sant'Elia's Citta Nuova, 1914

competing to bring down prices and increase the volume of sales in the manner of simpler manufactured goods through the previous two centuries. The most successful early mass producer of the automobile was to be Henry Ford, who increased production sharply in his new plant in Highland Park, Michigan. There, in 1913, Ford's team pioneered a continuous moving assembly line to carry the car chassis past each worker in turn, each of whom would add the same part again and again, becoming faster and more reliable than multi-skilled workers carrying out lots of different procedures – the same principle seen with the adoption of *opus latericium* on ancient Roman construction sites, but now powered by fossil-fuelled machines. Ford and his engineers built on a century or more of innovations in manufacturing. Around 15 million Model T cars were made in the nineteen years of its production, bringing down prices by more than 77 per cent between 1910 and 1921, and opening the new technology of petrol-driven private transport to millions more people.[6]

Coming after more than a century of efficiency contests between rival factory designers and industrialists, the assembly line seems like a very obvious improvement to make. Why had so many previous factories resigned themselves to their employees' wasted work humping goods around in an irrational order, increasing production time and cost?

Much of the answer lay in the power source. Factories powered by a watermill or a steam engine had a single source for all the energy used anywhere in the factory. Power was carried by an elaborate apparatus of rotating wheels and shafts, connected by spinning leather belts. Moving goods was easier than moving power, so the factory was designed around getting power to machines rather than around the logical movement of materials. Factories kept all machines as close as possible to the central power source, with lots of storeys stacked tightly together to be near the main belts, generating the classic image of the Victorian industrial area.[7]

In addition to awkward planning problems, these line-shaft-driven factories were inflexible, inefficient, high maintenance, very noisy and dangerous, with workers apt to lose body parts or their lives if a moment's inattention caused them to be caught in the powerful spinning belts.[8] Line shafts also made the machines they served wobble, contributing to

the difficulty of producing genuinely identical parts which would fit together to the precise tolerances required for car engineering.[9]

The revolutionary technology which enabled Ford's plant to escape this clattering, lethal, inefficient nightmare of an energy distribution system was electricity. The main power source of the Highland Park factory was a group of turbines which burned natural gas (at this period a by-product of the oil-extraction industry that powered the cars Ford made) to produce electricity.[10] The electricity was carried round the plant by narrow, silent, energy-efficient, safe, low-maintenance cables. Wherever needed, it could power the motors which ran machines placed usefully close together in a sequence determined by the logic of production. Electric motors were quieter and cleaner than steam engines and

line shafts, and wobbled much less. They offered better conversion of the energy source into useful work. So good was electricity that by 1925 89.5 per cent of all American industry was using electric-powered machinery.[11] The provision, thanks to electricity, of powerful lighting twenty-four hours a day, as well as dust-free ventilation, made a further contribution to the improved performance of factory workers and machines.[12]

In the USA, where large and comparatively accessible oil reserves came together with a vast industrial base and very substantial land resources, Ford's cheaper cars changed the patterns of where people lived and worked remarkably fast. The late twentieth-century English architect Cedric Price proposed the model of the city as an egg. He suggested that the ancient walled city had been like a boiled egg, tightly bound by the hard shell of its encircling defensive walls. The coal-powered city, extending into the countryside and bringing rapid urban sprawl, was like a fried egg, the yoke representing the city centre, the white the dense but irregularly shaped suburbs. The car-based city, Price proposed, was like scrambled eggs: there were bits of employment and industry dotted around amid housing and commercial or entertainment facilities, and strange pockets of little-used or informally used space.[13] Anything could be anywhere, provided it had plenty of car-parking space.

Price was writing about the cities of the 1950s and 1960s, but the tendency he described so memorably was starting to take shape between the world wars, both for real in the American suburbs and in the mind of Frank Lloyd Wright. From the early 1930s Wright had been hoping for the advent of 'the disappearing city', arguing that the petrol station was likely to become an important 'city equivalent' for each area as the car enabled the city to spread out into a sprawling indeterminate area of low-density suburb. He eagerly forecast that soon the only reason to visit the city centre rather than stopping at roadside shops would be 'to view the ruins'.[14] Cars, forward-looking architects could see, were starting to change everything about the built environment.

The 'engineer's aesthetic'

The changes of the twentieth century did not inevitably require radical changes in architectural aesthetics. Victorian Liverpool and New York's early skyscrapers showed that major technological transformations did not have to result in the abandonment of older architectural styles. Yet in the forward-looking, technically advanced factories of car production and other new industries there emerged a new aesthetic even more distinctive and free of historical quotations than Frank Lloyd Wright's houses. Ford's Highland Park plant was designed by the large and expert team of Albert Kahn (1869–1942), an architect whose own office borrowed from the industrial ideas of his clients to become a 'plan factory' – an efficient and speedy production centre for architectural designs.[15]

The factory Kahn built for Ford used a similar structure to the Liver Building in Liverpool – a system of slim concrete columns supporting wide, open floors, the concrete reinforced by steel bars. Reinforced concrete was fire-resistant and strong, fairly quick to build and capable of the kinds of orderly repetition and economies of scale that appealed to the great industrialists. Between the columns, as much of the wall as possible was given over to windows to produce good daylighting in the working areas.

Most office buildings of this period still had decoration carefully added to the structure, sometimes, as we have seen, at the expense of useful window area. The Ford plant, by contrast, was stark in its lack of historically influenced ornament or materials chosen for their historical associations. This probably reflects a genuine prioritization of function over appearance: people choose what office to rent partly on grounds of appearance and feeling; a factory's appearance has much less impact on car-purchasing decisions than the price and quality of the vehicles it makes. However, the remorseless repetition of identical windows over such a very large building was also architecturally spot-on for the

ethos immortalized in Ford's famous joke: 'Any customer can have a car painted any colour that he wants so long as it is black.'[16] The stunning publicity photos with which Ford's company advertised its scale of production make a feature of both the thousands of identical cars and the hundreds of identical factory windows. In Kahn's factories, a compelling visual shorthand had been found for the new age of unprecedentedly huge and efficient mass production. Over the following decades, its influence was to extend into all other building types: for some young architects, the shock and thrill expressed so strongly by Marinetti contributed to a rethinking of what architecture should look like as well as how it should perform. The muscularity, brightness and unsentimental clarity of the new factories embodied the look of progress.

A young Swiss architect, Edouard Jeanneret (1887–1965), was to become one of the most influential of these young challengers to architecture's old order. After an episode of poor mental health he renamed himself Le Corbusier (derived from his grandmother's surname) and wrote articles and books that explored how architecture and cities might look once designers had learned fully to exploit the wonders of electricity, oil-powered cars and aeroplanes, and the materials made affordably available by cheap coal heat: concrete, steel, plate glass and aluminium. He pored admiringly over pictures of industrial and engineering achievements, and contrasted them favourably with what he saw as the encrusted, heavy-handed historical fakery published approvingly in architectural publications. He extolled the 'engineer's aesthetic' – the beauty to be found in machines designed not to look like their distant precursors, as buildings so often were, but to be as efficient and powerful as possible.

Bridges, dock gates and liners showed Le Corbusier what might be achievable with concrete, steel, oil and electricity, as did plants like the Ford-influenced Fiat Lingotto factory on the outskirts of Turin (built 1916–23, to the designs of

engineer Giacomo Matté-Trucco, 1869–1934). The assembly line ran up through the building, starting at ground level with raw materials and ending with finished cars on the top floor. The supreme *coup de théâtre* was the testing track on the roof, where newly finished cars could race around before descending a spiral ramp back to ground level and off to be sold.

The Lingotto factory illustrated for Le Corbusier how revolutionary concrete and steel could be for cities. For a brick structure to allow heavy vehicles to move over it safely required very bulky, solid arches – weight to resist the shaking and twisting effects of the moving load. At the Lingotto

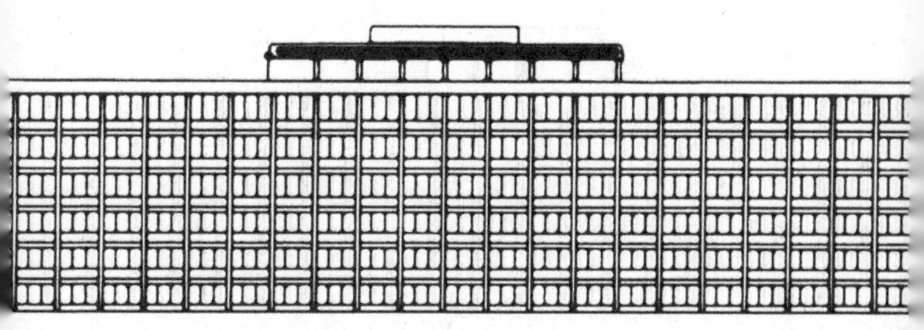

factory this resistance was provided by the strength of the slender steel bars embedded in the concrete. Fiat's factory showed that concrete structures could safely provide abundant open floor area with widely spaced columns beneath a road with rushing cars on it.

Suddenly, many long-established truths about the city were overthrown. Sant'Elia was right: the road no longer had to run along the earth's surface, with buildings standing squarely on the ground between the roads. Why not have a road running along the roof or the building continuing under the road? Why did gardens have to be at ground level? Why did pedestrians have to walk alongside the road, slowing down the cars when they wanted to cross (and risking their lives in the process given the iffy brakes, skinny tyres and

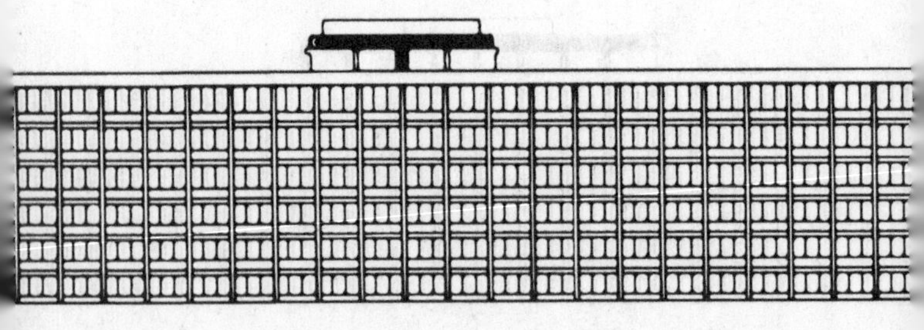

heavy weight of early automobiles, and an unapologetic culture of speeding and drink-driving)?

In 1923 Le Corbusier published one of the most famous architecture books of all time, *Vers une architecture* (*Towards an Architecture*).[17] In it, the young architect and an artist friend of his ambitiously attacked the existing norms of architecture under a series of headings like 'Eyes which do not see'. Preoccupied by the challenge and the opportunities presented to architects by the need to reconstruct the vast amount of destroyed housing in the former battlefields of the Great War, Le Corbusier felt that France faced a choice: 'architecture or revolution'.

The book begged and bullied its readers to look away from the historical ornament which still filled architectural publications of the period ('the architecture of horrors') and instead consider the great engineering marvels of the age, including the Lingotto factory. Le Corbusier compared four of Paris's most admired buildings with a large ocean liner to show how much bigger the liner was – and, with its vast coal-fired boilers, the liner could rush thousands of people from Liverpool to New York in a matter of days, weathering the heaviest storms, while buildings just stand still, passive and fragile.

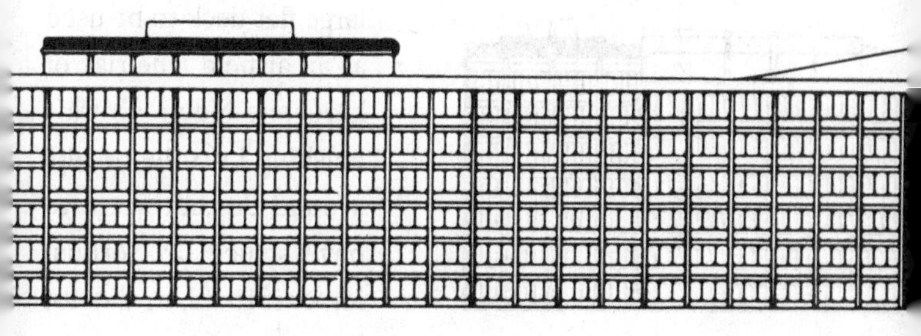

One of the young technophile's greatest enthusiasms was for the thrilling potential of the internal combustion engine. He compared the evolution of ancient Greek temples over centuries with the rapid evolution of sports cars and aeroplanes over the previous two decades.[18] At the time this must have seemed a deliberately outrageous debasement of Western architecture's most revered structures, but Le Corbusier was right: the development of oil-fuelled engines and the machines to exploit them was to do infinitely more to change the world than the careful entasis of the ancients.

By 1925 Le Corbusier's ideas on the future of the fossil-fuelled city were coming together. He proposed a plan for razing and replacing much of central Paris, including medieval churches and a large area of the characteristic nineteenth-century boulevards that still define the city today. The replacement buildings and roads would benefit to the full, Le Corbusier felt, from reinforced concrete, glass and the oil-fuelled internal combustion engine.[19] Huge concrete-framed blocks would provide abundant living and work space, while freeing as much of the ground level as possible for parkland. Over the top of this parkland layer the fastest direct roads would dash safely and without interruptions. Above this, in the city centre, would be a large flat deck to be used as an airfield (the risk of kerosene-laden aircraft crashing into the towers around doesn't seem to have struck Le Corbusier, or if it did, he did not

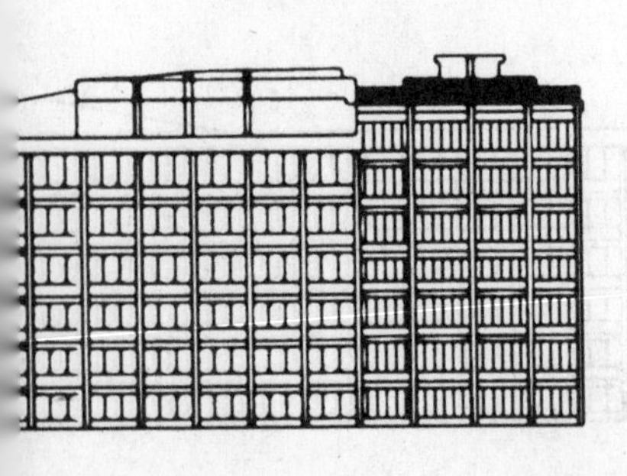

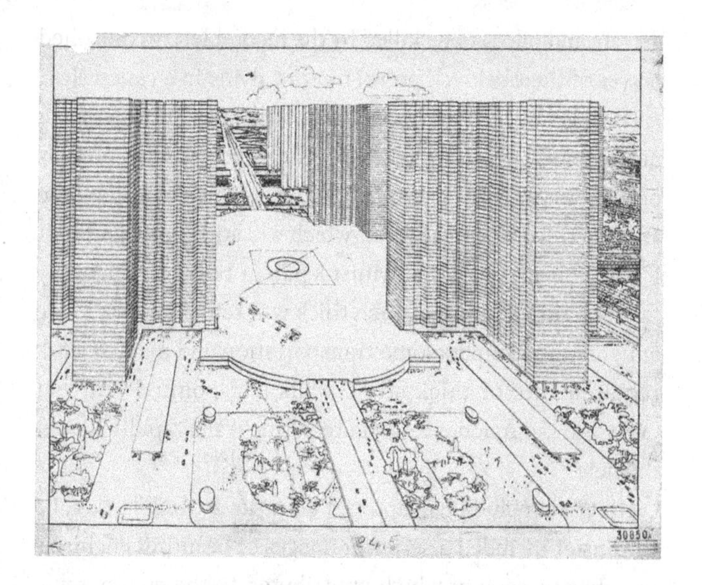

feel it was problematic). Although Le Corbusier held out the promise of immense financial profits from his proposed redevelopment, it seems unlikely that he thought it would actually happen. He was, however, clearly eager to show that it could happen if the will were there. His dreamed-of town would, compared with the grimy, congested Paris of the time, have been paradise for speed-loving admirers of green space.

Le Corbusier's enthusiastic embrace of oil and the internal combustion engine brought him into contact with other passionate promoters of these transformative powers. Le Corbusier's provocative scheme for Paris is known as the 'Plan Voisin', after its sponsor Gabriel Voisin, a pioneering aviator and manufacturer of cars and aeroplanes, including for military purposes in the recent war. In 1912 Voisin's younger brother had become one of many early technophile drivers and pilots to be killed by the incompletely controlled power of their beloved petrol motors, dying in a car crash.

Dazzled by the new and the industrial, Le Corbusier felt a profound mistrust of the heavy, unreliable natural materials of earlier ages, like 'massive timbers, as thick as you please and heavy for all eternity, but which will still spring and split if placed near a radiator, whilst a patent board [industrially manufactured panel] ⅛ inch thick will remain intact'.[20] He lamented the cumbersome transportation and manual work requirements of thick stone walls and pointed out that even after so much effort the resulting walls rapidly let out the heat.

In the concluding chapter of *Towards an Architecture*, Le Corbusier included a series of images of beautiful engineering objects, each of which contributed to the energy revolution of the early twentieth century: ventilating fans, a car engine, an aeroplane, a warship, a crane and a vast coal dock, among others. He was mesmerized by the new energy technologies, by the beauty of 'the poetry of facts'.[21]

Freedom for Architects

The key materials of this architectural revolution were steel, concrete and ever-larger sheets of plate glass. Walter Gropius (1883–1969), a leading figure in the various emerging movements that came to be grouped under the name 'modernism', was to write in 1925 of steel, concrete and glass as 'new synthetic substances'.[22] In fact, as we have seen, steel and glass had each been known and used by humans for thousands of years, and the cement that was the new component in modern concrete was only different by a few technical tweaks (above all, through hotter firing) from the lime that had again been in widespread use across many societies for tens of centuries. Nevertheless, Gropius's sense that these were new substances is, in architectural terms, broadly accurate: they had not been available on anything close to their twentieth-century scale until the advent of cheap fossil fuels.

In each case the 'new' substances cost a lot of intense energy to manufacture, on a scale that would have been absurdly uneconomical without fossil fuel heat. The energy consumption required to make one tonne of cement (the binding ingredient that typically made up around 20 per cent of concrete) by the 1960s was 2,614kWh, equivalent to the muscle work done by a manual labourer working steadily, ten hours per day, six days per week, for more than eleven years.[23] Put another way, without fossil fuels it would have taken a year's wood supply from an area of nearly 3,000 square metres of sustainably coppiced woodland to make enough charcoal to generate the heat for each tonne of cement.

Steel, too, guzzles heat energy in its production. Each tonne of structural steel today costs an amount of energy that an agrarian economy would never have squandered on concealed structural materials: a year's production from around 6,850 square metres of charcoal coppice.[24] In the twentieth century, the energy capability opened up by fossil fuels led to the manufacture of millions of tonnes of concrete and steel for hundreds of thousands of new buildings.

With fast-rising industrial production (France, for example, used 86 per cent more energy per person in 1910 than it had done forty years earlier) came economies of scale and very rapid development of expertise.[25] New technologies often suffered from glitches, and sometimes met powerful opposition from traditionalists, but the economic benefits of the new methods won them increasingly wide adoption.

As technical competence and engineering understanding grew, steel, concrete and plate glass offered almost incalculable improvements on agrarian construction norms. Twentieth-century technologies even improved performance and saved construction labour relative to the brick walls, wooden floors and more modest windows which had already improved cities in the previous phases of the fossil fuel revolution.

At the forefront of enthusiasm for the liberating potential of concrete, steel and cheap glass were neophiles like Le Corbusier, who felt the new freedoms acutely. Masonry buildings had always had to have uninterrupted vertical walls, pierced only where doors or windows were necessary, each brick or stone standing squarely on the one below, down to the foundations. This had meant that each storey had to have the structural walls in the same position, irrespective of the ideal shape and size of the rooms. The outer envelope of a masonry building was generally made up of structural walls, so their shapes tended to be comparatively flat and samey, decorated superficially by their architects according to passing fashions, but in their true essence the same from the Georgian terrace to the houses of Sefton Park or the hotel at St Pancras. The all-cloaking walls almost always landed heavily on the ground, in order to transfer their load to the foundations.

Le Corbusier experimented throughout the 1920s with the ways the structural miracles he had seen reinforced concrete perform at the Lingotto factory could be used to free architects from what he saw as deadening restrictions. His 'five points of architecture' in *Towards an Architecture*, which spread round the world's young, modernizing architects like a new Ten Commandments, made into dogma some of the thrilling new capabilities of steel-reinforced concrete: columns (for which he invented the term *pilotis*) beneath could raise the house above the street, freeing the ground level for other uses; slender columns were all that was needed to hold up each floor, enabling clients and architects to put rooms where they were wanted rather than within the straitjacket of structural brick walls; the concrete structure allowed the façade to be similarly free, dropping back for balconies or gardens where wanted; now that walls were free of load-bearing obligations, windows could run in long strips along the façade rather than being vertical slits cut out of the load-bearing masonry; and the solidity of a concrete roof could easily support the weight of a roof garden, saving the best views and brightest light of the house from being wasted on an inaccessible pitched roof.

The structural freedoms which Le Corbusier celebrated have shaped much architecture ever since, and as sustainability-conscious architects try to move away from the fossil-fuel-dependent production of cement and steel, they are discovering once again, this time in mourning rather than celebration, how life-enrichingly versatile the great materials of the twentieth century were.

The Flick of a Switch

The importance of concrete, steel and glass to twentieth-century architecture is inescapable at a glance. Equally transformative, though much less discussed by many commentators today, was electricity. We are so used to it that lighting is more often discussed by design historians in terms of the aesthetics of fittings rather than in terms of the revolution it wrought in the nature of buildings. Electric lighting was only one of the transformative gifts that electricity was to bestow on architecture. Early, cumbersome experiments with making buildings active rather than passive, like the steam engines which heated and cooled St George's Hall, were to be made economic in all scales of building by electricity, with its ability to be switched on and off rapidly, and to be delivered to small, almost maintenance-free motors wherever they were needed. Steam engines, by contrast, provided a gross oversupply of very local power and only after a substantial warm-up period overseen by a trained technician.

Once again, Le Corbusier was at the forefront of thinking on how fundamentally electricity would transform architecture; his selection of beautiful machines in the final chapter of *Towards an Architecture* includes two turbines for generating electricity. Just as Le Corbusier had, through his admiration of cars and aeroplanes, become close enough to Gabriel Voisin to be sponsored by him, so his love of new electrical technologies won him another relationship with industry. In this case it was a leading French glazing manufacturer who worked with Le Corbusier to test his ideas on how the developing technologies could produce entirely new forms of architecture.[26] Le Corbusier dreamed that, with electric motors, it would soon be possible to design buildings sealed against the noise and health-destroying filth of the city. Clean air could be pumped, at the right temperature, into the interior, improving health and comfort, while the external cold or heat trying to get through the glass façade could be controlled by having inner and outer walls of glass, between which air could be pumped at a temperature that would counterbalance the outdoor temperature to make the interior an even eighteen degrees Celsius all year round, anywhere on earth. He contrasted his dream of the fully controlled interior with the age-old reality: 'Every country builds its houses in response to climate. At this moment of general diffusion, of international scientific techniques, I propose only one house for all countries, the house of exact breathing.'[27] This, Le Corbusier had read, was the ideal temperature to ensure healthy lungs.[28]

A century later, some of these ideas read as weird pseudoscience. To claim that eighteen degrees Celsius is a worldwide ideal has ugly echoes of the Western belief that Europe's success since the Renaissance arose from the intrinsic superiority of its climate over parts of the world too hot for human health or productive work. The imposition of such norms since has created a huge demand for unsustainable air conditioning in the tropics.[29]

In the 1930s there was no awareness of the potential for fossil fuel consumption to cause devastating climate change. Nearly a century on, the wastefulness of Le Corbusier's servicing proposals has retrospectively taken on a nightmarish quality. Even at the time, his ideas were challenged by an American ventilation company, to whom they were sent for review. Not only were they inefficient and fuel-hungry (requiring four times more steam and twice the mechanical power of normal air-conditioning methods), but Le Corbusier's desire to introduce ozone into the air supply – far from being health-giving, as he believed – was actively dangerous to the lungs.[30]

Le Corbusier's belief in his new ideas was so great that he sought to install largely artificial ventilation in a hostel for homeless people that he designed in Paris. When estimates for the ventilation equipment came in too pricey, costing up to half the budget of the entire building, the architect continued to specify almost entirely sealed windows, with the result that the building was uninhabitably airless and foul when occupied.[31]

Machines for Living

More legitimate guinea pigs for Le Corbusier's half-brilliant, half-mad ideas than the homeless of Paris were the arty, avant-garde rich. Throughout the 1920s his main clients for built commissions were well-off Parisians wanting suburban villas which used and celebrated the very latest architectural materials and ideas, the most momentous artistic thinking and the best new technologies. Le Corbusier had a genius for attracting clients who, even if eventually disgruntled by his experiments at their expense, nevertheless allowed him to produce extraordinary buildings. He was ruthless in pushing through his ideas. For example, he asked contractors to price the building using materials and installations considerably cheaper than those he actually intended to use, so that his clients would agree to the outline designs and could then be talked up to pricier options one at a time.[32] It was in these villas that Le Corbusier explored his famous notion that a house was a '*machine à habiter*' – a machine for living.[33]

The reality of his building sites was much more craft-based than the resulting buildings might suggest. The design sketches and the photos of the completed houses seemed like what Le Corbusier described in *Towards an Architecture* as a 'pure creation of the mind' – industrial-looking windows giving rhythm to smooth white walls, hovering on the slimmest of white legs. Yet in construction they involved rather old-fashioned craft, from the joinery of the concrete's shuttering (the moulds into which the concrete was poured) to the brick-like concrete blocks that made up most of the walls, carefully plastered smooth, and the crude wooden scaffolding the builders erected.

For their dream client, every innovative architect would have chosen the Swiss banker Raoul La Roche, whose firm commitment to modernity, deep pockets and robust belief in the genius of Le Corbusier allowed the architect and his team to experiment with solving some of the problems produced by the new materials and technologies: condensation running down from metal window frames, new boiler systems that did not perform as advertised, inadequate new lighting systems and so on.[34]

Even La Roche sounded gently desperate in a few of his pleas to Le Corbusier to fix urgent problems:

> I understand perfectly your hesitancy over the way to light my house. But until you find something really good, it is essential at least that I should be able to see clearly in my home. It's six months since I moved in. [...] It is becoming clear that your various pieces of equipment, however ingenious they might be, do show certain drawbacks and, since they are also very dear, I hesitate to proceed any further with them.[35]

It was to be three years before the final lighting installation was in place. The original coal-powered boiler was replaced in 1931–2 with an oil-fired one, but after a few years of this second smelly, noisy, ineffective heating mechanism it was again replaced in 1939.[36] The leading historian of Le Corbusier's villas, Tim Benton, has calculated that, after hefty initial construction costs, between 1929 and 1938 La Roche spent 10,000 French francs per year on maintaining and upgrading his house.[37] Elsewhere in Paris, in 1929, George Orwell was struggling by on just six francs per day – less than a quarter of

La Roche's average daily costs for post-completion building work.[38]

Rather unfairly to the endlessly supportive La Roche, Le Corbusier gave the title of 'the nicest client we have ever had' to Emilie and Pierre Savoye.[39] Their wealth came from an insurance firm and their country house was the sixteenth private house Le Corbusier was to design. It has become perhaps the most famous of them all, its pristine white body floating on slender concrete legs above a broad swathe of lawn – a kind of Parthenon of the high modernist movement.

Some of the Villa Savoye's features were inspired by the couple's car – it is an exhilarating thirty-kilometre drive from Paris, a pleasurable run for the Savoyes and their wealthy guests in their large, powerful automobiles. The ground floor of the villa is curved to allow the owners' Citroën to drive round it in an elegant sweep rather than negotiating awkward corners. An early scheme even brought cars up to a first-floor garage on a ramp, but better sense prevailed and the three-car garage remained on the ground floor.[40] There is something glamorously dangerous about the slender concrete columns with heavy, powerful steel motorcars negotiating them, the drivers on occasion reeling home after cocktails. For Le Corbusier, Sant'Elia and Marinetti, safety concerns seem to have been for the backward-looking technophobe.

Emilie Savoye wanted modern facilities in her modernist villa. She asked Le Corbusier for electric lighting and power points, central heating and hot and cold running water.[41] The villa was built in the early years of the spread of electrification beyond city centres. High-voltage transmission systems, rolled out from the 1920s, allowed electricity to travel tens and eventually hundreds of kilometres without prohibitive loss of energy through heating the cables.[42]

So enamoured was Le Corbusier of electricity at this period that he investigated the possibility of heating the house electrically: not only would an underfloor electric heating system have produced draught-free, evenly distributed warmth, it would also have freed the architecture of the house from visually intrusive radiator installations. The idea was rejected because, in these early years of rural electrification, the increased power requirements would have called for a special transformer station a lot more expensive than the system itself.[43]

The villa was and remains utterly beautiful, but once again its habitability relied on technologies that did not reach the level that Le Corbusier seemed to hope they would. A house designed to optimise light and views, and to float elegantly on slender columns,

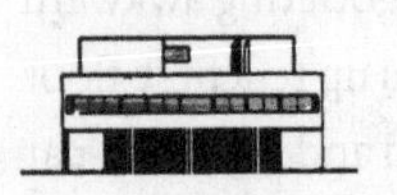

made for an awful lot of outside walls when compared to the compact huddle of rooms in traditional northern-European houses. The salon shows this at its most extreme: the best view was to the north, but Le Corbusier's love of sunlight made him reluctant to leave the principle living room devoid of south-facing windows, so he incorporated a substantial open terrace to let the sun into its fully-glazed south wall.[44] The light is beautiful in the resulting room, but thin concrete walls, together with very large single-glazed, metal-framed windows that were able to lose huge amounts of heat on a cold day, set the heating system a challenge to which it was never equal. The temperature difference between cold walls and warm air inside encouraged condensation throughout the house, dampening every surface and stimulating mould. Problems of rain penetration through experimental flat roofs redoubled the problem. The Savoyes' son, whose health was problematic, ended up spending 1933 in a sanatorium (they had moved into the house in 1930).[45] Even the architect felt that there was a need for the heating system to produce 'ample – even superabundant – heat' and that 'this may cost a lot'.[46] The clients were still pursuing Le Corbusier through the courts shortly before the Second World War, when they fled the forthcoming Nazi invasion. Their house was used first as a hay store by local farmers and then occupied successively by the Nazi and American armies, being more or less a complete wreck by the end of hostilities.

This elite group of early clients spent a lot to get sensationally beautiful houses with functional problems. They also bought immortality. Their names echo still among the world's architects: La Roche, Savoye, Cook, Stein and de Monzie.

If the practical experience of Le Corbusier's individual clients varied between problematic and disastrous, the ability of the high-profile architect to survive such hitches was boosted. Architecture had almost always been a relatively local activity until the rise of the steam engine – even for Gothic masons, whose itinerant careers could see them build for church or royal clients hundreds of kilometres apart, the client group was limited and well networked. If these earlier designers had treated the practicalities of real buildings with as cavalier an attitude as Le Corbusier, they could have expected their mistakes to catch up with them. In the twentieth century, however, the world had changed. Powerful but hard to use steam-engine printing presses had become available in the 1840s–50s, and by 1900 most presses were moving to electrical machines that were easier to work.[47] The volume of printed material increased proportionately. Particularly important for architecture were improved techniques for printing photographs and other illustrations. Le Corbusier's charismatic photographs and sketches were reproduced hundreds of thousands of times in his books and in architectural journals worldwide, accompanied by his aphoristic, prophetic-sounding dictums about architecture. The lure of his fame and his beautiful architecture was greater than the power of individual disgruntled clients to hold him back.

The publications did not show that, as Emilie Savoye complained in September 1936, 'it's raining in the hall, it's raining on the ramp and the wall of the garage is absolutely soaked. What's more, it's still raining in my bathroom, which floods in bad weather, as the water comes in through the skylight. The gardener's walls are also wet through.'[48] The pristine visual material that Le Corbusier published set architects and clients on fire with excitement. They saw perfect, thrilling, enviable images of a new world brought about through concrete, cars, industrial plate glass and electricity.

Le Corbusier's publications of his work and ideas contributed heavily to the 'system cultures' of electricity and the internal combustion engine: once a new technology starts to be widely adopted, economic and practical arguments expand the market further. But some brave or foolhardy early adopters are required, to take on and popularize the new technologies even while they are glitch-prone and expensive. The excitability that leads to people camping outside Apple Stores at each new release is a pale version of the neophilia that spread among a self-selected group of young architects and clients worldwide in the 1920s and 1930s.

Doing the Right Thing

In Britain and America, the internal combustion engine and electricity seemed like the latest advances in a well-established pattern of developing fossil fuel exploitation. In much of the rest of Europe the exhilaration of young architects came in the context of later, faster industrialization. At the turn of the twentieth century, as the first generation of modernists was growing up, countries across Western Europe were industrializing headlong, fuelled by indigenous or imported coal, sometimes supplemented by hydroelectric schemes.

Germany was emerging rapidly as Britain's great industrial rival, Austria and the Netherlands were changing radically, and France, Italy, Switzerland and others were each experiencing their own accelerated industrial revolutions. The excitable young architects who matured in these thrilling circumstances, Le Corbusier prominent among them, tore up the architecture of the past. In the Netherlands from 1917 a movement of artists and architects known from their journal's title as *De Stijl* (The Style) experimented with simple compositions of vertical and horizontal, made vibrant by blocks of primary colour. Mondrian was the most famous painter of the group, and one of its architects, Gerrit Rietveld (1888–1964), built a house in 1924 that seemed like a Mondrian you could live in. Its upper floor had no internal walls that could not be folded away – a new level of flexibility in house architecture.

The country which saw the most spectacularly rapid change was Germany. After a century of headlong industrial growth, England and Wales saw energy use per head rise only 14 per cent between 1880 and 1910 (to a notably high 42,400 kWh per person per year), whereas Germany's energy use doubled in the same period. The end figure for Germany was only just over half the English and Welsh figure (23,900 kWh per person per year), but the speed of change in Germany set many architects ablaze with excitement and provoked all-encompassing social, political and technological transformations that seemed to clamour for novel architectural solutions.[49]

At the Bauhaus school of architecture and design in Weimar, a group of gifted, self-consciously radical artists, designers and architects gathered to argue, collaborate, experiment and teach. Founded in 1919 by Walter Gropius, the school placed a major emphasis on finding appropriate, functional, aesthetically satisfying shapes for mass-produced goods like lamps and chairs. In architecture too Gropius saw mass production as having untapped potential to provide good housing and facilities for Germany's new industrial working classes: if the moving electric assembly line had enabled Ford to drop prices and increase technical quality in his cars, why should buildings not benefit from a similar process? Gropius was already looking into this possibility as early as 1909.[50]

When Gropius designed a purpose-built new home for the Bauhaus in Dessau (1925–7), he made it a showcase of his ideas for the new architecture of the twentieth century. Its plan has the clarity and simplicity of a good diagram, with design workshops, offices and residential accommodation provided in clearly distinct blocks, interlocking elegantly or connected by bridges. The elevations are equally carefully

composed, with each simple rectangular façade articulated by windows appropriate in size and shape to the functions of the rooms they light. The most memorable and striking block, appropriately, was the workshop block which housed most of the design work and artistic experimentation. Its open floors consisted of concrete decks standing on columns, enclosed only by a grid of metal-framed glazing. The lightness and transparency of the building thrilled visitors, and the much larger number of people who got to know it only through published plans and high-contrast black-and-white photographs. This, surely, was the future of architecture: democratically open, flexible and stunning to look at.

Yet a recent piece of research by Daniel Barber has uncovered the extraordinary price that the Bauhaus has paid over its first century for this level of transparency and clarity. The glass and iron 'curtain wall' of the workshops was single-glazed and uninsulated, resulting in very intense loss

of heat in Dessau's cold winters. To counteract this, the new building was given a substantial array of radiators in front of the glazing. Even so, the heating has proved consistently incapable, despite eleven phases of upgrading and replacement, of providing adequate warmth to the block. By 1972 the one original boilerman had become three, while the coal supply they shovelled incessantly into the overstretched boilers started out in a built-in space but sprawled to occupy ever more room, eventually forming a large, ugly mound behind the building. The boilers had to run night and day throughout cold weather, even when everyone was away for the holidays, to prevent frost damage to furnishings and fittings, and, even with the heating on low, pipes and radiators froze and burst on New Year's Eve 1978. Two of the earliest boilers cracked from overheating as a result of boilermen's frantic attempts to provide enough warmth, and coal smoke stained the crisp white paintwork, inside and out. The building's long-standing dependence on coal finally gave way to a district heating scheme in 1998 and at last, in 2011, a refurbishment sought to upgrade the Bauhaus's insulation and reduce heat loss.[51]

With a building of such worldwide influence and significance, the restoration was suitably sensitive, balancing improved performance with conservation of original details. The hardest balance to strike was in the workshops – the heart and soul of the building, and its most important architectural expression, but also the part that lost the most heat. In the end, the restoration architects concluded that the only way to avoid very radical intervention in the architecture was to abandon the workshops, leaving them as empty monuments to the architectural ideas of 1920s modernism. Ironically, the failure in energy terms of this beautiful building might be the most valuable way it can contribute to contemporary architectural education. At present architecture schools still tend to treat modernism as the foundation of today's architecture, but as the evacuated Bauhaus workshops show, there is no other architectural style in history that offers today's students such a bad model for approaching the relationships between energy and architecture.

Barber suggests that modernism's profligacy with heat and electricity was not negligent; it was, many believed, a matter of doing the right thing. Until more recently, when people became aware of the changes that fossil fuel use was causing to the Earth's atmosphere, the use of energy had seemed like 'a positive contribution to economic activity'.[52]

Ample glass was not just a good way of stimulating demand for coal, it was a promise of a brave new world. Ever since London's 563 metres-long Crystal Palace exhibition building of 1851 had shown the potential of glass and metal enclosures, dreamers and radicals around the industrializing world had proposed glass as the transformative modern material that would make human life literally and metaphorically transparent, and change our relationship with nature. A new glass architecture was needed, 'which lets in the light of the sun, the moon, and the stars, not merely through a few windows, but through every possible wall, which will be made entirely of glass'.[53] In 1920 in the USSR a prominent sculptor proposed a vast occupied

monument 400 metres high, straddling a river. Its huge double-helix iron structure would incorporate the meeting places for the central organs of state, each of which would be walled with glass to encourage openness and revolve at a different speed: the legislature at the bottom, in a glass cube, would complete a full revolution in a year; the executive, in a glass pyramid above, in a month; the cylindrical press bureau over that in twenty-four hours; and the hemispherical radio station at the top every hour.[54] The early years of the Soviet Union saw many such radical proposals. The Constructivist architects behind them were thrilled by the potential of high-energy materials and machinery, but frequently unable in reality to muster enough of either to make their schemes come to fruition in the challenging economic and political conditions of the early USSR.

The ideas and imagery that De Stijl in the Netherlands, the Constructivists in the USSR, Le Corbusier in France, the Bauhaus in Germany and the Futurists in Italy were developing in the early decades of the twentieth century have all come to be seen as threads in the complex fabric of modernism. Many of its leading figures met up from 1928 under the auspices of CIAM, the French acronym for International Congresses of Modern Architecture. Together they discussed the great challenges of the changing world and issued proclamations on the future of architecture and the city.

With their fanatical celebration of the changes they saw coming from new forms of energy and of the boundless possibilities these seemed to promise for buildings and cities, modernists have had an enduring influence over architectural thinking and architectural aesthetics.[55] Modernists had in common a rejection of the ornament beloved of nineteenth-century architects. Now that it was possible for a factory to churn out any quantity and style of decoration at lower and lower costs, it was losing its age-old cachet. What had once been a sign of the wealth of the client and the skill of the craftspeople who made it was fast becoming a sign instead of the poor taste of everyone involved – like an adult wearing plastic jewellery out of a Christmas cracker. The Austrian architect Adolf Loos opined in a famous article 'Ornament and Crime', first published in 1913, that the tendency to decoration was improper by his time in Western cultures, and so were other expressions of the longing for ornament:

> the modern man who tattoos himself is a criminal or a degenerate. There are prisons in which eighty per cent of the prisoners are tattooed. Tattooed men who are not behind bars are either latent criminals or degenerate aristocrats. If someone who is tattooed dies in freedom, then he does so a few years before he would have committed murder.[56]

To many, the pristine white surfaces of the new buildings, influenced by the aesthetics and easy-to-clean materials of medical architecture, as well as the unashamed exposed structure and lack of ornament of factory architecture, seemed a welcome relief from the visual and physical clutter and the omnipresent coal soot of the bourgeois house.

The Second World War

While the avant-garde was taking this rejection of ornament to extremes from the 1910s, most architecture in the industrialized world before the Second World War showed a total indifference to Le Corbusier's five points. Some modernist motifs were softened by the art deco style, often used for forward-looking buildings like cinemas and car showrooms, but for most new houses, pitched roofs and strong links to local architectural habits remained the norm, whether in the form of the Arts and Crafts-influenced suburban houses that spread out around London, into any area that the tube or the car now made accessible; or in Reims, France, which was being extensively reconstructed with clear art nouveau influences after devastation in the Great War; or in the US, where most houses continued in an established American tradition of wooden structure and facing, and generous stoop.

London, Reims, Columbus, GA, all inter-war

The tipping point for the adoption of modernist ideas came after another cataclysmic war. The energy systems of oil and the internal combustion engine experienced further rapid development under the pressure of combat, as did electrical networks struggling to meet the unprecedented energy demands of the arms race, and the refinement of weapons-grade nuclear fissile material. With large bomber planes gaining ever greater range and destructive power, many cities were damaged and a few devastated, making the 1930s modernist dreams of new cities suddenly seem like realistic reconstruction schemes.

In munitions factories, the rapid production of countless engines for everything from motorbikes to lorries, tanks and aeroplanes gave a stimulus to new production techniques and heightened the existing emphasis on speed and efficiency. More than three decades on from his first Ford plant,

Albert Kahn headed the team that produced a very large new tank factory for the Chrysler Corporation and the US government. Even while the rapid construction of the factory was proceeding, the first tanks were beginning production in the completed portion of the building.[57]

The plant produced more than 25,000 tanks between 1941 and 1945, over a quarter of all those built in the USA.[58] In factories across America air-conditioning systems were stripped out of civilian commercial buildings as a contribution to the war effort. Steady, cool temperatures in munitions factories were helping to improve the accuracy of bomb sights and other precision parts, and many other industrial processes that were affected by atmospheric conditions.[59] By the end of the Second World War the daring early adopters of modernist architecture had been proved right: mass production, electricity, concrete, steel and plate glass really were ready to change the built world beyond recognition.

CHAPTER 11

'Too cheap to meter'

The Post-war Boom, 1939–90

In September 1954, in a Manhattan hotel, America's National Association of Science Writers met for a celebratory dinner. They were addressed by Lewis Strauss, the chairman of the USA's Atomic Energy Commission, who was in bullish mood:

> It is not too much to expect that our children will enjoy in their homes electrical energy too cheap to meter, will know of great periodic regional famines in the world only as matters of history, will travel effortlessly over the seas and under them and through the air with a minimum of danger and at great speeds, and will experience a lifespan far longer than ours, as disease yields.

The predictions of improved health care, reduced famine and fast travel each proved broadly accurate for the following six decades or more. The speech has become famous, however, for its less accurate prediction that energy would become 'too cheap to meter'. Strauss's agency was at the time working on America's programme of nuclear power plants, and exploring longer-range possibilities for even greater levels of power generation from nuclear fusion.[1]

Nuclear energy was, many hoped, to become the powerful

new technology that would grow to dwarf fossil fuels in electricity generation. At the height of nuclear optimism, the *Eagle* comic imagined that the 1990s would see 'Dan Dare: Pilot of the Future' dashing round the galaxy in spacecraft and across London by monorail. Nuclear has, in the event, turned out to be disappointing. The danger of devastating accidents, the significant difficulties involved in safely disposing of radioactive material produced by fission and the expense of belt-and-braces safety procedures to minimize the risk of radiation leaks have made it uneconomic and politically unattractive, with relatively few countries opting to have it as more than a relatively modest proportion of their electricity generation. Despite the absence of a transformative new energy source in this period, however, much of the world's population saw substantial increases in their energy access over the three decades following the end of the Second World War from ever-rising fossil fuel extraction.

Britain was the first country in the world to achieve a national electricity grid, just before the Second World War. Networking electricity generation and use made it far more efficient. In the 1890s, when Britain's few power stations were small and local, 90 per cent of their capacity went to waste: supply needed to be reliable, but demand was unpredictable, so the generators kept turning whether their output was used or not. Even in 1929, with much larger supply, 84 per cent was going to waste. A decade later, once the National Grid was set up, only 16 per cent of capacity went unused.

In the post-war years British electricity supply capacity

rose sharply. In 1955 London's Battersea Power Station, designed and half built before the war, was one of the biggest producers of electricity in the UK, burning 10,000 tonnes of coal each week to supply up to 509,000kW of electricity, around a fifth of London's demand.[2] It was to be overtaken by ever-larger fossil-fuelled power plants, no longer built in cities, the largest being Drax power station, opened in 1973, and since much expanded and modified. Drax can produce up to 4 million kW – the equivalent power of 50 million pre-industrial labourers working steadily.[3] Even that is only around 7 per cent of the UK's electricity requirements.

The very large power stations of the post-war period became major features of the UK landscape. Their enormous, gently curving cooling towers, in particular, were familiar objects on the horizon for much of the population. Excitingly, the UK is succeeding in significantly reducing its fossil-fuelled power generation, but the resulting systematic demolition of cooling towers – these monuments to the industrial past and the decarbonization of the grid – is short-sighted and vandalistic.[4]

In the early 1960s oil overtook coal in providing the biggest proportion of the world's fossil energy: in 1965 coal provided 16 billion kWh of energy, but oil over 18 billion. With natural gas rising to produce well over another 6 billion, the world's energy supply that year was a record-breaking 40.5 billion kWh – the equivalent of more than half a trillion hours of steady pre-industrial labour.[5]

Ever-larger oil tankers carried the world's new leading energy source around the globe with relatively low overheads, powering engines for a huge variety of purposes.

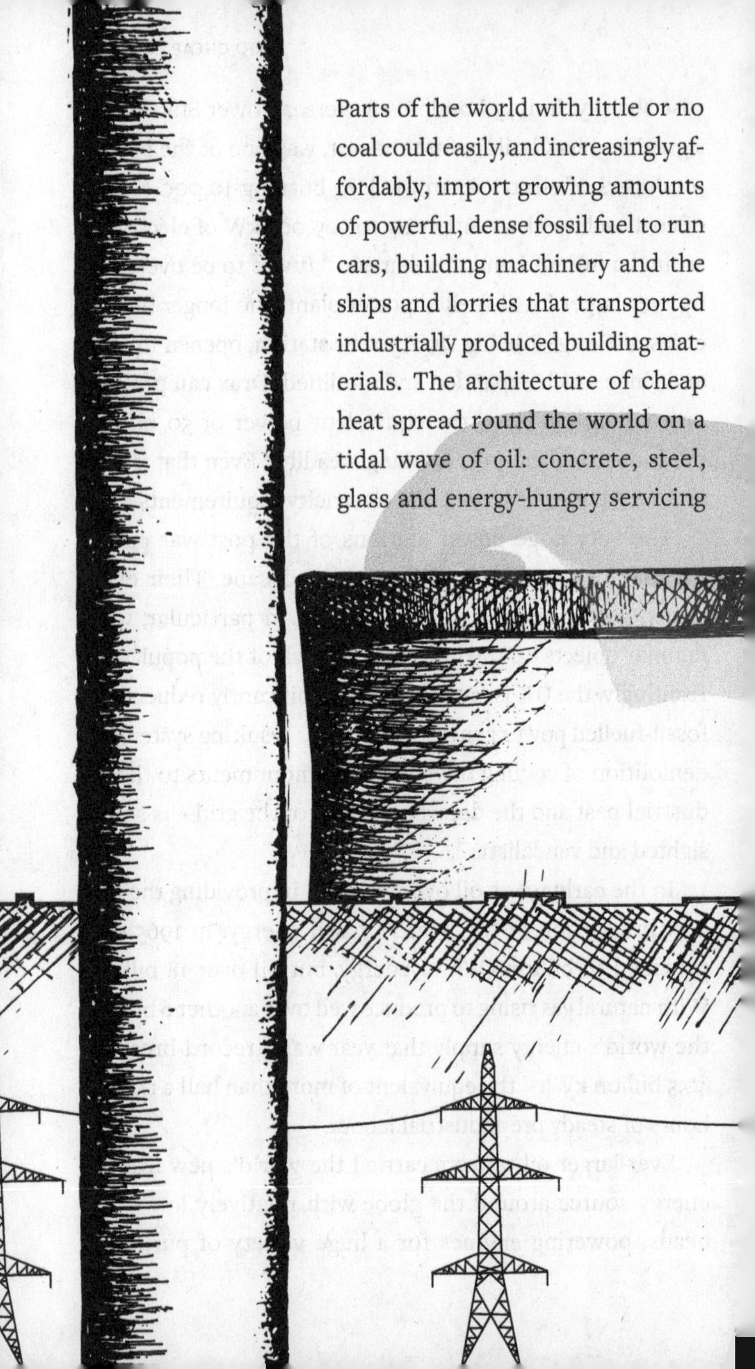
Parts of the world with little or no
coal could easily, and increasingly affordably, import growing amounts
of powerful, dense fossil fuel to run
cars, building machinery and the
ships and lorries that transported
industrially produced building materials. The architecture of cheap
heat spread round the world on a
tidal wave of oil: concrete, steel,
glass and energy-hungry servicing

propagated across the globe accompanying new fossil fuel patterns of business and industry.

If Le Corbusier and his generation of modernists had in some instances tried to run before they could walk, their fundamental insights into the radical implications of oil, the internal combustion engine and electricity tended to be borne out in the decades following the Second World War, as the technologies with which they had experimented became increasingly universal realities in rich countries, and a form of conspicuous consumption for the powerful in many poorer countries.

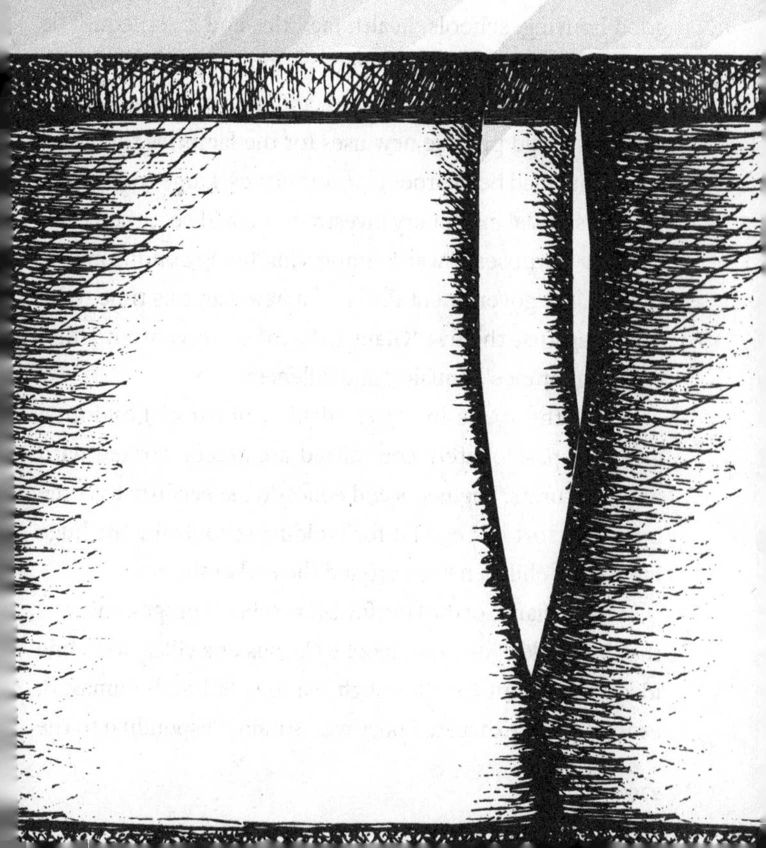

Any Colour, So Long as It is Grey

Gropius's dream of buildings mass-produced cheaply and quickly in factories came much closer to widespread reality after the Second World War. The military's need for rapidly erected factories, barracks, aircraft hangars and so on had driven rapid developments in prefabrication, and when peace came, idealistic architects, engineers and industrial designers sought to harness the power of factory provision to increase the quantity, quality and construction speed of vital social welfare facilities. With prefabrication, they hoped, good housing, schools, health facilities and so on could be built fast, to predictable budgets, on limited public finances. Governments in many countries, and industrialists too, saw that this could provide new uses for the factories that until so recently had been producing war planes. Large workforces and substantial machinery investments could be turned from military purposes towards improving living conditions. In the UK, the government declared a new war, this time to be waged against the five 'Giant Evils' of society: 'want', 'disease', 'ignorance', 'squalor' and 'idleness'.[6]

From the 1940s in Hertfordshire, north of London, a group of passionately committed architects worked with manufacturers, engineers and education specialists to come up with a sort of Lego kit for building schools for the huge number of children born around the end of the war.

The brilliance of the Hertfordshire schools programme was that it was designed not, like Le Corbusier's villas, for some ideal future, but for the tough realities of bomb-damaged, economically exhausted post-war Britain. Responding to the shortage of labour and machinery in a country with much urgent repair work to do and many of its building labourers still overseas in uniform, the designers produced a structural kit of components so light and slender that they could be slotted into place by men without much training, standing on ordinary wooden ladders, rather than having to wait, and pay, for specialist skilled workers or scarce cranes.[7] The schools had a pleasing simple aesthetic, made child-friendly by cheerful murals painted at weekends by designers coming in for free to do work they believed in.

In the immediate post-war years in Britain, materials and the energy to produce them were in shorter supply than designers and engineers, so the team did very precise calculations to minimize the amount of metal used in the structural components. American structures, by contrast, were built in a context of greater energy wealth: abundant cheap steel meant it cost less to use more steel than needed instead of paying an engineer to spend the extra time required to shave a little off the dimensions.[8]

Experiments were also made in mass-producing concrete buildings. When the British government decided to build seven new universities to expand the national provision of higher education, leading architectural practices used this as an opportunity to set out their stalls on the future of education, architecture and cities. The University of York used a prefabrication system that had originated in the Hertfordshire schools programme. At the University of East Anglia, Denys Lasdun & Partners designed their own prefabrication systems for the magnificent continuous wall of teaching and research space, and again for distinctive 'ziggurat' student residences in front of the Teaching Wall. Some of the large components for the university's buildings were manufactured in a factory and trucked to the site in powerful diesel vehicles. Others were cast on the ground near

their final location. Both types were dropped into place by a tall steel crane that travelled round the site on train tracks, effortlessly lifting components far larger and heavier than the modest human workforce of the project could have shifted. It is not the first time in this book that very heavy pieces have been used in construction, but whereas the movement of column shafts had been a major form of ostentation in ancient Rome, similar-sized steel-reinforced concrete beams were driven round Britain's fast-expanding road network as a matter of course by the 1950s and 1960s, only to end up hidden within structures. The prestige that once attached to using large components – the prestige of heavy labour requirements – was now reversed. With diesel-burning lorries and cranes, larger prefabricated components could bring down expensive labour costs by reducing the need for on-site assembly.

Prefabricated concrete, and concrete cast into moulds in its final position ('*in situ* concrete') both featured heavily in the single most influential building of the post-war period. It was, once again, by Le Corbusier, now in his sixties but no less creative and charismatic as an older firebrand than he had been as a young one. His Unité d'Habitation in Marseille (1947–52) was a block of 337 flats, with a row of shops part-way up. It still followed some of his pre-war points (standing on columns and with a roof garden), but now in a completely different language. His 1920s villas had seemed to want the new industrial materials to look smooth, bright white, hygienic – expressions of factory-made perfection. After years of experience of the messy realities of concrete building sites, Le Corbusier's Unité d'Habitation revels in the chance marks left by construction processes. The formwork moulds into which the *in situ* concrete was poured were made of wood. Looking at the finished concrete you can see at once how the imprint of the wood grain and the irregularities of rough, construction-site carpentry are lastingly preserved. The concrete elements that were prefabricated elsewhere and then installed in the building are equally recognizable, with their more precise craft and visible joint lines where separate pieces meet. Even the fire escape staircase becomes a kind of modernist totem pole, the stair itself winding curvaceously round a continuous central slab of concrete up to the shopping level.

Although locals for a long time called the unusual building 'the house of the crackpot', for visiting architects and artists the lure of Le Corbusier's muscular, well-controlled crudity was irresistible. On the roof terrace, Le Corbusier placed facilities for a children's playgroup and made space for residents to exercise and relax, all in a fantasy landscape

of abstract shapes that would have made a good set for an avant-garde staging of *Alice in Wonderland*. Yet in fact these apparently self-indulgent formal games are the housings for modern infrastructure: a boiler-house chimney, roof vents for air extraction, a water tank and lift machinery.

Concrete, the wonder material of the post-war decades, was very widespread and, under the influence of the Unité d'Habitation, very often exposed. While Le Corbusier's buildings tended to be accented with brightly coloured paint, their beauty was broadcast to the world overwhelmingly in black-and-white photos, and architects fell in love with, and reproduced, this monochrome appearance in their work.

Brick continued to appear in many buildings, but now, in a strange historical reversal, it had gone from being a cheap structural workhorse, often concealed behind stone or plaster, to becoming the more prestigious surface, increasingly often applied in the factory as thin brick tiles – 'slips' – on a concrete backing, before the entire panel was trucked and craned into place. Not even in this most optimistic and fast-improving moment in the history of architecture did the backward-looking tendency that was so dominant in agrarian architecture entirely disappear, and by the 1950s many seem to have found brick an attractive material for its evocation of architecture from earlier in the fossil fuel centuries – architecture that was now just about old enough to seem 'traditional'.

Over the post-war decades the consumption of energy-hungry building materials soared. In the USA, the use of cement (the powder that makes up about 20 per cent of most concrete) more than quadrupled from under 20 million tonnes in the year after the Second World War to nearly 82 million tonnes in 1973.[9] The energy required to make so much cement was equivalent to the charcoal production of an area of coppiced woodland the size of Michigan, or put in terms of human labour it would amount to the full year's useful work output of a group of manual workers more than five times the total population of the USA at the time.[10]

Concrete, steel and glass, which had come to dominate the post-war lexicon, had been referred to by Gropius as 'new synthetic substances'. In the post-war decades two genuinely new materials were to add to the options available to architects.

Aluminium

If a number of building materials grew fast in the post-war decades, a newcomer to architecture grew faster than most. Aluminium is very abundant, making up around 8 per cent of the earth's crust, but it had been impossible to extract even in laboratory conditions until the 1820s.

When it first came to prominence in the 1850s, it was as a new precious metal. It was shiny when polished, like older precious metals, but thrillingly lightweight compared to clunky silver and gold. In 1884 a 2.7-kilogram pyramid of it was used in a position of great honour, as the tip of the Washington Monument on the National Mall in Washington, DC.[11] The following year, a small amount of aluminium was displayed as a precious novelty alongside the crown jewels of France at an international exhibition in Paris.[12] Shortly afterwards, a new chemical solution considerably improved the extraction process, but it continued to cost up to ten times as much energy to make as steel.[13]

Strong, light, rigid and not flammable, aluminium was much used for making military aircraft during the Second World War. The American architect Pietro Belluschi was approached during the war by the head of a dam administration whose electricity fuelled the aluminium industry.

The businessman, who was worried that the huge demand for aircraft aluminium would collapse after the war, asked Belluschi for advice on whether there might be architectural uses for the metal. Belluschi initially hoped to devise an aluminium structure for an office tower he designed, the Commonwealth Building, in Portland, Oregon (1946–8), but when this proved not to be achievable he was nevertheless able to use aluminium for a shining façade and numerous other elements.[14] Wartime fears of a peacetime collapse in demand for the new metal were successfully avoided: between 1939 and 1956 US aluminium production rose by a factor of ten, although the boom was driven by Cold War rearmament as well as new architectural applications.

Internationally prizes were awarded for clever new uses of aluminium, like Denys Lasdun's 1957 prefabricated roof, made in a British aluminium factory and test-erected in a field before being packed flat and shipped to Ghana, where it still serves as the roof of the National Museum. It embodies

the Liverpool iron-founder Cragg's imperialist dream of flat-pack architecture for export, but in an even lighter material.[15] The 1951 Festival of Britain, a national exhibition intended to perk up and advertise a country still in the grip of food and goods rationing after the war, included a magnificent dome 110 metres across. With its skin and structure entirely of aluminium, it was the largest aluminium dome there had ever been.[16]

In Pittsburgh, the USA's leading city for heavy industry, the producers of different materials made their headquarters into shopfronts for the architectural use of their products: US Steel put up a sixty-four-storey tower of a special steel called Corten, which rusts but does not flake, so never needs painting; Pittsburgh Plate Glass built a complex that emphasized shiny plate glass; and Alcoa, the great aluminium corporation, built a tower that not only used aluminium internally wherever possible, but was also clad in panels of aluminium, the windows built in, all of a piece, in the factory.[17]

Plastics

An even more transformative group of materials that took off in the post-war decades tends to be overlooked, or seen as having failed in architecture: plastics. Plastics not only required high energy inputs (in the region of four times the energy input per kilogram of steel), but were made with crude oil.[18] Plastics had originated in the first decade of the century with Bakelite, which was invented as a synthetic insulator for the growing production of electrical cables, but became much more abundant and varied in the 1950s–80s. In the 1950s–70s plastics were one of the great hopes of some of the most radical architects, who believed that these materials would soon be revolutionizing architecture as fundamentally as they were revolutionizing household goods and packaging. Plastic enclosures could be light, quick to build in big pieces, easy to transport and assemble, waterproof, brightly coloured and easy to wipe clean. One of Britain's leading post-war architects, James Stirling (1926-92), experimented with architectural plastics in a training centre he designed for Olivetti (1969–72). Despite the fact that the building was largely hidden from the road, planners vetoed the very loud colours he originally hoped for, but even so its rounded shapes bore a distinct resemblance to the stylish, fun typewriter designs for which Olivetti was famous.[19]

In the longer run, though, plastics proved underwhelming for external architectural use. The technical performance of most plastics under ultraviolet sunlight turned out to be disappointing, scuffs and damage were very visible and hard to repair, and plastic soon got a negative reputation as cheap and unattractively artificial. Despite this, plastics are in standard use in contemporary architecture for a huge range of floor coverings, suspended ceilings, pipes, electrical fittings, kitchen surfaces, window surrounds and doors, including the very widespread retrofitted PVCu double-glazing installations that have improved the insulation of so many buildings over recent decades, usually at the cost of an ugly appearance. Perhaps 21 per cent of all plastic use is for architectural applications.[20]

The most important current role of plastics in architecture might well be as new thermal insulation layers in buildings, though here too the story is complex. Despite fairly heavy energy requirements for their manufacture, their versatility is valuable in finding ways of reducing heating loss in existing buildings and saving operational energy. Yet irresponsible use in some instances has led to a series of fires internationally, most catastrophically at Grenfell Tower in London in June 2017, when flammable insulation and cladding panels spread fire from flat to flat, emitting poisonous gases as they did so. Seventy-two people died. Plastics can be made with good fire-retardant properties, but in a context of excessive deregulation and inadequate funding, together causing an aggressive race to the bottom on price, a much more dangerous option was selected for Grenfell Tower. Despite appropriate grief and anger at this entirely avoidable tragedy, by June 2020 more than 600,000 people in the UK are thought to have been living in buildings still clad in flammable materials, some without even an upgrade or replacement plan in place.[21]

Blowing Hot and Cold: Heating, Ventilation, Air Conditioning and Lighting

If 'new synthetic substances' were coming of age after 1945, so were the transformative mechanical services (heating, cooling, lighting, electrical power points and so on) whose potential had been foreseen by Le Corbusier and his generation.

As the power and quantity of pipes, cables, ducts and machinery increased with growing technical capabilities, architects needed to find ways of handling these potentially intrusive systems so as not to detract from the quality of the architecture. The architect who did most to resolve complicated servicing demands into poetic, beautiful buildings was Louis Kahn (1901–73). Kahn practised and taught architecture in Philadelphia, where he won a commission for a university laboratory, the Richards Medical Research Laboratories (1957–65), and it was here that he first developed his key idea for controlling the growing technical complexity of buildings – especially scientific research buildings like this. He divided the brief into two types of space: 'served' spaces, which were the key places for humans to work, and 'servant' spaces, which were there to maximize the functionality and pleasantness of the 'served' spaces. This translated, in practice, into open, simple, flexible rooms of 'served' space, with substantial conduits for services running up the outside, and unobtrusively through the floor slabs, to provide whatever was needed, wherever it was needed: water, electricity, air, specialist gas supplies for scientific experiments and so on. The attached service tower was to become a major feature of architecture for two decades, both as a practical measure and as a powerful aesthetic element to give excitement to the shape of potentially boxy buildings.

Louis Kahn used another variation of the idea of servant and served spaces at another new laboratory, the Salk Institute (1962–5). Jonas Salk, who had led the team that produced a successful vaccine against the widespread and serious childhood disease polio, gained funding to build a super-lab for other leading scientists and their research teams. The site was a ridge overlooking the Pacific in La Jolla, California, and Kahn was selected as the architect by Salk, who believed strongly in the value of beautiful architecture and art (he was married to the distinguished artist Françoise

Gilot). Salk expressed a desire that Kahn should 'create a facility worthy of a visit from Picasso' – an interesting choice, given that Gilot had lived with Picasso for ten years.

The simplicity and timeless beauty of the Salk Institute, which Kahn himself compared to a 'ruin', seem a long way from the messy complexities of lighting cables and laboratory air pumps. Kahn tamed all the technical apparatus of the high-specification laboratories by spanning their open spaces with long, deep concrete trusses a bit like the criss-crossing ironwork of St Pancras's train shed. The 2.74-metre depth of these trusses was then treated as an extra floor into which all the services could be built, keeping the laboratories flexible and adaptably well serviced, and the architectural expression clean and pure.

The laboratories focused on life sciences. In order to avoid any risk of recirculating experimental disease agents through the ventilation system, they were to have enough air flowing through them to change all the air inside twelve times every hour. All this air was heated in cold weather and cooled in hot in order to make the building's users comfortable. So indispensable was this air supply, and the lighting of sometimes windowless interiors, that the institute had two separate connections to the local electricity mains, and its own generator, along with two back-up generators to provide emergency power, just in case all these failed. An electricity supply was now an indispensable basic requirement for human life within parts of many buildings.

It was not only specialist scientific contexts like the Salk Institute that saw the creation of sealed buildings with entirely artificial ventilation, like those Le Corbusier had foreseen between the wars. Sealed façades were to become normal for office buildings, beginning in the USA in the 1940s and spreading worldwide. They remain today a common mode of design for office buildings, despite their high energy demands.

The first example of this new type was Belluschi's Commonwealth Building. Sealed façades were in fact needed for mechanical ventilation systems to be effective (one open window could destroy the pressure gradients needed to move the air around predictably and evenly). In Belluschi's building, innovative technology produced glass that reduced the overheating caused by the summer sun hitting so many large windows.[22] The designs of some of the fixed-façade buildings that followed, however, seem to take the same pride in consuming large amounts of cheap energy as the big, heavy 'gas-guzzler' Cadillacs and Thunderbirds of the period. Engineers and manufacturers tweaked, rethought and improved ventilation systems with each successive building, minimizing the potentially lettable space they wasted, moving from space-hungry air ducts to smaller water ducts to carry the cool or warm temperature from the plant rooms to ventilation units all over the building, making diffusers better and generally competing to provide the most space-efficient and powerful systems.[23]

Tall office buildings with uninsulated glass façades, metal window frames and no shade from the sun looked sleek, but maximized the requirement for conditioned and pumped air to be sent to every space. In large plant rooms, powerful machines worked constantly to keep up with the demand, heating the air during winters and cooling it when the bright summer sun pounded in through so much glass. Air vents were often near windows in order to keep users comfortable despite the extensive heat transfer through the building envelope.

The dream of the modernist architects who had come up with such ideas was that if you poured enough energy into it, modern servicing could eliminate the age-old local differences between buildings around the world: a glass tower in the tropics would simply use more cooling plant, while one in the Arctic Circle would use more heating. The office worker in each would be equally comfortable in the conventional Western business clothes that were spreading like modernist architectural façades through the commercial centres of the world. For these architects the new energy-hungry technologies were a great unifier, allowing what they saw as 'ideal' conditions to exist everywhere – a truly international style.

As night fell, America's office buildings became even more striking. The new bright, cool lighting technology of the post-war years, fluorescent tubes, had starter mechanisms which wore out periodically when lights were turned on and off daily. Office managers found it cheaper to run the lights twenty-four hours a day, even when the building was empty.[24] The first building to show this effect, Lever House in New York City, shone like

a beacon, making surrounding streets that had previously been thought bright look gloomy by comparison.

The sleek image of the new blocks was indeed copied in other energy contexts. In London, buildings that looked like America's slickest new office towers went up within years of the originals, but as London's planning environment was much more cautious about height (for fear of fire and out of respect for historical urban skylines) than the typical US city, and as the cost of air-conditioning equipment was initially prohibitive in the UK, neither air conditioning nor space for it was provided. The resulting blocks have often proved uncomfortable to use and hard to retrofit, resulting in early demolition.[25]

Not only the façade systems but also the volumes possible in new office blocks were transformed by the improving mechanical services of the mid-century. The relationship between window height and room depth had long determined how useful a room was: as we have seen, more than 8.5 metres deep and US offices were too dark to attract high rents, six metres being more typical. Gas lighting was hot, pricey and degraded the air quality. Early electrical fittings did not produce gases but were hot, and required regular replacement as well as an expensive electrical supply. By the 1940s, however, fluorescent tubes could provide bright, even lighting with almost no heat and less use of electricity, which was in any case much cheaper by then.

Whole new office layouts became possible, with very large areas of open space divided as desired by management into smaller and larger offices, laid out – like a Ford factory

– according to working process rather than, as always before, to meet the need for access to precious window light.

It was not only office buildings that were transformed by cheap, reliable electricity, and the ever-improving machinery to exploit it. As the Salk Institute showed, some kinds of rooms now no longer needed windows at all. And if they did not need windows, their walls did not have to have direct contact with the outside air. Buildings could be wider, with some rooms entirely surrounded by other rooms; they could go down deep into the ground and still provide decent light and air quality to every room. Le Corbusier had realized in considering the Lingotto factory that cities needed no longer to be built on the ground, with roads running along the earth's surface. This was now expanded by post-war architects: buildings could potentially be any size and shape desired, with electricity making spaces that had previously been impractical into habitable ones.

Suddenly, a dense urban site could pack in many more facilities and much more useful space than ever before. The Barbican Estate and Arts Centre in the City of London, designed by Chamberlin, Powell and Bon from the late 1950s, was built through the 1960s and 1970s, with the Arts Centre being finally completed in the 1980s.[26] In it, a concert hall with nearly 2,000 seats was entirely buried under a large courtyard intended for exhibiting sculpture. Its foyers folded over on themselves between a library to one side and an art gallery above. Roads, car parking and even an underground railway threaded their way through different levels of the building's subterranean areas. Beneath it all was a plant room with an array of powerful ventilation fans, boilers and machinery that would have put many ocean liners to shame.

On all sides of the arts centre, and indeed above it, the Barbican Estate houses around 4,000 people in over 2,000 flats built in large blocks supported on slender concrete and steel legs. The entire site is only a little over sixteen hectares, yet thanks to the density of the flat planning and the versatility lent by steel-reinforced concrete and electrically powered services, the estate has a remarkable atmosphere of tranquillity, openness, greenery, water and quiet, car-free space. It is a peaceful spot in the middle of noisy London. The monumentality of the buildings – aesthetic successors to ancient Egyptian and Mayan religious architecture in their appearance of chunky permanence –

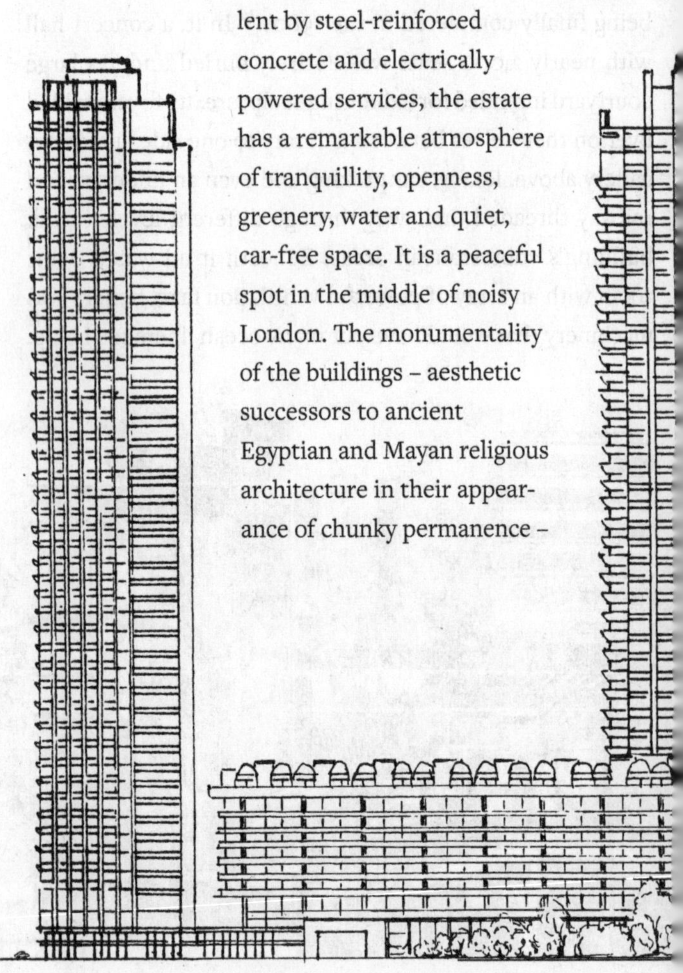

seems to rejoice in the new architectural possibilities. The architects' practice Chamberlin, Powell and Bon was very interested in the history of architecture and compared aspects of the Barbican to Venice and Georgian London. Yet they felt no need to tiptoe gently, or echo the styles of the past. Their design seems instead to derive its strength and confidence from the self-evident technical superiority and artistic richness they found in their sudden ability, thanks to structural and servicing engineers, to build in almost any shape and to provide flats and communal facilities of such life-enhancing loveliness. The Barbican is the architecture of fossil fuels at its glorious best, built in guilt-free celebration before the knowledge spread of the environmental harm oil, coal and gas were doing.

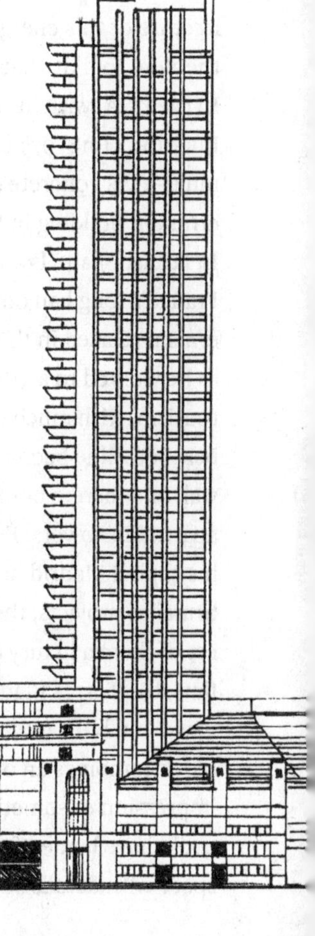

International Brutalism

In England, the term 'brutalism' came to be used to describe this new concrete style. It originated from the French *béton brut* – exposed concrete – but it may have stuck because of the almost violent animal vigour or 'brutality' of some of the best brutalist buildings.[27] Over much of the world, the architecture of this energy boom period shows similar elation at the new toolkit that its designers were able to work with. In the USA, with his trademark idiosyncratic brilliance, Paul Rudolph (1918–97) explored the curves of the baroque in a multi-level concrete government building in Boston. In fascist Spain, Francisco Franco's long ban on modernism relaxed in the 1960s as he worked to attract more tourism, with much of the best coastline becoming a wall of concrete hotels and apartment towers. Even in his capital Madrid he allowed brutalist projects, the greatest of them a luxury block of flats, the Torres Blancas (designed and built 1961–9). The block combines an overall impression of something that could have landed from outer space with details suggesting

that its architect, Francisco Javier Sáenz de Oiza (1918–2000), was working out the best concrete treatment with the contractors as the building rose, trying different solutions on successive floors to the tricky challenge of getting wooden moulds off the set concrete without damaging the concrete or leaving splinters of wood trapped in it.

The moods of this flowering of concrete architecture, typically having in common a bullish confidence, could range from the curvaceous expressionism of Giovanni Michelucci (1891–1990)'s Chiesa dell'Autostrada (1960–64) on the outskirts of Florence to the heavily modelled façades of repeated elements used for office buildings by Marcel Breuer (1902–81).

All over the world, brutalist architecture was enthusiastically taken up as visual shorthand for buildings that were new and benefited from the huge technical advances of recent years: offices, banks, flats, shopping centres, multi-storey car parks for fast-growing numbers of vehicles, schools and universities, hospitals and motorway service stations. Social housing in cities dotted around the planet took inspiration from Le Corbusier, resulting in millions of new flats in big, tough-looking blocks. Their political and architectural promoters had generally taken it

for granted that the new residents would feel delighted to move from crowded, mouldy slums with bad sanitation into modern flats with hot and cold water in their bathrooms and fitted kitchens.[28]

For the growing global middle classes, the tourism industry leapt enthusiastically to provide modern hotels from Tunis to Kiev. New resorts in the Alps opened to cater for a

boom in skiing – a sport of the energy boom if ever there was one, with its high demands for heating, transport to remote snowy places, constantly churning ski lifts and, abhorrent innovation of recent years, outdoor hot tubs in sub-zero temperatures.

The infrastructure of sport scaled up too, with ever more ambitious provision for the Olympic Games becoming a major form of cultural competition, from a masterly pair of stadiums by Kenzo Tange (1913–2005) for the 1964 Tokyo

Olympics to Montreal's 1976 Olympic stadium. The swirling shapes and cable-hung roofs of Tange's buildings simultaneously bring to mind the roofs of traditional Japanese architecture and convey the technical achievements and aspirations of high-technology post-war Japan. Japan, with its limited geological stocks of fossil fuels, was a surprising country to thrive in the post-war years in economic, technological and architectural terms. However, the mobility of oil enabled it to achieve a boom based on manufacturing. By 1965 Japan was already using more oil (almost all of it imported) than the UK, and by 1973 it was importing nearly three times as much as the UK, supporting an ongoing 'economic miracle'.[29]

Cheap oil also allowed Brazil to build a new capital in an unpromising location in the middle of arid, remote terrain with a hostile climate, without a navigable river or significant existing transport or industry. Diesel vehicles could bring the concrete, steel and glass, the food and the workforce, and aeroplanes could fly in the politicians. Brasília, the new modernist dream city founded in 1960, was built in just a few years on a site apparently foreseen in a vision by a nineteenth-century Italian monk. Don

Bosco's distinctly coal-age vision had a celestial being show him round South America by train, revealing the sites of future fossil fuel extraction, mineral mines and railways, furnishing a new 'land of milk and honey' on what was to become the site of Brasília.[30]

Redevelopment and Conservation

If new cities were called into being at will on the unprecedented energy surge of oil, the implications for existing cities were also all-changing. Cities had always developed and altered over time, but the speed and scale of change brought about by post-war fossil fuels represented a new phenomenon: soaring populations in the millions in large cities, worsening vehicle traffic and suburban sprawl enabled by easy transport for all appeared to call for new thinking on the entire nature of cities. Suddenly, very dense offices could stack floor after floor of clerical workers more tightly than had been possible under the limitations of daylight; housing could rise tens of storeys above other buildings wherever it was needed; roads could be at a different level from pedestrians, rather than crowding people on to narrow pavements while the cars chugged in traffic next to them. Le Corbusier's Plan Voisin – shocking science fiction in the 1920s – looked old-fashioned and unambitious by the 1960s, largely based as it was on an acceptance of the idea of ground level. With the new megastructures of concrete and steel, roads and ventilation ducts, the only restriction on what could be done seemed to be cowardice. In Japan young architects responded to Tokyo's size and growth by proposing vast city extensions in the bay or giant inhabited

tree-like structures bursting up out of the dense streets.

Japanese architects may have pushed such dreams further than most, but in Western Europe and America too the 1950s and early 1960s were characterized by grand plans for technological modernity. Few tried to argue for preserving the shabby, soot-blackened Victorian city. It suddenly seemed clear that the dirty windows and oppressive valley streets of Victorian towns denied light and views; their offices and living spaces were the wrong shapes, dark, chilly and obsolete. To the taste of the time, the shabby ornamentation of these older buildings was a hideous anachronism, hoarding dirt as if it were precious treasure and preserving the now-comical pretensions and snobberies of the stuffy Victorians. The social revolution of the 1960s called for clean surfaces and the egalitarian, unpretentious plain lines of the new architecture.

Hand in hand with idealistic excitement at the new possibilities for cities went the commercial realities of the property market. As seen in New York in Chapter 9, if a developer could significantly increase the lettable space of a city block by demolishing and replacing its buildings, the profits of redevelopment in areas with high land values could be enormous. One London property developer in the 1960s is said to have made £7.3 million from the well-timed Drapers Gardens office block in the City of London – enough to buy nearly 2,000 average-priced houses.[31] With so much money to be made, developers became very ingenious in exploiting whatever tools got their developments through, whether that was clever legal loopholes, charm offensives, corruption or arson.[32]

With cities changing so fast – not just buildings but whole areas of familiar townscape disappeared – a movement to preserve existing buildings became prominent in the late 1960s in Britain, America, France and elsewhere. The conservation movement sought not just to protect the most important architectural classics, but also to save ordinary decent buildings dating from the eighteenth and nineteenth centuries. Local and national groups protested against demolitions, and Victorian and Edwardian houses that had been seen as irretrievably technologically outdated came to be appreciated and carefully refurbished. A large and energy-hungry industry producing central heating and bathroom and kitchen fittings developed, exploiting improving plastic technologies that could turn a chilly, damp, squalid slum into a distinctive and comfortable family home.[33]

During the agrarian millennia there had been a strong international tendency to reuse buildings where possible and to recycle their components when demolition was unavoidable. Any of the energy cost of a new building that could be saved by using existing fabric brought expenses down sharply. The conservation movement since the 1960s has been different. A rare pre-Victorian survival in Manchester shows the new pattern. Shambles Square, a row of half-timbered Tudor houses and shops, stood in an area which was slated in the 1970s for comprehensive redevelopment into multi-storey parking, offices and retail. The Tudor row was retained as a result of conservationist sentiment, but in a way that consumed far more energy than it must originally have cost to build: it was floated on a deep concrete and steel raft, from beneath which the earth was then dug by diesel-engined earth-movers. The concrete raft and the buildings

perched on it were then raised to the new street level that had been chosen for the redevelopment of the area, and incorporated on a new artificial ground level as set dressing for the multi-storey megastructure around it. It has subsequently been moved again, and rearranged, when the 1970s shopping centre was demolished and replaced after only a quarter of a century.

Walking through the centre of a contemporary rich-world city usually feels like walking on the ground, between buildings that stand on the ground, just as was the case in most medieval cities. In reality, you are very often walking on an invented ground level floating above service roads, transport infrastructure, basement levels, sewers and drainage, service ducts for utilities and so on. Behind the façades are often very deep buildings, just as dependent on artificial lighting and ventilation as the basements beneath. Our city centres are composed of the sorts of multi-layered, multipurpose megastructures predicted in the 1960s, with heavy energy demands for ventilation, heating, cooling and lighting. However, through cultural preference we choose that they should be sculpted to have the appearance, from the front at least, of conventional older streets.

Airports

One pressure towards a new appreciation of the historical fabric of cities came from growing levels of tourism. Between 1950 and 1984 the worldwide value of the tourism industry rose by a factor of thirty-three. Aviation fuel was the key source of power behind this increase, with scheduled flights rising almost as sharply over the same period.[34] In the post-war decades aircraft got bigger, faster and longer-range as competition between manufacturers and airlines converted wartime experimentation with jet engines into a viable consumer service. The huge expansion in numbers of flights required proportionate increases in airport size, with major cities like Paris, London and New York building more than one very big airport to service ballooning traveller numbers.

If city centres were increasingly fraught battlegrounds

between conservationists and the promoters of new developments, airports operated in architecturally freer circumstances. Just as with trains in the nineteenth century, air travel in the twentieth century was to bring with it its own architecture. The best of the new airport terminals were among the most exciting and forward-looking buildings of their periods, built in vast out-of-town empty spaces, liberated from messy interaction with older neighbours or awkward-shaped sites. Their architects could produce joyous expressions of functional organization, handled in unapologetically contemporary materials and futuristic-looking architectural language.

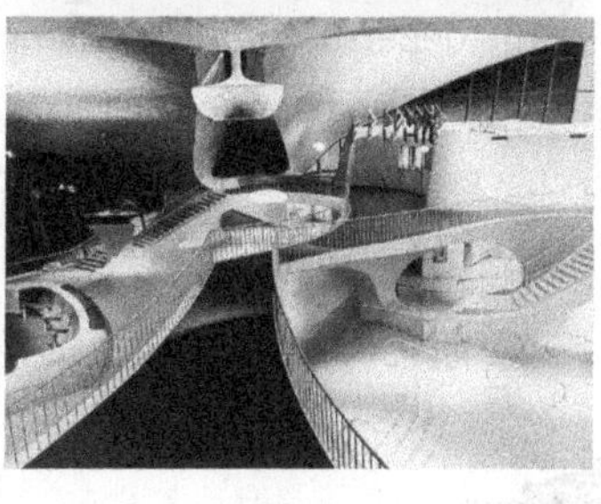

The consistently exciting Finnish-American architect Eero Saarinen (1910-61) produced one of the most beautiful terminal buildings, at New York's Idlewild Airport (now JFK, designed and built 1955–62), its swirling interiors sheltering under giant bird-like wings of concrete – 'functional, comfortable and dramatic', in the architect's words.[35]

An ambitious attempt to integrate the entire system of travel was seen in the first terminal of Paris Charles de Gaulle, commissioned in 1964 and designed by the architect

Paul Andreu (1938–2018). The core building is a vast concrete doughnut whose functions are divided up not by making separate rooms on the same floor but by making each floor have its own function – departures, arrivals, parking, baggage handling and so on. Everywhere, fossil fuels take the strain for the pampered passengers: cars drive in along a bridge, then up an external ramp to the parking floors at the top, from which passengers can then descend by lift to the departures level. There they could check in and go through the minimal security checks thought necessary in the early 1970s in a large volume of stylish space, protected from aircraft noise by sealed glazing and pumped air at an agreeable temperature. Their bags would then go down to the basement by electric conveyor belt, from which further conveyor belts would carry them up to the plane. The passengers, meanwhile, would not even have to stroll under their own muscle strength to the aeroplane, but could take a travellator out through an artificially lit and ventilated tunnel that looked like a science-fiction set, running under the plane taxiway to the departure pod. To change levels within the terminal itself, escalators in sealed tubes run constantly, carrying passengers upwards and downwards through the air across a

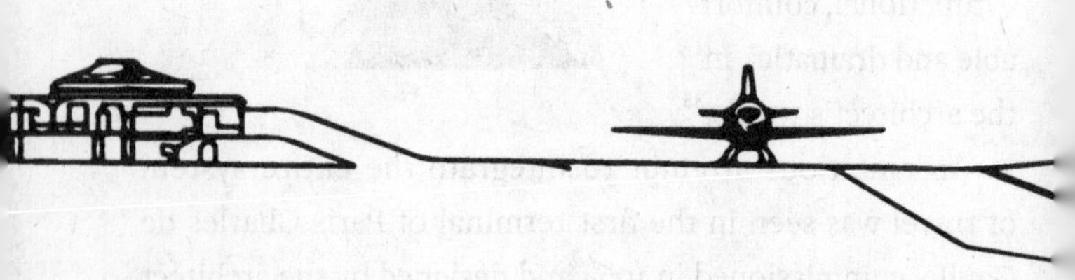

central open courtyard. Pedestrians need never meet a vehicle, planes and cars can circulate on their own levels without ever slowing for pedestrians or other vehicles, and the whole process of arrival and departure was free of sweaty physical effort: smooth, comfortable, safe and breathtakingly stylish – a step into a glamorous future where robots, it was widely hoped, might take over all our menial tasks.

Even as the terminal was opening for business in 1974, however, its design was creaking under the pressure of higher passenger numbers and larger aircraft.[36] The original intention to build several such terminals was quickly abandoned in favour of a new design that seemed to offer better versatility in the face of continually escalating levels of air travel.

That design too has continued to be developed to cope with repeated enlargement.

'Fancy nozzles'

This kind of instant obsolescence was a major concern for architects and their clients in the post-war years. Building types like science laboratories, hospitals, universities and shops cost a substantial amount to build, but could be very rapidly overcome by changes in the scale and nature of the operations they housed: bigger scientific machines, new surgical equipment, soaring student numbers, the advent and growth of supermarkets. It seemed increasingly impossible that even brutalist architecture, with all its technical innovations and versatility, could keep up with the changes in the world that were being driven by more abundant and affordable energy. Even as early as the 1910s some had been warning that concrete was inflexible and hard to remove.[37]

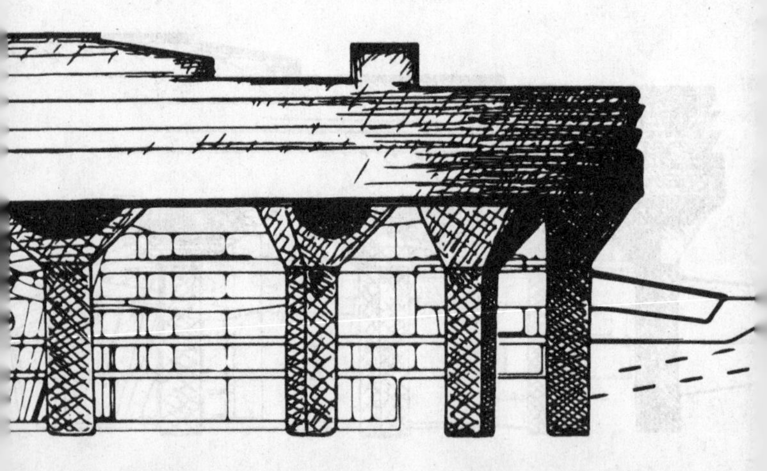

From the 1950s, a loose international grouping of architects began trying to rethink architecture in order to cope with the demands that the pace of change was placing upon it. These architects, who looked for technological solutions to ever-accelerating change, came to be known under the style name 'high tech'. In England in particular, young architects such as Norman Foster had, as children, pored obsessively over science-fiction comic strips like Dan Dare, and carried their enthusiasm for the technologies and ideas of space travel into their architectural careers.[38]

One group began their own comic-book, a home-made architectural magazine that they called *Archigram*, whose pages through the 1960s incorporated both sensible ideas on how architects might produce buildings that could adapt to change and mad-seeming fantasies about future worlds in which entire cities might float in the sky on balloons, or make their way round the globe accommodated in interlinked walking neighbourhoods, like so many giant, inhabited insects.[39] These ideas might seem fantastical with sixty years' hindsight, but in their context in the 1960s these attempts to imagine the future appear less wild. Western Europe saw its levels of energy consumption more than triple between the end of the Second World War and 1970, on what looked over most of the previous hundred years like a roughly exponential curve. If nuclear and fusion power, and other as yet unforeseen power sources, could keep that curve rising, the pace of change and development would surely increase with it. In the context of such rapid and accelerating change, walking cities were not self-evidently inconceivable. A sober British report on the implications of increasing traffic for city planning felt the need to discuss whether personal jet-pack flight was a probable form of transport in the foreseeable future.[40]

While some dreamed futuristically, other high-tech architects focused more tightly on the practicalities of what could be done now. Cedric Price (1934–2003) built very little, partly because he tended to offer radical challenges to the brief rather than acquiescently fulfil it. Yet his ambitious schemes for world-changing cultural institutions were carefully designed to be achievable straight away if someone would put up the money. His Fun Palace scheme, designed with theatre director Joan Littlewood over the course of the 1960s, tried to escape the tendency for buildings to shape and limit the activities that happen in them. It was more like a shipyard scaffold than a conventional building – big gantries of

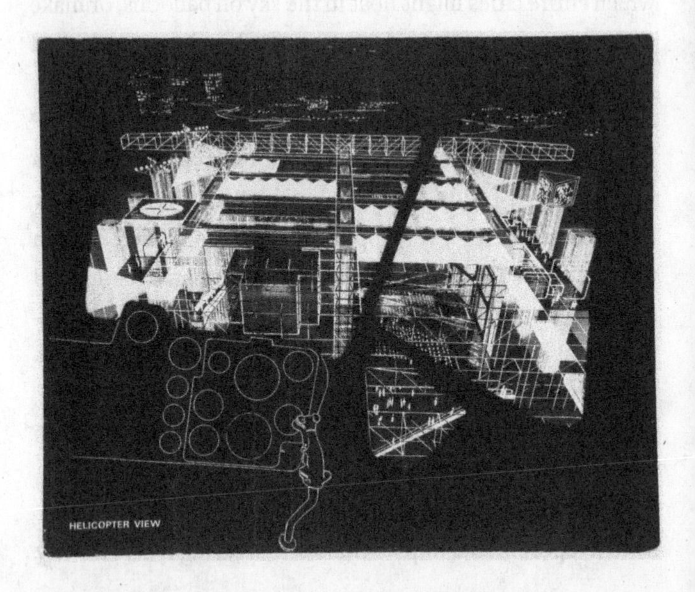

steel from which any facility that the users wanted could be suspended, built quickly as canvas enclosures, with warmth, light, sound and images all being provided by advanced electrical systems.[41] Its architecture was a candid expression of the view given by the leading American high-tech thinker, Richard Buckminster Fuller (1895–1983), that buildings were increasingly coming to be 'fancy nozzles' whose purpose was to distribute infrastructural services to users.[42]

Price's opposition to the 'medieval piles with power points' that he saw as making up most 1960s architecture was so great that when he specified the steel for the Fun Palace's only lasting element, the framework, he chose one that would rust dangerously within ten years, then need to be demolished. He hated the idea that his guesses about the future of architecture would come to constrain future change.[43] If people still wanted a Fun Palace in ten years, he felt, they could build a better one.

High-tech architects (often now preferring the term 'high performance') have mostly emphasized in recent years the ways in which their technology-driven approach can allow buildings to consume less operational energy, but in its pre-1970s form most high-tech architecture seems more like a celebration of massive consumption of cheap energy. At the time, hardly anyone had any suspicion how much harm might be done by the way fossil fuels were changing the atmosphere. The freedoms and social revolutions being brought in by bigger and better energy supplies were self-evidently delightful – easy travel and entirely new levels of information; personal choice about ways of life, forms of entertainment, sexual partners and so on.

High-tech architects dreamed of architecture that would be as freeing as a VW camper van, but more comfortable: one drawing shows a kind of blow-up sack whose separate lower inflatable compartment could allow it to adapt to any surface beneath. Within the main bag the inhabitant could relax in comfortably conditioned air, freed even from the need to wear clothes. The chilled-out occupant could be entertained and kept in touch by advanced devices, and could simply sit cross-legged on the bouncy-castle floor and smoke weed. The man shown doing so in a famous drawing was Peter Reyner Banham (1922–88), an English historian of architecture who had become an increasingly radical advocate of high-tech ideals, writing the first history of architecture as servicing as opposed to aesthetics, *The Architecture of the Well-Tempered Environment* (1969).[44] Surely, thought Banham and the high-tech architects, the heavy, unchangeable reinforced concrete of the Barbican or Charles de Gaulle Terminal One was the very worst possible technology for buildings facing such a rapidly evolving world. Rather, with plastics to build it from and helicopters to drop it into place, the Futuro House – a sort of habitable flying saucer, complete with a down-folding door with built-in stairs – was the right way for architecture to be going. Buckminster Fuller frightened architecture students with the alarmingly challenging question 'how much does your building weigh?'

Oil Crisis

Yet this future-gazing was to prove short-lived. Only a few years after the Futuro House went into small-scale early production, the map of world energy changed radically once again. In 1973, in retaliation against international support for Israel in the Yom Kippur War that year, the leading oil exporters of the Middle East stopped shipping oil to the USA, the UK, the Netherlands and a number of other countries. By the end of the embargo the following year the affected countries had experienced painful shortages, oil prices had quadrupled and demand and prices for other energy sources had risen sharply too.[45]

High-tech architecture was quick to adapt, turning its mechanical ingenuity to the problem of reducing energy consumption. Richard Rogers's practice was the unlikely winner of a competition to design a new office building for the long-standing insurance institution Lloyd's of London, in the heart of the City. The building they produced (1978–86) was the ultimate exploration of Kahn's idea of servant and served spaces: large rectangular floors round a central light well were uninterrupted by columns or service risers. On the outside of this simple core were all the staircases, lifts, air ducts, water pipes, vertical electrical cables and so on that were needed to serve the main spaces. From the exterior, these servant spaces form a beautiful expressionist sculpture, their stainless-steel weather protection glinting brightly even on a hemmed-in site under London's often-overcast skies.

Designed just after the 1973 oil crisis, the building is triple-glazed to improve insulation, with the aluminium glazing bars

insulated, unlike Gropius's highly heat-conducting window frames at the Bauhaus. Lloyd's also exploits the heat energy in extracted warm air sucked out through ducts built into the light fittings, carrying away much of the heat from the lights themselves and that from the room below. This warm air is drawn down between two layers of the triple-glazed façade to prevent the work stations near windows from becoming uncomfortably cold. It is a version of Le Corbusier's early 1930s aspiration to control internal temperature through air circulating between glazing, but this iteration profits from existing waste heat to reduce energy bills.

Yet although Rogers's team tweaked their design to reduce energy hunger, it is clearly closely related to the ideas of Price and Archigram in the energy-profligate 1960s. Even the energy-saving air circulation is achieved by powerful pumps, and by the time it opened the building as a whole was dealing with the new conditions of the high-energy 1980s. Early computer technology of the sort that was becoming part of the brief here doubled the electrical capacity required by the building, at the same time generating unwanted heat that was constantly pumped into the air of the office by the gently humming fans that have provided the sonic backdrop of so many workplaces since the 1980s. The unwanted heat generated by computers and lighting (not to mention the warmth from the hundreds of people in the building) was substantial in Lloyd's. Despite the relatively cool climate of London, the new building needed cooling most of the day, even in cold weather, with large cooling fans at the tops of the service towers and a chiller plant in the basement with a maximum capacity of 4,200kW. This was needed for a heat load more than four times that of typical office buildings at the time.[46] The resulting building is magnificent in the ingenuity of its detail and in the overall punchiness of its architectural presence. The budget was high and the architects used the money to wonderful effect.

At the other end of the budgetary spectrum, the sudden spiralling of energy costs was devastating for many buildings and their users. With unemployment and recession spreading through much of the industrialized world, governments, businesses and individuals struggled to meet the increased energy costs of lifts, lighting and ventilation systems now widespread in housing and public buildings. Building programmes ground to a halt irrespective of the viability of what had been built so far, leaving countless housing estates without their community facilities and numerous half-completed school and university expansion schemes forced to cram staff and students awkwardly into a mix of old, new and supposedly temporary accommodation.

Most disastrously, the large populations accommodated in the newish social housing estates of the industrial world hit socio-economic problems which for many turned an unfamiliar but potentially thrilling new way of life into a nightmare: very high unemployment; creaking and increasingly expensive transport networks; new illegal drug trends that temporarily dulled the pain of worsening living standards; lurches of housing policy that suddenly dumped newly released prisoners in large numbers into already unpopular blocks without adequate employment or support; collapsing levels of building maintenance. Once-valued estates could descend rapidly to be seen as the roughest parts of town, hotbeds of suffering from which those who could escape did, steepening the decline.[47]

These practical consequences of the oil crisis contributed to a collapse in the Western world's confidence in the viability and value of the modernist project. A rapid swirl of ideas came and went, driven in part by a substantial quantity of underemployed architects vying for limited work, abetted by architectural theorists seeking their moment of prominence. The heady building boom and optimistic technophilia of the 1950s and 1960s seemed to many regrettable, crazy or even, among the most hysterical voices, actively evil.

New buildings, rarer in any case in the face of economic difficulties, often diverged ostentatiously from the aesthetic norms of the 1960s. Some architects fell back on older styles of architecture, hoping that the new-found unpopularity of architecture could be overcome by reviving the look of familiar and now popular older buildings and materials. Some architects quoted the past as directly as they could, seeming to wish away two centuries of architectural change, while others were more playful, mixing old motifs with modern technologies and imagery in a sophisticated and humorous mix that they branded as 'postmodernism'.[48] Even before the

oil crisis, the leading early theorist of postmodern architecture, Robert Venturi, had proposed that Mies van der Rohe's famous minimalist mantra 'less is more' was wrong: Venturi felt that 'less is a bore'.[49]

For the architecture and construction industries in general, the effects of the oil crisis tended to be more intellectual and economic than ecological. Only limited effort was put into reducing energy consumption in most buildings, particularly in terms of the embodied energy it cost to make them in the first place. The architectural world fought intense internal battles over what constituted good and appropriate architecture, but for mainstream practice the habits of energy wealth were not much challenged until the 1990s. Only on the fringes of the mainstream were a small number of not yet influential figures starting to question the relationship between architecture, humanity and nature.[50] In 1987 this group got their rallying cry from an influential report drawn up by a commission chaired by Norwegian politician and diplomat Gro Harlem Brundtland (born 1939), which drew the attention of the public and the professions to the interlocking environmental crises that human industry was provoking and provided the lasting definition of 'sustainability': 'development that meets the needs of the present without compromising the ability of future generations to meet their own needs'.[51]

CHAPTER 12

Today's Great Energy Revolution

When I set out to research this book I confidently expected that the final chapter would be about sustainable architecture – the architectural response to the overwhelming scientific consensus that human energy use is producing a climate crisis that may soon be catastrophic for life on earth. According to a UN agency report, constructing and running buildings is responsible for 39 per cent of the human population's energy-related greenhouse gas emissions.[1]

Sure enough, interest in sustainability has gone from a small niche in the 1970s to a dominant, sector-wide preoccupation now. There is hardly any writing about contemporary architecture and cities that does not refer to sustainability as the central challenge of our time. Technical research on architecture and construction consistently refers to sustainability. Materials manufacturers and construction companies make noisy competing claims for green credentials. A seemingly endless stream of books on sustainable architecture and sustainable cities pours out of presses the world over, while blogs, social media feeds and community groups supply and exchange information on the complex decisions involved in trying to make buildings old and new as sustainable as possible. A large and ever increasing number of architects, engineers, materials specialists, architectural educators and clients have thrown themselves into the challenging and serious business of informing themselves about the great challenge of taking fossil fuels out of our built environment.

Both the ethical avant-garde and the mainstream of big commercial architecture practices devote real attention and design ingenuity to reducing aspects of their buildings' energy hunger, and seek external environmental assessment to corroborate their claims. The 'Architects Declare' and 'Architectural Education Declares' movements have seen thousands of practitioners and educators signing up to honourable goals which would improve the sustainability of the built environment.

Yet mainstream practice around most of the world remains stubbornly dependent on heavy fossil fuel inputs. The carbon emissions of construction and buildings have risen around 1 per cent every year for the past decade.[2] Almost everyone talks about sustainability and makes gestures towards it, but, despite heroic efforts from so many, so far it has been less fundamental in transforming the world of mainstream architecture than two other energy changes of very recent decades: first, the growing impact of computers in shaping the making of buildings, and second, the industrialization of much of the previously largely agrarian, developing world. I will return at the end of the chapter to the challenge posed to architecture by the need to substantially reduce our dependence on fossil fuels and to the faltering first steps that have been taken so far.

Computers

In 1957, the young Danish architect Jørn Utzon was the surprise winner of an architectural competition for a new opera house overlooking the bay in Sydney, Australia. His proposal broke with the self-restrained boxiness of many of the other entrants, suggesting a series of sail-like concrete vaults arching high over the auditoriums. He won, of course, but it rapidly transpired that he did not in fact have a structural solution for making these graceful shapes solid enough to survive gravity and battering by maritime storms.

The outstanding London-based firm of structural engineers Ove Arup & Partners was called in to work out how to make complicated, flimsy shapes into robust engineering. Despite significant disagreements, they managed it in the end, resulting in the internationally famous, widely loved roof shells of the Sydney Opera House.

Engineers had since the nineteenth century been working out mathematical calculations to tell architects and builders how big a beam would need to be for a given span between columns. The complexity of engineering involved in good reinforced concrete had made their role all the more crucial.[3] The complex geometry of the Sydney roof sails pushed their existing techniques to the limits of what was achievable, and Ove Arup & Partners built very large models of the proposed vaults, hanging weights from hundreds of different points on the shells to monitor how the structure responded. The complexity of the mathematical calculations that were required to process the resulting measurements, however, remained too great. The sheer number of sums was getting beyond what the engineers themselves could manage in a realistic time frame. Instead, they took a bold step and paid for a series of twelve- to fourteen-hour sessions in the offices of Ferranti, one of the earliest of all commercial computer manufacturers. There they fed their figures on a long tape punched with holes that represented numbers into a Pegasus computer which took up most of a room. Its array of big, expensive vacuum tubes did the many calculations required considerably faster than humans could and the engineers carried away another tape punched with holes that could then be translated back into numbers.[4]

The arrival of computers was as revolutionary an energy change as the arrival of the steam engine. Just as the latter replaced limited human capabilities in physical work, allowing vast scaling up in the quantity and types of work achievable, so computers have allowed electrical energy to replace much human mental work. And as with steam engines so with computers; once tasks were freed from the inflexible limitations of human capabilities, they could be scaled up with no apparent bounds. The growth in computing power since the invention of transistors has become legendary. The computer that landed Apollo 11's astronauts on the moon in 1969 was a hundred thousand times less powerful in terms of processing than the one a teenager's phone in 2019 used to operate their social media.[5]

By the 1990s computing power had risen to a level where the most advanced systems were capable of running software that could help with architectural design. The architect whose work showed the world the potential of this development was Frank Gehry (born 1929). Gehry's practice produced a large sculpture of a fish for the 1992 Barcelona Olympic Games. It is a pleasant enough object in its own right, but its importance lies in the fact that it was the first high-profile project in which the computer design model

the architects used to finalize its form was then also used to inform the fabricators and contractors what to make and how to assemble it.

Gehry's team was to apply similar software – software originally created to meet the much more precise requirements of aircraft engineering – to help design and build the Guggenheim Museum in Bilbao, which opened in 1997. The famous exterior of Gehry's masterpiece would have been very difficult to represent using the traditional two-dimensional paper drawings that had dominated European and North American architectural production since the Renaissance at least, but with the best new software a precise set of measurements and construction details could be produced and communicated to everyone involved. The complex curved stone parts of the building were cut by computerized lathes following the model's dimensions and contours – almost a sort of cumbersome digital printing process.[6]

The Bilbao Guggenheim was an instant worldwide smash hit with the public. Its striking appearance, totally different from anything built before – so shiny, so fantastical – resulted in a new word of praise sought by architects internationally for a decade or more: iconic. Its stirring shapes were not produced exclusively on screen, but originated with physical models, which were then scanned and tweaked on computer, printed in three dimensions, tweaked in the physical version, rescanned and retweaked on computer, until the results were not only buildable but, to the best of the architect's considerable abilities, beautiful.[7] The building created a partially mythical 'Bilbao effect', where substantial tourism revenue could allegedly be generated by a single eye-catching building. This purported effect was used to justify lavish spending on funny-shaped buildings in hundreds of cities over the world throughout a decade or more of fast-diminishing returns.[8]

In the following years, architects like Zaha Hadid (1950–2016) and Daniel Libeskind (born 1946) made their names internationally with their own variations of the expressionistic shapes and complex geometries now possible thanks to the new design software. It is clear from the plans that the dynamic shapes of the building are driven by aesthetic rather than functional considerations. In both Gehry's Guggenheim and Hadid's Guangzhou Opera House (2005–10), for instance, many of the rooms hidden behind the

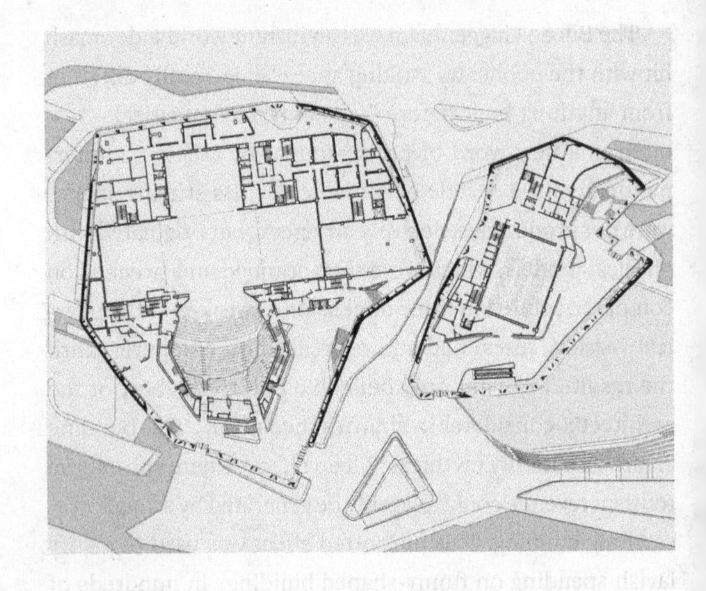

swooping and bending façades and foyers are conventional rectangles of the sort that are generally the best at accommodating normal human uses, furniture and standard, affordable fittings. Better three-dimensional printing may over time make unusual shapes easier and cheaper to construct, allowing greater freedom to designers and clients, but for now most components are produced using older methods. The fifty-nine irregular-shaped steel joints for the façade of Guangzhou Opera House, for example, were made using an ancient technique of casting metal into moulded sand – an extreme version of the twenty-first century tendency for digital technologies to move far faster than those in the world beyond the screen.[9]

While complex irregular shapes are the most visible expression of computers in architecture, the real transformative powers of IT are less obvious but more profound. The improved communications and smarter manufacturing and transportation which computers have brought about have made a modernist dream into a reality: a substantial proportion of contemporary architecture is manufactured in factories elsewhere and brought to the building site ready for a relatively simple, Lego-like assembly. The designers have at their fingertips software which can download the details of a given window type and click it into place with far less design time and need for personal knowledge about window detailing than were required until recent decades. The computer-aided supply chain then allows the windows to be made and transported with a seamlessness that reduces delays and cuts the need for expensive and unreliable on-site specialism and craft.

The latest generations of Building Information Modelling (BIM) aim to create a digital model of all aspects of the building, not only to accelerate the design and construction process but also to help with management and maintenance. Because much of the writing on such technologies is by enthusiastic adherents or by the companies who are selling the software, it is very hard to know how consistently reality meets aspiration, and how often contractors and subcontractors, in the privacy of the building site, improvise and tweak. What is certainly the case, however, is that the manufacture and deployment of prefabricated building components have grown very sharply over recent decades and will almost certainly continue to do so.

Megacities

Zaha Hadid's Guangzhou Opera House (see above) is one of a great many prominent buildings of recent decades where the named architects are based half a world away from the construction site. This is made feasible partly by quick, affordable long-haul flights and partly by the speed and reliability of computer-based communications. The international transfer of very large amounts of computerized information is extraordinarily easy. As recently as the 1990s an international collaboration involved blotchy-looking faxed drawings being sent to and fro. The original drawings could move only as fast as a courier service could transfer and fly them.[10] Now architects sitting in an office in Sydney, London or New York can exchange drawings and photographs instantaneously with people on a building site anywhere.

Another thing that has made prominent architectural practices so very international in recent years is the abrupt appearance of extremely wealthy client institutions and individuals in the fast-industrializing cities of Asia, the Middle East, South and Central America and, increasingly, Africa too. Thrilling possibilities are offered to architects by clients wanting to make a splash with an exciting building. In some of these contexts, perhaps there is also exhilaration in escaping the well-intentioned planning limitations of many Western countries.

The global 'starchitect' is the product of this internationalization. Within the stylish offices of many of this high-profile elite are often to be found the children of the upper middle classes, working in unpaid 'internships', helping the named architect to enter international competitions with minimal cost and maximum publicity. The young architects who work unpaid in a sequence of prestigious practices then have a far higher chance of establishing a prominent career of their own, entrenching a pattern of exclusion for the children of the less affluent in the world of architecture.[11]

International teams propose projects for cities they sometimes know little about, thousands of kilometres away, parachuting similar buildings into diverse cityscapes and architectural cultures modified by only superficial gestures plucked from local architectural traditions. When the famous Dutch architect Rem Koolhaas (born 1944) cited the influence of a Japanese castle on his choice of cladding for a Japanese housing project, Osamu Ishiyama (born 1944), a prominent Japanese architect, parodied what he saw as a superficial and condescending response to local architecture by offering to go to the Netherlands and 'stick windmills on all the houses'.[12]

It is not just the architects who have to adapt fast to the new environments of the urbanizing world. No one at all – not even a resident – is deeply familiar with these megacities; they change so fast, with such a complex blending of technological, economic and cultural practices, that the whirlwind development of Western industrial cities over the nineteenth and twentieth centuries looks gradual, and their final size modest, by comparison. In China, for example, the inhabitants of the new conurbations are predominantly from the first or second generations of their families to live in cities rather than on farms. Many are official residents, but there is also a substantial undocumented population of migrants from poorer areas all over China. Many towns also include villages swallowed by the growing urban area, where villagers have shown impressive entrepreneurial spirit in redeveloping their family homes into ad hoc mass housing for industrial workers.[13]

Shenzhen, 1980

Shenzhen, 2013

One of the most famous of the new megacities is Shenzhen in China. The region was impoverished and largely rural, with Shenzhen a town of under 60,000 people, until, in 1980, the Chinese government declared it a Special Economic Zone. Since then it has urbanized at an astonishing speed, to a current official urban population of 12.53 million. After only twenty years of very speedy coal-fuelled industrialization its residents were, on average, seventy-two times better off than inhabitants of the area had been in 1980.[14] Some of this raised average comes from a new elite of people grown rapidly rich in the booming industrial economy, but the level of increased material wealth for ordinary people is an even more remarkable story, and with bigger implications for the growth and development of the city, including soaring levels of car ownership and use, fast-rising demands on infrastructure, building booms of a scale and speed never seen before in human history, very rapid changes in land values and challenging levels of industrial and transport pollution.

China's architectural patrons have thrown themselves into intense competition to produce the most exciting, the most eye-catching, the best or the biggest. The largest building in the world (in terms of usable area), the New Century Global Center, is in Chengdu, China. Half a kilometre long

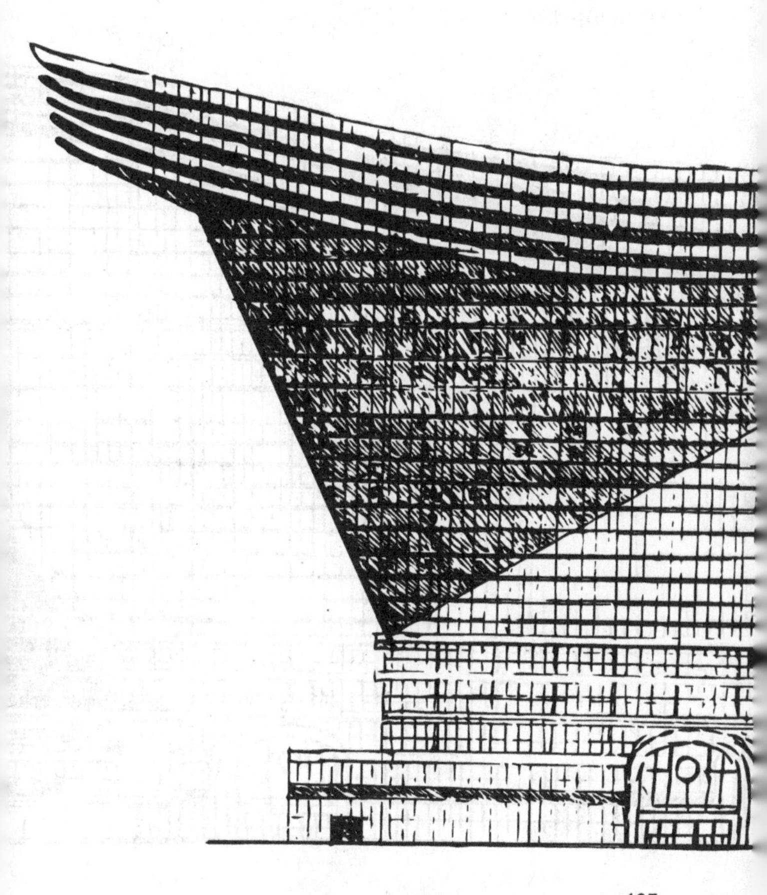

by 400m deep, and 100m high, it houses shopping, offic-es, a swimming pool with an artificial beach lit by tropical sunsets from an enormous LED screen, a skating rink, a cinema, and facilities for culture, university education and conferences. The ancestry of its steel and glass lies in mid-century American office buildings and in British high tech, while its undulating roof refers back to thousands of years of Chinese architectural history. Its size, however, is entirely contemporary.

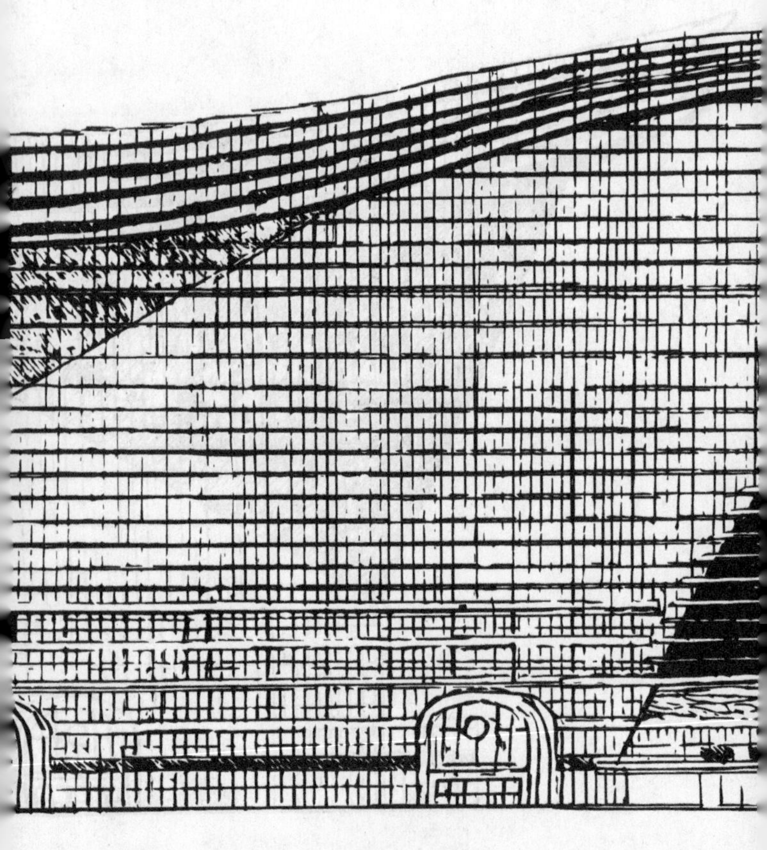

The scale of China's building boom over the past decade is highlighted by a remarkable statistic. The USA, the world's biggest economy throughout the twentieth century, used 4.4 billion tonnes of cement over the course of the entire twentieth century. In just three years, 2011–13, China used 6.4 billion tonnes.[15] Cement typically makes up only around 20 per cent of concrete. Its intensity rises and falls with the world

economy, but China's building boom of recent years has been by far the largest the world has ever seen.

The rapid development of China and other world economies is driving a demand for concrete which makes it the most used material on earth after water. The chemical reaction which turns limestone into clinker, the predominant ingredient in almost all cement, releases substantial carbon dioxide: for every tonne of limestone that goes into the process, 560 kilograms of clinker is produced. The remaining

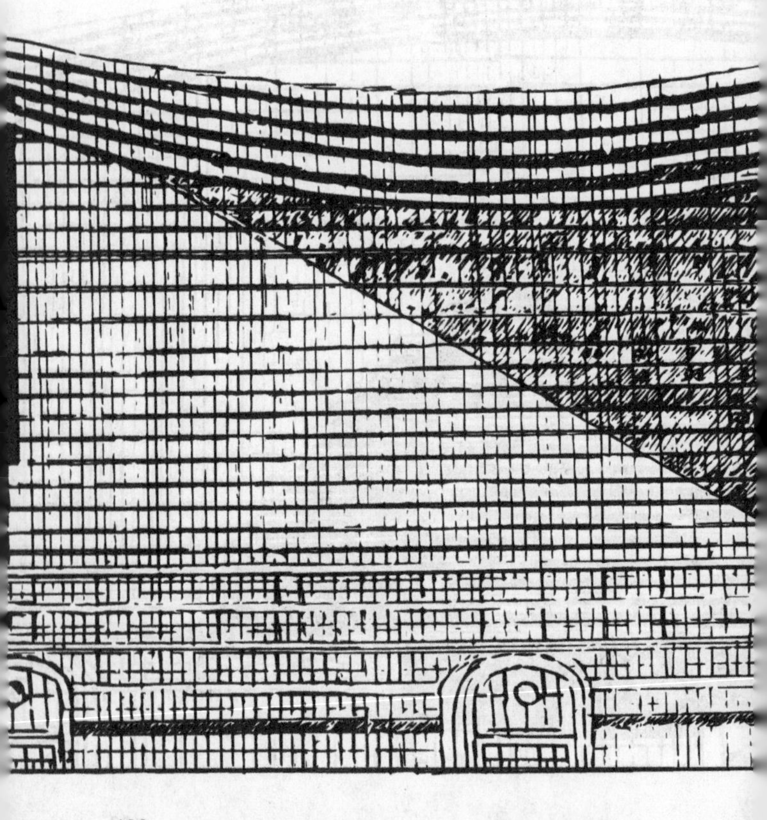

440 kilograms are emitted as CO_2.[16] Once the carbon cost of the heat required in this process is included, the cement industry emerges as the second biggest industrial contributor to climate change (18 per cent of all industrial carbon emissions) after the iron and steel industry (24 per cent), around half of whose output is also used in construction.[17] The energy requirements of the world's cement industry at present are immense: around 6.7 trillion (million million) kWh. In terms of pre-industrial energy sources the charcoal demand of today's cement industry might involve covering every metre of an area of land larger than Australia in coppiced timber to meet demand sustainably. That same amount of power, expressed in human labour, would require eight hours a day, six days a week, every week of the year, of serious physical work from a population of adult male labourers 4.7 times the size of the entire human population.[18]

The Chinese government is taking steps to reduce the pollution problems of so much coal-fired industry and oil-fired transport, and is attempting to reduce the greenhouse gas emissions of the world's fastest-growing economy, with a target of radically reducing carbon emissions and offsetting the remainder (known as net zero carbon) by 2060.[19] The proportion of China's energy to come from coal fell by 14 per cent over the decade to 2018, but even so it remained at 58 per cent, with the use of other fossil fuels rising sharply, and even coal use is still rising in absolute terms. China's building boom is, so far, a fossil fuel energy boom just as much as England's Industrial Revolution was. Renewables, though growing fast, provided less than 4.4 per cent of China's primary energy in 2018.[20] The headline figures for many Western countries are more flattering on the face of it. Yet in reality a significant proportion of China's emissions is produced making goods for export, including much of the new renewable energy infrastructure that allows Western countries to improve their local energy mix.

The Greenest Building

While sincere efforts are being made by many individuals and organizations around the world, and a few countries are managing to decarbonize their local energy supplies fairly effectively, the overall picture worldwide is one of continuing fossil fuel dependency. In 2018 carbon emissions rose by 2 per cent, with natural gas use rising sharply, and coal and oil rising more slowly, but still on the up. Renewable power grew considerably faster, but over 85 per cent of the world's primary energy continued to come from fossil fuels.[21]

The world of architecture and construction, as usual throughout this book's story, remains very closely interlocked with the main energy story of its time. In an unsustainable world, almost all architecture remains heavily dependent on fossil fuels, whatever green claims developers and designers might publicize. On building sites everywhere, new buildings are still rising in the unsustainable concrete and steel that characterized the gleeful energy frenzy of the 1960s.

Many architects are conscious of the poor performance of the construction industry so far and are distressed by it. A group of leading UK architects came together in summer 2019 to make a common declaration on the need for 'a paradigm shift', signed at the time of writing by 1,071 practices large and small.[22] Yet change so far feels incremental rather than revolutionary. The concrete mixers continue to churn and buildings that could have been repurposed continue to be demolished and replaced in service of property market profits.

Part of the problem may lie in fuzzy thinking. Anthropogenic climate change is largely a problem of human energy use: over 73 per cent of greenhouse gas emissions are produced by burning fossil fuels for energy.[23] A further 3 per cent is from CO_2 released by cement-making. Reducing these emissions to net zero by 2050 is the only way to meet targets without which climate change will reach catastrophic levels. The challenge is the biggest humanity has yet faced, but at least it is clear. In the words of the Intergovernmental Panel on Climate Change: 'The remaining equivalent CO_2 budget available for emissions is very small, which implies that large, immediate and unprecedented global efforts to mitigate greenhouse gases are required.'[24]

Yet, as the term 'sustainability' has gained increasing acceptance and importance, other good causes have staked their claims for being part of what it means to be sustainable. Social justice, the avoidance of pollution, conservation of water supplies, appropriate labour conditions and species diversity are all admirable goals, and are all important considerations in building projects. They are, however, only tangentially related to the question of how much a new building contributes to anthropogenic climate change. The critical measure in terms of climate change is the level of greenhouse gas emissions produced anywhere in the world for a given project.

As things stand, this diversity of goals embodied in definitions of 'sustainability' risks fatally undermining the usefulness of environmental assessment systems as tools for reducing the greenhouse gas impacts of constructing and operating buildings. The UK's leading sustainability assessment scheme, known as BREEAM, encourages all sorts of good behaviour from architects submitting new buildings to it, but only 31 per cent of the credits available relate to either the materials used in the building or the energy the building is expected to use in heating, lighting, cooling, lifts and so on. The other 69 per cent of credits are spread across a wide variety of topics considered to be relevant to 'environmental, social and economic sustainability'. While many or all of the goals encouraged by BREEAM are good things in themselves, they allow projects to get very high ratings while still contributing heavily to our unsustainable levels of greenhouse gas emissions.

Take, for example, the headquarters for the financial news organization Bloomberg, opened in 2017. Admirably, the founder wanted his new office building at the heart of London's financial district to focus on sustainability. He appointed one of the world's most experienced and technically advanced architecture practices, Foster + Partners, led by the prominent high-tech architect Norman Foster, to head the design team. They spared no expense exploring innovative

designs for lighting, water handling, ventilation and a 'living wall' of plants in the dining room. It was duly awarded the highest environmental rating ever given to a major office building by BREEAM – 98.5 per cent. This all sounds encouraging, and extensive press coverage reported this breakthrough in sustainable architecture.[25]

The headline sustainability claims of the new building were a 73 per cent reduction in water use and a 35 per cent reduction in the amount of energy it takes to run the building.[26] Yet the publicity for the Bloomberg HQ was much less prone to discuss the energy costs of constructing the new building. The abundant concrete, steel, bronze and stone – even if the most sustainable versions of those materials now available – will have produced substantial carbon emissions to manufacture, transport and erect. Nor was this building on a virgin site. A large 1950s concrete-framed office building which previously occupied the site had been demolished in 2010 to make way for new development. The amount of greenhouse gas emissions generated by the production of materials for the demolished building, added to the energy costs of demolishing it and disposing of the waste, added to the energy costs of the new building's materials and construction, are, as far as I know, uncalculated, but they must be substantial. As the advocates of reuse and adaptation of existing buildings argue, 'The greenest building is the one that already exists.'[27]

In the view of a leading specialist in sustainability, Simon Sturgis, the Bloomberg HQ is 'not a truly sustainable building itself nor is it a model to others for the future'.[28]

Foster + Partners and their consultants are impressive teams. Set them a target and they stand as good a chance as anyone of hitting it. If BREEAM's targets for embodied and operational carbon had been more stringent it seems probable that they would have made more progress with this difficult challenge and still earned a very high rating.

LEED, the US counterpart of BREEAM, has come under mounting criticism, including a claim by one study that, in 2011, buildings certified by LEED showed, on average, no reduction in operational energy over buildings not certified by it. Critics suggest that some of its criteria (bike racks, room cards in hotels requesting that guests reuse towels, and marking out preferential parking spaces for greener vehicles, whether or not they are then policed) are 'greenwashing'.[29]

A true climate change audit of any project ought to measure the total carbon cost of replacing what is there now and to weigh that against the possible savings of energy that could be made by upgrading the existing building. This is attempted by academics working in the important and fast-growing field of carbon life cycle assessment, but, if we are to reduce our carbon footprint as much as we need to, it needs to become part of the planning process in all countries with a substantial existing stock of robust buildings. Unless a new project can either be proved to offer a carbon saving within a fairly short timescale or be shown to be overwhelmingly necessary, the presumption should be that the existing building is kept and upgraded. For now, though, the case for doing less is argued for by few and easily shouted down by well-funded marketing departments selling a succession of financially profitable interventions whose noisy claims of improving environmental performance are unproven.

Under the current UK and US legislation, it is hard to see how the main environmental assessments could be more muscular in calling out unsustainable new buildings: they are largely voluntary schemes that succeed or fail on their ability to attract architects and their clients. If one scheme gave out lots of bad marks, or gave high marks only to a tiny number of buildings whose costs were substantially increased to meet stringent standards, it is hard to imagine the environmental assessor getting the number of new applications they need to grow.

Any voluntary system, with rival schemes in competition with each other, risks encouraging a race to the bottom. Rather than relying on voluntary assessments, we urgently need transparent compulsory regulation of embodied and operational carbon. Many architects and engineers who want to do the right thing are blocked by cost-shy clients or timid insurers. Net zero carbon, rigorously measured, needs to become as normal a minimum standard as structural resilience or fire escapes. Only then will the vast economic power and ingenuity of the construction industry focus on the most challenging and important aspects of cutting fossil fuel use and carbon emissions.

The current often-heard excuse for continuing to use energy-hungry concrete, steel and brick – that their permanence offers a long enough building life to justify the high original carbon cost of these materials – is unconvincing in most circumstances. If Norman Shaw's Dawpool, built on a rural site, for a rich family and intended to last for centuries, was gone in scant decades, surely many of today's urban buildings have a similarly unpredictable and limited lifespan. Even if they are not demolished before their time, if carbon savings are only to be realized over the course of decades then many of the buildings for which this argument is made may have been washed away by the rising seas before they have time to recover their initial carbon investment.

New buildings, meanwhile, need to measure their environmental impact not through peripheral if worthwhile considerations, but through rigorous calculation of the embodied energy costs of their materials and construction, and their running and maintenance, adaptation and eventual demolition. This is not easy – given the economic threat it poses to construction industry profits, it will be powerfully obstructed and lobbied against – but the long story of architecture set out in this book shows that humanity has an astounding ability to adapt its architecture to changing energy contexts rapidly, effectively and creatively.

The full implications of climate crisis response in architecture require a revolution as sweeping and total as that seen in the nineteenth and twentieth centuries as architecture adapted to fossil fuel energy. An unusually complete attempt to design for the new energy conditions has recently been completed on a riverbank near London. Cork House, Eton, completed 2019, was the result of years of research collaboration between the architects who designed it and live in it, and two leading schools of architecture, whose investigations into materials have allowed the resulting house to claim that its materials have taken more CO_2 out of the atmosphere than they have put in.[30]

The house is built of expanded cork – low-quality cork bark rejected by the wine industry and heated to produce a coarse, foam-like material which provides excellent thermal insulation. The cork was cut very precisely into identical blocks that could be fitted together closely enough to be made airtight with only the use of foam inserts which sit between blocks. The cork makes up the structure of walls and roofs, tethered in structurally vulnerable places by wood treated in as sustainable a way as possible to delay decay and avoid the distortion that natural wood suffers with changing moisture levels.

Excellent insulation and airtightness minimize the need for heating from the stove, which burns renewable fuel, and

in summer the openable skylights carry heat out efficiently from the funnel-like roofs. In terms of its original construction and its running, then, Cork House goes to impressive lengths to minimize energy consumption. Even beyond this, a further layer of thought has gone into the house's design: the team has very carefully thought through the end of the house's life. In 2016 the UK produced 66.2 million tonnes of waste from demolition and construction. As a result of significant efforts to improve recycling, only 9 per cent of that was put into landfill, which costs energy and space, and offers no reuse.[31] The rest was recycled – glass, steel and aluminium melted back down, plastics reprocessed, concrete crushed into new aggregate. Recycling, however, is no magic bullet. The processes of transport and recycling cost a lot of intense heat energy and in some cases, particularly with plastics and concrete, the result is a less good material than the original – 'downcycling'. The outcomes from a conventional mix of downcycling and recycling of demolition waste save around 36 per cent of emissions compared with landfill.[32] Cork House is designed to avoid recycling in favour of reuse, which offers much the best low-energy afterlife for materials. Accordingly, each element of the building has been designed for easy, non-destructive disassembly: glues and nails are rejected in favour of screws for floorboards and other carpentry, and simple gravity for holding wall blocks and foam insulating strips in place. Coatings to materials are avoided, as they make recycling difficult or degrade the quality of the recycled product. Instead of paint or varnish, the outside of the cork is protected by running rainwater clear of it with well-designed drainage details, and internally the cork is fireproofed by sprinklers rather than fire-retardant coatings. Woodwork within is all left exposed and untreated.

In its loving attention to detail, Cork House offers rich inspiration to designers and clients seeking genuinely sustainable architecture. It also presents them with a major challenge: the amount of work that has gone into researching and designing this modest-sized, simple house is immense. The design team explicitly acknowledge that it is a one-off, not a scalable solution even for other houses, let alone office buildings, sports halls or factories. Yet the rigorous thinking about the whole life of the building, and the honesty about what sustainability requires of architects, make Cork House a revolutionary building – sustainable architecture's closest counterpart so far to the early villas of Le Corbusier, exemplifying the principles that must come to dominate the architecture of the coming decades, even while the technologies are in their infancy. Le Corbusier's villas were themselves small but had a big influence. Let's hope Cork House proves similarly inspiring.

Just as important as its technical accomplishments and intellectual appeal is the fact that Cork House has found a wonderful architectural language within the technologies and materials used. Sustainability claims often seem to rely on additions like solar panels on roofs or external sunshade louvres on large modernistic windows. Cork House seems instead to

derive a beautiful architectural language from the materials and structures that arise logically from its sustainability principles. There is a comforting atavistic feeling of home that comes from the monumentality and primitivism of the coarse, dark walls, and the funnel roofs with their ancient echoes of Neolithic Scottish brochs or corbelled Mayan palaces. This is a reassuring and uplifting place in which to spend a cold winter's evening, warmed by the log burner, or to relax in the shady cool on a summer afternoon, looking on to its lovely garden.

The very earliest buildings in this book, the mammoth-bone houses discussed in Chapter 1, showed a desire to make buildings visually satisfying or meaningful. Cork House brings this ancient and distinguished tradition into the age of sustainable architecture. It is an early demonstration of the potential of the latest change in architecture's energy rules to generate another crop of ingenious, intelligent, beautiful buildings. May there be many more.

Conclusion

It is a strong recurrent theme in the history of architecture that the designers of buildings understand their practice with reference to the past. Understandings of the architectural past have helped to shape the contemporary ambitions of architecture for millennia. As architects and technicians come to consider the great energy change that faces us all – decarbonizing our built environment – architectural history needs to lead the discussion. As we have seen throughout this book, changing energy conditions have time and again given designers and builders new tools, new limitations, new challenges and new sources of artistic and intellectual stimulation.

In the quest to reach net zero carbon, human conservatism and inertia pose significant threats to the achievement of genuinely sustainable architecture. At present, most new institutional and commercial buildings seem to me to be based largely on leftover modernist and postmodernist habits and assumptions about what buildings are made of, what they look like and how they function: unnecessarily large amounts of glazing, massive concrete foundations, height achieved by steel, and exteriors covered in brick or other facings rather than rain-proofed by good detailing.

Their attempts to improve their energy performance have about them something of the awkwardness of the early historicist skyscrapers of New York, the solar photovoltaic panels on their roofs only highlighting the contradiction with their wasteful and destructive materials. Zero carbon architecture needs to rethink every assumption. It is not enough to say, like the leading modernist architect Ludwig Mies van der Rohe, 'It's up to the engineers to find some way to stop the heat from coming in or going out.'[1]

Architectural education has a role to play, with much design teaching and architectural history still profoundly shaped by modernism.[2] Beautiful, thrilling twentieth-century buildings by Le Corbusier, Mies van der Rohe and Louis Kahn haunt the design studios and architects' offices of the twenty-first century just as the classical orders ran through centuries of architecture like a golden chain, guiding and constraining. Yet, as we have seen, the wonderful buildings of modernism were the very antithesis of everything that sustainable architecture needs to become: they gloried in profligate heating, cooling, ventilation and lighting systems, in new energy-hungry materials, in car-based cities and unlimited international air travel.

While the social-democratic values embodied in much modernism retain their appeal, in technical terms today's architects might learn better attitudes from more surprising precedents like the monuments of Rome. Ancient Rome explored with exceptional effectiveness the efficiently systematized use of construction technologies that were able to exploit locally occurring materials and to minimize heat inputs. The specific details of agrarian construction will generally be impractical or undesirable, but approaches to design and engineering from earlier ages of net zero carbon architecture offer a fascinating and inspiring challenge to today's norms. Sure enough, a number of engineers and architects are exploring returning to the use of stone as a structural material, with engineer Steve Webb concluding that in terms of its ratio of strength to carbon footprint, stone performs ten times better than concrete.[3] In France the engineer turned architect Gilles Perraudin (born 1949) has designed a mixed-use tower of twenty storeys in structural stone.[4]

Generally, as with the fabric-influenced structures of Uruk or the wooden origins of the classical orders, architectural change has developed from what is already around. It is probable that the major technologies and ideas that will lie at the heart of sustainable architecture are already in use somewhere, and that their successful combination and scaling up, rather than a completely revolutionary new technology, will hold the key to producing attractive, good-quality buildings which do not require heavy fossil fuel inputs.

New buildings are only one part of the challenge. Most of the world's current building stock is incapable of being run at zero carbon, and retrofitting hundreds of millions of houses and flats, offices and industrial buildings, with such a wide body of owners, to require much smaller energy inputs is a daunting task.

Changing the patterns of transport, commerce and industry in our cities to minimize carbon inputs will also require extensive evidence-based effort and thought, before careful implementation. It will necessarily be driven not only by the design and construction industry but also by governments and institutions, and by consumer demand.

There is already a great deal of good-quality research taking place at universities and elsewhere on ways to reduce fossil fuel dependence in buildings. If the economic pressure towards business as usual is to be overcome, concerted public support will be needed for the implementation of these findings. We must hold clients to account when individuals or organizations are proposing concrete-heavy new buildings, and we must demand better regulation and legislation.

The Passive House standard has established consistent and clear parameters for genuinely sustainable building performance, with homes that meet the standard requiring almost no heating beyond the warmth of bodies and electrical appliances, slashing the biggest energy cost of most houses in temperate and cold regions. It should be made mandatory for new buildings, but at the moment UK building regulations demand much lower levels of insulation and allow a lot more precious hot air to leak out. The resulting buildings frequently underperform their unambitious sustainability targets. When measured in terms of actual heating consumption as opposed to target heat consumption the design was aiming for, homes built to the Passive House standard use around 89 per cent less energy for heating than other new buildings in Britain.[5] Many of the new buildings going up now will require substantial adaptation in a matter of years if we are to achieve net zero carbon by 2050.

A long perspective on architecture highlights the extent to which little nudging improvements on the fossil fuel dependence of the typical architecture of ten years ago are dwarfed by the vast fossil fuel dependence that had grown up over the previous two centuries. Nor, when looked at in a long historical time frame, is our reliance on fossil fuels a matter of bad habits and self-indulgence. While there are plenty of examples of both in our built environment (patio heaters should simply be illegal), fossil fuels have provided us with miraculous improvements in our quality of life over the past century or two. In some respects we will miss them.

From the radical changes that are required over the coming decades we will see new beauties emerge: new expressions of different materials and ideas, and a hybrid, varied cityscape, based on the creative reuse and retention of more ordinary existing buildings, with modest infilling and extension to accommodate necessary change and expansion. There needs to be much less work for concrete mixers and the steel industry, and much more for designers who can help improve what is already there with as knowledgeably light a touch as possible.

There are lots of things that individuals can do to improve the situation. Architects and other construction professionals can inform themselves and their clients, acting as advocates for good practice on embodied energy and energy performance. Home owners can improve the performance of their own property enormously by insulating the walls and windows, making the envelope airtight, and installing pumped ventilation with a heat exchanger to save the heat energy while changing the air. A leaky UK house can use 30,000kWh of heating each year, the equivalent energy of 400,000 hours of manual labour.[6] This can be improved by a factor of twenty by good retrofitted insulation and airtightness. With solar panels on the roofs and a sustainable choice of heating technology (a heat pump rather than a gas boiler), any house can get close to net zero carbon.[7] There are government grants to help with this, but to reach high standards will still cost the owner money. The ethical choice for comfortably off people should be to upgrade the energy performance of the home they have rather than moving to a larger one with even higher energy demands. Insulation values and annual heating requirements need to become central to the social one-upmanship of the years ahead, as car engine sizes or exotic holidays were in previous decades.

Every tool, big and small, heroic and expedient, needs to be brought to bear on the urgent race to decarbonize. Net zero carbon by 2050 is possible, but it is not going to happen without everyone making efforts and sacrifices, informing themselves and taking a clear ethical decision to live and work as sustainably as they can. The consequences of failure would be catastrophic.

Acknowledgements

This book is the outcome of the most exciting and mind-expanding research project of my life. There are many people to thank for the support, guidance and enthusiasm that have given me the courage to stray so far from my normal comfort zone of post-war British architectural history. These acknowledgements only cover the most salient debts I've incurred since starting my research in 2015 into the links between architecture and energy.

I began working on ideas for an architecture primer after a conversation between Jon Elek, my agent at the time, and Laura Stickney at Pelican. I've been well steered through the project since by Jon, Jim Gill, who succeeded him, and Pelican's editorial team, especially Laura, Rowan Cope and designer Matthew Young, who has made the book such a pleasure to look at.

After idly wondering what the embodied energy of the Pyramid of Khufu was, I started reading energy history by Kander, Malanima and Warde, and Vaclav Smil, and my world view changed for ever.

Research over such a diversity of periods and regions was made possible by an invaluable year in the British Library, thanks to a British Academy Mid-Career Fellowship.

Students at Liverpool School of Architecture have contributed enormously to my work; I've fed off their enthusiasm and tried out on them different ways of telling the story. My colleagues have been generously supportive and patient, in particular successive heads of the Liverpool School of Architecture: André Brown, Andrew Crompton and Soumyen Bandyopadhyay. University of Liverpool funding – the David Foster Wicks Endowment and the School of the Arts Research Development Initiative Fund – have provided invaluable help with image costs.

I have profited immensely from the advice of the distinguished scholars who read a full draft – Adrian Forty, Simon Pepper, Mark Swenarton and Florian Urban – and those who read one or more chapters: Daniel Barber, Alex Bremner, Fei Chen, Janet DeLaine, Fuchsia Hart, Gail Fenske, Elizabeth McKellar, Anarkali Musgrave, Joseph Sharples, Steve Sharples, Nancy Steinhardt, and Patrick Zamarian. Those of their suggestions which I couldn't squeeze into this volume will enrich my future research projects.

I have had valuable help, too, from Benjamin Akrigg, Mark Crinson, Richard Dod, staff and partners at Feilden Clegg Bradley Studios, Wolfgang Feist, Steve Finnegan, Michael Kenneth Lawson, Hentie Louw, Helia Neto, Sarah Nichols, Miles Pearson, Otto Saumarez Smith, Erica Stoller, Simon Sturgis, Rosa Urbano Gutierrez and Paul Warde.

In the final stages, Ewan Harrison lent his impressive research skills to the project, filling in gaps and improving precision with brilliant acuity and efficiency.

Working with my illustrators has been a joy, and I'm thrilled with the results. Elisabeth Gouldbourne and Clare Shields at Gingerhead produced the wonderful same-scale drawings throughout the book, the complex project brilliantly overseen by Christina Carter and Caroline Butler. Chris Dove drew the superb Liverpool maps, and Tim Rodber and Dominic Walker (@GatheringPlans on Instagram) made the beautiful comparative plans on p. 39.

Photographers, professional and amateur, have also been critical to the success of this research, both by providing such abundance and variety of photographs online for me to be able to get to know buildings I have been unable to visit, and by kindly permitting me to use their images in the book. I'm also grateful to café staff, especially at Polidor 68 in Liverpool, for letting me linger and write with my daily coffee.

In the difficult months of 2020, when Covid and Long Covid hit my family, Paul Garner helped steer us back to health and good spirits.

Helen, my wife, has been involved in this project from the start, encouraging me, letting me try out ideas in discussion with her, and helping me to improve my working patterns to speed up a project that felt increasingly urgent as its implications for the present became clearer. My wonderful parents have been a constant support in this project as always, and they, Cheryl, Peter, Katy, Michael, Ellen, Julian, Dominic and Sebastian have provided uncountable hours of loving attention to my children, allowing me to sneak extra time on the book. My daughters themselves have kindly adapted to my obsession. Harriet, at two, points out 'big, tall buildings' to me, and Lottie, six, castigates the latest energy-guzzling speculative high-rises of Liverpool every time we pass. I'm very proud.

Barnabas Calder, St Michael's, Liverpool, March 2021
@BarnabasCalder #ArchitectureAndEnergy

List of Illustrations

457

Further Reading

Energy History:

— Astrid Kander, Paolo Malanima and Paul Warde, *Power to the People: Energy in Europe over the Last Five Centuries* (Princeton University Press, 2014)
— Cutler Cleveland (ed.), *Encyclopedia of Energy* (Elsevier, 2004)
— Vaclav Smil, *Energy and Civilization: A History* (MIT Press, 2018)
— E. A. Wrigley, *Energy and the English Industrial Revolution* (Cambridge University Press, 2010)
— Ian Morris, *Foragers, Farmers, and Fossil Fuels: How Human Values Evolve* (Princeton University Press, 2015)

Books and articles that shed significant light on architectural history and energy:

Plenty of architectural history research worldwide has potential to contribute to the understanding of architectural history and its relationship with energy. The following list is not exclusive, but highlights some of the best research that contributes to this reading.

— Daniel A. Barber, 'Heating the Bauhaus: Understanding the History of Architecture in the Context of Energy Policy and Energy Transition', report published by Kleinman Center for Energy Policy, October 2019, available at https://kleinmanenergy.upenn.edu/sites/default/files/proceedingsreports/KCEP-Heating-the-Bauhaus-Singles.pdf

- Daniel Barber, *Modern Architecture and Climate: Design before Air Conditioning* (Princeton University Press, 2020)
- Tim Benton, *The Villas of Le Corbusier, 1920–1930* (Yale University Press, 1987)
- G. A. Bremner, '"In bright tints ... nature's own formation": The Uses and Meaning of Marble in Victorian Building Culture', in N. J. Napoli and W. Tronzo (eds.), *Radical Marble: Architectural Innovation from Antiquity to the Present* (Routledge, 2018)
- G. A. Bremner, 'Material, Movement and Memory: Some Thoughts on Architecture and Experience in the Age of Mechanisation', in E. Gillin and H. Joyce (eds.), *Experiencing Architecture in the Nineteenth Century: Buildings and Society in the Modern Age* (Bloomsbury, 2018), pp. 175–91 (p. 185)
- Jiat-Hwee Chang and Tim Winter, 'Thermal Modernity and Architecture', *Journal of Architecture*, Vol. 20, 2015, pp. 92-121
- Janet DeLaine, 'The Baths of Caracalla: A Study in the Design, Construction, and Economics of Large-scale Building Projects in Imperial Rome', *Journal of Roman Archaeology*, Supplementary Series Number 25 (1997)
- Cecil D. Elliott, *Technics and Architecture: The Development of Materials and Systems for Buildings* (MIT Press, 1992)
- Richard A. Goldthwaite, *The Building of Renaissance Florence* (Johns Hopkins University Press, 1980)
- Dean Hawkes, *The Environmental Imagination: Technics and Poetics of the Architectural Environment* (Routledge, 2007)
- Dean Hawkes, *Architecture and Climate: An Environmental History of British Architecture* 1600–2000 (Routledge, 2012)
- Dieter Kimpel, 'Le Développement de la taille en série dans l'architecture médiévale et son rôle dans l'histoire économique', *Bulletin Monumental*, Vol. 135 (1977)
- Ranald Lawrence, *The Victorian Art School: Architecture, History, Environment* (Routledge, 2020)
- Hentie Louw, 'Window-glass Making in Britain c. 1660–c. 1860

and Its Architectural Impact', *Construction History*, Vol. 7 (1991), pp. 47–68
— David Martlew, 'History and Development of Glass', in Michael Tutton and Elizabeth Hirst (eds.), *Windows: History, Repair and Conservation* (Donhead, 2007), pp. 121–58
— Elizabeth McKellar, *The Birth of Modern London: The Development and Design of the City 1660–1720* (Manchester University Press, 1999)
— Elizabeth McKellar, *Landscapes of London: The City, the Country and the Suburbs, 1660–1840* (Yale University Press, 2014)
— Ben Russell, *The Economics of the Roman Stone Trade* (Oxford University Press, 2013)
— L. F. Salzman, *Building in England down to 1540: A Documentary History* (Clarendon Press, 1952)
— Robert A. Scott, *The Gothic Enterprise: A Guide to Understanding the Medieval Cathedral* (University of California Press, 2003)
— Joseph M. Siry, *Air Conditioning in Modern American Architecture, 1890–1970* (Penn State University Press, 2021)

For the wider history of architecture globally, the best single source and starting point is:

— Murray Fraser (ed.), *Sir Banister Fletcher's Global History of Architecture* (21st edition, Bloomsbury, 2019)

Notes

A note on the notes: instead of providing a comprehensive bibliography, each chapter will be annotated like an academic article – the full bibliographical details will be given the first time a source is referenced in that chapter, with later mentions given more briefly.

INTRODUCTION

1. IPCC Special Report, *Global Warming of 1.5°C*, 2018, https://www.ipcc.ch/sr15/ (retrieved 21 February 2020).
2. UN Environment and International Energy Agency, *Global Status Report 2017: Towards a Zero-emission, Efficient, and Resilient Buildings and Construction Sector* (2017), pp. 6, 2.
3. Estimate taken as the midpoint of the estimated 69–87 million days produced by Stuart Kirkland Weir, 'Insight from Geometry and Physics into the Construction of Egyptian Old Kingdom Pyramids', *Cambridge Archaeological Journal*, Vol. 6 (1996), pp. 150–63 (p. 162).
4. 78 million eight-hour days working at 75W makes 46,800,000kWh for the Pyramid of Khufu. US per capita energy consumption 2018: 90,559kWh (2018 EIA figure for annual per capita energy consumption, 309,000,000 Btu, https://www.eia.gov/tools/faqs/faq.php?id=85&t=1 retrieved 17/02/2020); 2017 US life expectancy https://www.cdc.gov/nchs/fastats/life-expectancy.htm, retrieved 17/02/2020. So at present annual rate over a 78.6 year life, an average US citizen would use 7,117,937 kWh of energy: 6.57 people's lifetime energy consumption to make the pyramid. US Population in 2018 was 327.2 million. So the US population could build 49,802,131 pyramids the size of Khufu's, if they spent all their lifetime's consumption of energy on it. Khufu's pyramid is a square of 230m sides, 52,900 m², so total area of 49,802,131 pyramids that size would be 2,634,532,729,900 m², or

2,634,533km². The area of the USA is 9.834 million km², so the proportion of the USA that its citizens could cover in pyramids is 0.27.

5. Astrid Kander, Paolo Malanima and Paul Warde, *Power to the People: Energy in Europe over the Last Five Centuries* (Princeton University Press, 2014), p. 43; the figure for sustained effort is based on six eight-hour days per week for an average 35-year-old labourer, from Eugene A. Avallone and Theodore Baumeister III, *Marks' Standard Handbook for Mechanical Engineers* (10th edn, McGraw-Hill, 1996), Section 9, p. 4.
6. Mercedes A-class 2018 quoted as having up to 301 bhp (224kW), 2,987 times more power than steady human labour.
7. Figures for firewood and coal from Benoit Cushman-Roisin and Bruna Tanaka Cremonini, *Useful Numbers for Environmental Studies and Meaningful Comparisons*, available from http://www.dartmouth.edu/~cushman/books/Numbers/Chap3-Energy.pdf (retrieved 17 February 2020). For crude oil the figure given is the standard energy measurement the 'tonne of oil equivalent'.
8. Price per tonne of Brent crude on 17/08/2020, https://markets.businessinsider.com/commodities/oil-price.
9. Although a proportion of world oil consumption is used for things other than burning to supply energy. https://clockify.me/working-hours (retrieved 16 December 2019) quotes US working hours per year at 1,764.
10. Primary steel production (recycled steel is significantly less energy-hungry) could be achieved at a theoretical minimum energy input of 6.9GJ per tonne, but industrial reality is typically two or three times higher: Matthias Ruth, 'Steel Production and Energy', in Cutler J. Cleveland (ed.), *Encyclopedia of Energy* (Elsevier, 2004), pp. 695–706. Averaging that at 2.5 times higher, energy input of steel is around 17.25GJ, or 4,792kWh, per tonne. Averaged across a year, producing charcoal from coppiced hardwood gives a maximum of 0.1W/m of energy: Vaclav Smil, 'Land Requirements of Energy Systems', in Cleveland (ed.), *Encyclopedia of Energy*, pp. 613–22. So a year's production of energy from a metre of coppiced hardwood turned into charcoal is 0.876kWh. So a tonne of steel, produced with contemporary industrial efficiency using charcoal, would require the annual charcoal production of an area of coppiced woodland of 5,470m².

CHAPTER 1 – LIFE WITH LESS ENERGY

1. Lioudmila Iakovleva, 'The Architecture of Mammoth Bone Circular Dwellings of the Upper Palaeolithic Settlements in Central and Eastern Europe and Their Socio-Symbolic Meanings', *Quaternary International*, Vols. 359–60 (March 2015), pp. 324–34.
2. L. Marquer et al., 'Charcoal Scarcity in Epigravettian Settlements with Mammoth Bone Dwellings: The Taphonomic Evidence from Mezhyrich (Ukraine)', *Journal of Archaeological Science*, Vol. 39 (2012), pp. 109–20; Megan Glazewski, 'Experiments in Bone Burning', *Oshkosh Scholar*, Vol. I (April 2006), pp. 17–25.
3. Iakovleva, 'The Architecture of Mammoth Bone Circular Dwellings', p. 331.
4. Tiina Manne, 'Vale Boi: 10,000 Years of Upper Paleolithic Bone Boiling', in Sarah R. Graff and Enrique Rodriguez-Alegria (eds.), *The Menial Art of Cooking* (University Press of Colorado, 2012), pp. 173–99 (p. 185).
5. Michael W. Diehl, 'Architecture as a Material Correlate of Mobility Strategies: Some Implications for Archeological Interpretation', *Cross-Cultural Research*, Vol. 26 (1992), pp. 1–35.
6. 10,500kJ of energy per day for a modern adult man to retain a healthy weight, 365 days per year, for seventy years, is around 270GJ. 43GJ per 28,400km round trip (a bit less than London to LA and back), David J. C. MacKay, *Sustainable Energy – without the Hot Air* (Green Books, 2009), p. 35. This excellent book is available free online at https://www.withouthotair.com/.
7. E. B. Banning, 'So Fair a House: Göbekli Tepe and the Identification of Temples in the Pre-Pottery Neolithic of the Near East', *Current Anthropology*, Vol. 52, No. 5 (2011), p. 624.
8. Klaus Schmidt, 'Göbekli Tepe: A Neolithic Site in Southeastern Anatolia', in Gregory McMahon and Sharon Steadman (eds.), *The Oxford Handbook of Ancient Anatolia (10,000–323 BCE)* (Oxford University Press, 2011), pp. 917–33.
9. Banning, 'So Fair a House', pp. 619–60.
10. Schmidt, 'Göbekli Tepe'.
11. Astrid Kander, Paolo Malanima and Paul Warde, *Power to the People: Energy in Europe over the Last Five Centuries* (Princeton University Press, 2014), p. 44.
12. Jared Diamond, *Guns, Germs and Steel: A Short History of Everybody for the Last 13,000 Years* (Cape, 1997), p. 105.
13. Laurence Douny, *Living in a Landscape of Scarcity: Materiality and Cosmology in West Africa* (Routledge, 2016), p. 21.
14. Ibid., pp. 141–2.

15. Ibid., pp. 145–8, 21–2.
16. Ibid., pp. 75–7.
17. Nadine Wanono, 'Le Hogon d'Arou, chef sacré, chef sacrifié?', in R. Bedaux and J. D. van der Waals (eds.), *Regards sur les Dogon du Mali* (Leyden, Rijksmuseum voor Volkenkunde, 2003), pp. 104–9 (p. 104).

CHAPTER 2 – FARMING, THE CITY AND MONUMENTAL ARCHITECTURE

1. Children of the Hadza people in northern Tanzania spend five to six hours per day foraging, and gather around half their own calories by the age of five: Karen L. Kramer, 'The Cooperative Economy of Food: Implications for Human Life History and Physiology', *Physiology & Behaviour*, Vol. 193 (2018), pp. 196–204 (p. 198).
2. Mario Liverani, *Uruk: The First City* (London: Equinox, 2006). The standardized grain payments are hypothesized from the standard-sized bowls found in the excavations of Uruk. A rival theory proposes that they were used to bake standard-sized loaves of bread as payment: Jill Goulder, 'Administrators' Bread: An Experiment-based Re-assessment of the Functional and Cultural Role of the Uruk Bevel-rim Bowl', *Antiquity*, Vol. 84 (June 2010), pp. 351–62.
3. Liverani, *Uruk*, pp. 16–19.
4. Ibid.
5. Ibid., pp. 11–12.
6. Ibid., p. 22.
7. Ibid., p. 38.
8. Ricardo Eichmann, 'Uruk's Early Monumental Architecture', in Nicola Crüsemann, Margarete van Ess, Markus Hilgert and Beate Salje (eds., English-language edition edited by Timothy Potts), *Uruk: First City of the Ancient World* (Getty, 2019), pp. 96–107 (p. 101).
9. Ibid., p. 97.
10. Hans J. Nissen, 'Uruk and the Formation of the City', in Joan Aruz (ed.) *Art of the First Cities* (Yale, 2003), pp. 11–20.
11. David Reat, personal communication, August 2010.
12. See Virginia E. Miller, 'The Castillo-sub at Chichen Itza', in Linnea Wren, Cynthia Kristan-Graham, Travis Nygard and Kaylee Spencer (eds.), *Landscapes of the Itza: Archaeology and Art History at Chichen Itza and Neighbouring Sites* (University Press of Florida, 2018), for a synthesis of the

calendric interpretations of the temple complex at Chichen Itza. For an analysis of cosmic symbolism in iconographic and decorative programmes in Maya temples more broadly see Karl Taube, 'The Classic Maya Temple: Centrality, Cosmology, and Sacred Geography in Ancient Mesoamerica', in *Heaven on Earth: Temples, Ritual and Cosmic Symbolism in the Ancient World*, papers from the Oriental Institute Seminar held at the Oriental Institute of the University of Chicago, 2–3 March 2012.

13. Arthur Demarest, *Ancient Maya: The Rise and Fall of a Civilization* (Cambridge, 2004), p. 121.
14. Ibid., p. 117.
15. Ibid., p. 120.
16. Roland Fletcher, Damian Evans, Christophe Pottier and Chhay Rachna, 'Angkor Wat: An Introduction', *Antiquity*, Vol. 89 (2015), pp. 1,388–401 (p. 1,395).
17. Damian H. Evans et al., 'Uncovering Archaeological Landscapes at Angkor Using Lidar', *Proceedings of the National Academy of Sciences of the United States of America* (July 2013), pp. 12,595–600.
18. Etsuo Uchida and Ichita Shimoda, 'Quarries and Transportation Routes of Angkor Monument Sandstone Blocks', *Journal of Archaeological Science*, Vol. 40, No. 2 (February 2013), pp. 1,158–64. Earlier studies had conjectured a 90km route including travelling against the stream up the Siem Reap river, but this study uncovered further parts of the artificial waterway network that shorten and improve the route.
19. This is not to say that farming practices force just one urban pattern on their societies. The rice-growing south of China, for example, has a history of thousands of years of walled cities reminiscent of those of wheat- or barley-dependent societies.
20. Dieter Arnold, *Building in Egypt: Pharaonic Stone Masonry* (Oxford University Press, 1991), p. 3.
21. Ibid.
22. Herodotus, *Histories*, Book 2, 125, gives the figure of 1,600 talents of silver. An Egyptian talent was around 27kg.
23. Barry J. Kemp, *Ancient Egypt: Anatomy of a Civilization* (2nd edn, Routledge, 2006), p. 188.
24. Each labourer had a 60cm-wide stone face to work on and the trench they made as they dug deeper was only 75cm wide: Arnold, *Building in Egypt*, p. 37.
25. Ibid., p. 65.
26. J. Donald Hughes, 'Sustainable Agriculture in Ancient Egypt', *Agricultural History*, Vol. 66, No. 2, *History of Agriculture and the Environment* (Spring 1992), pp. 12–22 (pp. 17–18).

27. Arnold, *Building in Egypt*, p. 63.
28. Ibid., p. 3.
29. Ibid., p. 40.
30. Ibid., p. 37.
31. Ibid., p. 63.
32. The pyramid's sides are within a twentieth of one degree of due north–south: Kate Spence, 'Ancient Egyptian Chronology and the Astronomical Orientation of Pyramids', *Nature*, Vol. 408 (16 November 2000), pp. 320–24 (p. 320).
33. Robert C. Ellickson and Charles DiA. Thorland, 'Ancient Land Law: Mesopotamia, Egypt, Israel', *Yale Law School Faculty Scholarship Series*, Paper 410 (1995), p. 332.
34. Herodotus, *Histories*, Book 2, 124–6.
35. Kemp, *Ancient Egypt*, pp. 158–9.
36. Agrarian societies tended to place considerable emphasis on obedience in the education of their children, whereas hunter-gatherers typically emphasize self-reliance: François Nielsen, 'The Ecological–Evolutionary Typology of Human Societies and the Evolution of Social Inequality', *Sociological Theory*, Vol. 22 (2004), pp. 292–314 (p. 307).
37. Gülru Necipoğlu, *The Age of Sinan: Architectural Culture in the Ottoman Empire* (Princeton University Press, 2007), pp. 178, 185.
38. Jean-Louis Cohen, *Architecture in Uniform: Designing and Building for the Second World War* (Yale University Press, 2011).
39. R. J. Knecht, *Francis I* (Cambridge University Press, 1982), p. 264.
40. The neurological response to symmetry is strong and 'people reliably prefer symmetrical to abstract patterns': Marco Bertamini and Alexis D. J. Makin, 'Brain Activity in Response to Visual Symmetry', *Symmetry*, Vol. 6 (2014), pp. 975–96.
41. The influential nineteenth-century German architect Gottfried Semper argued that woven mats and fabrics were the origins of architecture in *The Four Elements of Architecture: A Contribution to the Comparative Study of Architecture* (1851): see Gottfried Semper, *The Four Elements of Architecture and Other Writings*, translated by Harry Francis Mallgrave and Wolfgang Herrmann (Cambridge, 1989), pp. 103–8.
42. Cassandra Adams, 'Japan's Ise Shrine and Its Thirteen-Hundred-Year-Old Reconstruction Tradition', *Journal of Architectural Education*, Vol. 52 (1998), pp. 49–60. For an account of the shrine's role as a cultural signifier see Jonathan M. Reynolds, 'Ise Shrine and a Modernist Construction of Japanese Tradition', *Art Bulletin*, Vol. 83 (2001), pp. 316–41.
43. Joan Oates, *Babylon* (rev. edn, Thames & Hudson, 1986), p. 47.
44. See Chapter 10.

CHAPTER 3 – US AND THEM: THE PARTHENON AND PARSA

1. Henry P. Colburn, 'Connectivity and Communication in the Achaemenid Empire', *Journal of the Economic and Social History of the Orient*, Vol. 56 (2013), pp. 29–52.
2. Barbara A. Barletta, 'The Architecture and Architects of the Classical Parthenon', in Jenifer Neils (ed.),.*The Parthenon from Antiquity to Present* (Cambridge University Press, 2005), pp. 66–99 (p. 95).
3. Alison Burford, 'The Builders of the Parthenon', *Greece & Rome*, Vol. 10, Supplement: Parthenos and Parthenon (1963), pp. 23–35 (p. 27); Léopold Migeotte, *The Economy of the Greek Cities: From the Archaic Period to the Early Roman Empire* (University of California Press, 2009), pp. 109–10.
4. Barletta, 'The Architecture and Architects of the Classical Parthenon', p. 88.
5. Lisa Kallet, 'Wealth, Power and Prestige: Athens at Home and Abroad', in Neils (ed.), pp. 34–65 (p. 56).
6. Burford, 'The Builders of the Parthenon', p. 29.
7. Murray Fraser (ed.), *Sir Banister Fletcher's Global History of Architecture* (21st edition, Bloomsbury, 2019), Vol 1, p. 67.
8. J. Coulton, 'The Toumba Building: Description and Analysis of the Architecture', in M. Popham, P. G. Calligas and L. H. Sackett (eds.), *Lefkandi II: The Excavation, Architecture and Finds* (British School of Archaeology at Athens, 1993), pp. 33–71. Coulton's interpretation has been challenged recently by Georg Herdt, 'On the Architecture of the Toumba Building at Lefkandi', *Annual of the British School at Athens*, Vol. 110 (2015), pp. 203–12. Herdt suggests a shorter building surrounded not by a veranda but by a fence, more like longhouses found elsewhere in Europe at this date. My thanks to Alan Greaves for drawing my attention to the building.
9. Ian Morris, *Foragers, Farmers and Fossil Fuels: How Human Values Evolve* (Princeton University Press, 2015), pp. 77–9.
10. Still the most charming and readable account of the long story of classicism is John Summerson, *The Classical Language of Architecture*, in any of its many editions.
11. Jeffrey M Hurwit, 'Space and Theme: The Setting of the Parthenon', in Neils (ed.), *The Parthenon from Antiquity to Present*, pp. 8–33 (p. 15).
12. Lothar Haselberger, 'Bending the Truth: Curvature and Other Refinements of the Parthenon', ibid., pp. 100–157 (p. 106).
13. Barletta, 'The Architecture and Architects of the Classical Parthenon', pp. 67, 73.

14. Burford, 'The Builders of the Parthenon', p. 30.
15. Manolis Korres, G. A. Panetsos and T. Seki, *The Parthenon: Architecture and Conservation*, exhibition catalogue (Athens: The Foundation of Hellenic Culture, 1996), Fig. 42, p. 69.
16. Barletta, 'The Architecture and Architects of the Classical Parthenon', p. 74.
17. Haselberger, 'Bending the Truth', p. 108.
18. Ibid., p. 103.
19. Robert C. Ellickson and Charles DiA. Thorland, 'Ancient Land Law: Mesopotamia, Egypt, Israel', *Yale Law School Faculty Scholarship Series*, Paper 410 (1995), pp. 336, 339–41.
20. Kallet, 'Wealth, Power and Prestige', p. 46; Josiah Ober, *The Rise and Fall of Classical Greece* (Princeton University Press, 2015), p. 92, Table 4.4.
21. Ibid., p. 46.
22. Christopher Tuplin, 'The Administration of the Achaemenid Empire', in Ian Carradice (ed.), *Coinage and Administration in the Athenian and Persian Empires* (BAR International Series 343, Oxford, 1987), pp. 109–67 (p. 133).
23. Ibid., pp. 153–5.
24. Ibid., pp. 137ff.
25. Jeanet Sinding Bentzen, Nicolai Kaarsen and Asger Moll Wingender, 'Irrigation and Autocracy', *Journal of the European Economic Association*, Vol. 15 (February 2017), pp. 1–53.
26. Alain Bresson, *The Making of the Ancient Greek Economy* (Princeton University Press, 2016), pp. 142–52. Bresson gives a fascinating and sophisticated analysis of the relationships between economic growth and energy supplies in ancient Greece.
27. Kallet, 'Wealth, Power and Prestige', pp. 51–2.
28. Donald N. Wilber, *Persepolis: The Archaeology of Parsa, Seat of the Persian Kings* (Darwin, 1989).
29. Rémy Boucharlat, Sébastien Gondet and Tijs De Schacht, 'Surface Reconnaissance in the Persepolis Plain (2005–2008). New Data on the City Organisation and Landscape Management', in Gian Pietro Basello and Adriano V. Rossi (eds.), *Dariosh Studies II: Persepolis and Its Settlements. Territorial System and Ideology in the Achaemenid State* (Naples, 2012), pp. 250, 278, 281.
30. Wilber, *Persepolis*, p. 36.
31. Ibid., p. 34.
32. Janett Morgan, 'Who Has the Biggest Bulls? Royal Power and the Persepolis Apadāna', *Iranian Studies*, Vol. 50, No. 6 (2017), pp. 787–817 (p. 790).
33. Wilber, *Persepolis*, p. 44.
34. Ibid.

35. A. Shapur Shahbazi, 'Persepolis', in *Encyclopaedia Iranica*, online edition, 2012, available at http://www.iranicaonline.org/articles/persepolis (accessed on 30 December 2018).
36. Plutarch, *Perikles* 14. While neat stories like this might seem unreliable as evidence, there is another case described in a stone inscription where the same politician's family asked permission to pay for a new spring house and were refused as important buildings should remain matters of collective rather than individual prestige: Kallet, 'Wealth, Power and Prestige', p. 58.
37. Burford, 'The Builders of the Parthenon', p. 25.
38. Katherine A. Schwab, 'Celebrations of Victory: The Metopes of the Parthenon', in Neils (ed.), *The Parthenon from Antiquity to Present*, pp. 158–97 (p. 161).
39. Jenifer Neils, '"With noblest images on all sides": The Ionic Frieze of the Parthenon', ibid., pp. 198–223.
40. Wilber, *Persepolis*, p. 42. There is debate as to how literally to take the sculpture cycle: did it show an actual event in which all these regional delegations turned up at Persepolis with their tribute, or was it a more stylized representation of something happening all over the empire, with tribute brought to the local rulers? In either case the apparent emphasis on energy supplies is clear.
41. Matt Waters, *Ancient Persia: A Concise History of the Achaemenid Empire, 550–330 BCE* (Cambridge University Press, 2014), pp. 98–100.
42. Herodotus, *Histories*, Book VII, 8C.
43. Margaret Cool Root, 'Reading Persepolis in Greek – Part Two: Marriage Metaphors and Unmanly Virtues', in Seyed Mohammad Reza Darbandi and Antigoni Zournatzi (eds.), *Ancient Greece and Ancient Iran: Cross-Cultural Encounters*, 1st International Conference, 11–13 November 2006, Athens (National Hellenic Research Foundation, 2008), pp. 195–221.
44. For the colouring of the Parthenon: Schwab, 'Celebration of Victory', pp. 159–60; for the first evidence of the presence of paint on the Parsa friezes: A. Askari Chaverdi, P. Callieri, M. Laurenzi Tabasso and L. Lazzarini, 'The Archaeological Site of Persepolis (Iran): Study of the Finishing Technique of the Bas Reliefs and Architectural Surfaces', *Archaeometry*, Vol. 58, No. 1 (2016), pp. 17–34; for the staircase at Parsa: Wilber, *Persepolis*, p. 36.
45. For the bronze Athena see Hurwit, 'Space and Theme', p. 12; for the weight of gold possibly being more than a full year's tribute from the Athenian Empire see Kenneth Lapatin, 'The Statue of Athena and Other Treasures in the Parthenon', in Neils (ed.), *The Parthenon from Antiquity to Present*, pp. 260–91; for Persepolis loot see Wilber, *Persepolis*, p. 11.
46. Neils (ed.), *The Parthenon from Antiquity to Present*, p. 220.

47. On Athenian colonies see Alfonso Moreno, *Feeding the Democracy: The Athenian Grain Supply in the Fifth and Fourth Centuries* BC (Oxford, 2007), p. 232. This is currently a subject of controversy amid economic historians, with a rival theory proposing that Athens was wholly or almost wholly self-sufficient in grain, but the balance of expert opinion seems to be in favour of the view that imports were extensive and important. On the Athenian Empire (never called this by the Athenians) and its role in paying for the Parthenon, see Kallet, 'Wealth, Power and Prestige', p. 57. On p. 59 Kallet suggests the idea of the Athenian citizenry as a kind of collective tyrant.
48. Plutarch, *Perikles*, 12.2–4.
49. Kallet, 'Wealth, Power and Prestige', p. 62.
50. For a detailed later history of the Parthenon see Robert Ousterhout, 'Bestride the Very Peak of Heaven: The Parthenon after Antiquity', in Neils (ed.), *The Parthenon from Antiquity to Present*, pp. 293–330: see pp. 302–17 for an account of the Parthenon's use as a church, pp. 317–20 for its use as a mosque and p. 321 for an account of its damage during the second Turkish-Venetian War.

CHAPTER 4 – ENERGY BOOMS: THE ROMAN EMPIRE AND SONG-DYNASTY CHINA

1. Both ancient commentators and modern historians have had arguments over the proportion of Rome's grain supply that came from Egypt. A reasonable estimate seems to sit somewhere around an increase of perhaps 25 per cent to the existing grain imports to Rome from other parts of the empire including Sicily and North Africa. Some important contributions to this debate are G. E. Rickman, 'The Grain Trade under the Roman Empire', in *Memoirs of the American Academy in Rome*, Vol. 36, *The Seaborne Commerce of Ancient Rome: Studies in Archaeology and History* (1980), pp. 261–75; Dominic Rathbone, 'Mediterranean Grain Prices c. 300 to 31 BC: The Impact of Rome', in Heather D. Baker and Michael Jursa (eds.), *Documentary Sources in Ancient Near Eastern and Greco-Roman Economic History: Methodology and Practice* (Oxbow, 2014), pp. 289–312; Dominic Rathbone, 'Mediterranean and Near Eastern Grain Prices c. 300 to 31 BC: Some Preliminary Conclusions', ibid., pp. 313–22; Paul Erdkamp, *The Grain Market in the Roman Empire: A Social, Political and Economic Study* (Cambridge, 2005).
2. Rathbone, 'Mediterranean Grain Prices', p. 307.

3. Michael McCormick et al., 'Climate Change during and after the Roman Empire: Reconstructing the Past from Scientific and Historical Evidence', *Journal of Interdisciplinary History*, Vol. XLIII, No. 2 (Autumn 2012), pp. 169–220 (p. 183): Egypt 'appears to have enjoyed exceptionally favorable conditions for cereal production during this period'; and best floods were consistent between 30 BCE and 155 CE.
4. Suetonius, *The Life of Augustus*, 29:3.
5. Fu Xinian, Guo Daiheng, Liu Xujie, Pan Guxi, Qiao Yun and Sun Dazhang; English text edited and expanded by Nancy S. Steinhardt, *Chinese Architecture* (Yale University Press, 2002), p. 136.
6. Laurence Sickman and Alexander Soper, *The Art and Architecture of China* (Pelican History of Art, integrated edition, 1971), p. 422.
7. Fu et al., *Chinese Architecture*, p. 185.
8. Ibid.
9. Ibid., pp. 136, 140.
10. Ibid., p. 139.
11. R. Barker, 'The Origin and Spread of Early-ripening Champa Rice: Its Impact on Song Dynasty China', *Rice*, Vol. 4 (2011), pp. 184–6 (p. 185).
12. Ibid., p. 185.
13. Francesca Bray, *The Rice Economies: Technology and Development in Asian Societies* (Basil Blackwell, 1986), pp. 203–4.
14. Ibid., pp. 204–5.
15. Rickman, 'The Grain Trade under the Roman Empire', p. 261.
16. Pierre Tallet and Gregory Marouard, 'The Harbor of Khufu on the Red Sea Coast at Wadi al-Jarf, Egypt', *Near Eastern Archaeology*, Vol. 77, No. 1 (March 2014), pp. 4–14
17. C. J. Brandon, R. L. Hohlfelder, M. D. Jackson and J. P. Oleson, *Building for Eternity: The History and Technology of Roman Concrete* (Oxbow, 2014), p. 57.
18. Erdkamp, *The Grain Market in the Roman Empire*, p. 244.
19. Annalisa Marzano, 'Trajanic Building Projects on Base-metal Denominations and Audience Targeting', *Papers of the British School at Rome*, Vol. 77 (2009), pp. 125–58 (p. 130).
20. Fu et al., *Chinese Architecture*, p. 146.
21. Pierre-Louis Viollet, *Water Engineering in Ancient Civilizations: 5,000 Years of History* (CRC Press, 2017), p. 251.
22. Peter Lorge, *The Reunification of China: Peace through War under the Song Dynasty* (Cambridge University Press, 2015), pp. 244–5.
23. Fu et al., *Chinese Architecture*, p. 136; Roman bridges survive at the Ponte Sant'Angelo in Rome and elsewhere.

24. Jan Theo Bakker (ed.), *The Mills-Bakeries of Ostia: Description and Interpretation* (Gieben, 1999), p. 9.
25. Ibid., p. 9.
26. Ibid., p. 10.
27. T. Ritti, K. Grewe and P. Kessener, 'A Relief of a Water-Powered Stone Saw Mill on a Sarcophagus at Hierapolis and Its Implications', *Journal of Roman Archaeology*, Vol. 20 (2007), pp. 139–63.
28. Joseph Needham's heroic multi-volume work of scholarship, *Science and Civilisation in China*, begun in 1954 and continued by others after his death in 1995, contains numerous examples of the sophistication of technical disciplines under the Song. See Joseph Needham and Wang Ling, *Science and Civilisation in China*, Vol. 4, *Physics and Physical Technology. Part II: Mechanical Engineering* (Cambridge University Press, 1965), p. 28. Needham dates the use of watermills to power metallurgical bellows to the first century and 'water power widely applied to textile machinery' from around the end of the Southern Song.
29. A. Wilson, 'The Economic Impact of Technological Advances in the Roman Construction Industry', in E. Lo Cascio (ed.), *Innovazione tecnica e progresso economico nel mondo romano* (Bari, 2006), pp. 225–36 (p. 227).
30. Ibid.
31. Ibid., p. 229.
32. Ibid., p. 230.
33. DeLaine, summarized ibid., p. 227.
34. Ibid., p. 228; H. Gerding, '*Later*, *Laterculus* and *Testa*: New Perspectives on Latin Brick Terminology', *Opuscula: Annual of the Swedish Institutes at Athens and Rome*, 9 (2016), pp. 7–31 (p. 11).
35. Andreas G. Heiss and Ursula Thanheiser, 'A Glimpse of Mediterraneanisation? First Analyses of Hellenistic and Roman Charcoal Remains from Terrace House 2 at Ephesos, and Their Possible Implications for Vegetation Change, Woodland Use, and Timber Trade', *Open PAGES* (2014), Focus 4 Workshop, pp. 28–9.
36. The standardization of pipe sizes is confirmed by Sextus Julius Frontinus, *The Aqueducts of Rome*, Book 1.
37. Rickman, 'The Grain Trade under the Roman Empire', p. 264.
38. The average time for news of an emperor's death to reach Egypt was fifty-seven days: Henry P. Colburn, 'Connectivity and Communication in the Achaemenid Empire', *Journal of the Economic and Social History of the Orient*, Vol. 56 (2013), pp. 29–52 (p. 48). The carving, transport and erection of one of the large obelisks at Karnak is recorded on the obelisk as having

taken seven years. Roman columns were smaller and plainer, but travelled further and were needed in much greater numbers.

39. Jean-Pierre Adam, *Roman Building: Materials & Techniques* (Batsford, 1994), p. 116, gives an estimate of fifty tonnes, and p. 26 an estimate of fifty-six tonnes. Paul Davies, David Helmsoll and Mark Wilson Jones, 'The Pantheon: Triumph of Rome or Triumph of Compromise?', *Art History* (June 1987), pp. 133–56, states that a shaft of fifty Roman feet was a little under twice the weight of one of forty Roman feet.
40. Ben Russell, *The Economics of the Roman Stone Trade* (Oxford University Press, 2013), p. 202.
41. Wilson, 'The Economic Impact of Technological Advances in the Roman Construction Industry', p. 228.
42. Qinghua Guo, '*Yingzao Fashi:* Twelfth-Century Chinese Building Manual', *Architectural History*, Vol. 41 (1998), pp. 1–13 (p. 4).
43. Ibid., p. 7.
44. 'Well building hath three Conditions. Commoditie, Firmeness, and Delight': Sir Henry Wotton, *The Elements of Architecture* (1624), p. 1.
45. Feng Jiren, 'Bracketing Likened to Flowers, Branches and Foliage: Architectural Metaphors and Conceptualization in Tenth to Twelfth-Century China as Reflected in the "Yingzao Fashi"', *T'oung Pao*, Second Series, Vol. 93, Fasc. 4/5 (2007), pp. 369–432 (p. 370).
46. Guo, '*Yingzao Fashi*', p. 4.
47. Wolfgang Drechsler, 'Wang Anshi and the Origins of Modern Public Management in Song Dynasty China', *Public Money and Management*, Vol. 33 (2013), pp. 353–60 (p. 355).
48. Guo, '*Yingzao Fashi*', p. 4.
49. Ibid.
50. Ibid.
51. Ibid.
52. Nancy Steinhardt, *Chinese Architecture* (Princeton University Press, 2019), pp. 162–3; Guo, '*Yingzao Fashi*', p. 1.
53. Steinhardt, *Chinese Architecture*, pp. 162–3.
54. Guo, '*Yingzao Fashi*', p. 1.
55. Fu et al., *Chinese Architecture*, p. 9.
56. Steinhardt, *Chinese Architecture*, p. 161.
57. Heping Liu, 'The Water Mill and Northern Song Imperial Patronage of Art, Commerce, and Science', *Art Bulletin*, Vol. 84 (2002), pp. 566–95.
58. Pliny the Elder, *Natural History*, Book XXXVI, 1.1.
59. All figures from Russell, *The Economics of the Roman Stone Trade*, pp. 30–31.

60. Ibid., p. 27.
61. Ibid., p. 355.
62. Lise M. Hetland, 'New Perspectives on the Dating of the Pantheon', in Tod A. Marder and Mark Wilson Jones (eds.), *The Pantheon: From Antiquity to the Present* (Cambridge University Press, 2015), pp. 79–98, convincingly argues that previous dating as a work entirely under Hadrian's rule is unsatisfactory. Janet DeLaine, 'The Pantheon Builders: Estimating Manpower for Construction', pp. 160–92 in the same volume, suggests a minimum construction period of around six years (p. 182), but perhaps more like nine (p. 190).
63. Emperor Vespasian and column shafts described by Dr Barbara Levick on *The Roman Way*, Episode 3, first broadcast on BBC Radio 4, February 2003.
64. Janet DeLaine, 'The Baths of Caracalla: A Study in the Design, Construction, and Economics of Large-scale Building Projects in Imperial Rome', *Journal of Roman Archaeology*, Supplementary Series Number 25 (1997), p. 99.
65. Adam, *Roman Building*, pp. 115–16.
66. Ibid., p. 184.
67. Wilson, 'The Economic Impact of Technological Advances in the Roman Construction Industry', p. 229.
68. Mark Wilson Jones, 'Building on Adversity: The Pantheon and Problems with Its Construction', in Marder and Wilson Jones (eds.), *The Pantheon*, pp. 193–230 (p. 194).
69. This 'compromise hypothesis' was first stated by Davies, Helmsoll and Jones, 'The Pantheon', and refined by Wilson Jones, 'Building on Adversity'.
70. Ibid., p. 211, confirms that the portico was the last part built, and a brick in the portico is stamped with information that dates it to having been made six years into Hadrian's rule (123 CE): Hetland, 'New Perspectives on the Dating of the Pantheon', p. 93.
71. Wilson Jones, 'Building on Adversity', p. 220.
72. DeLaine, 'The Pantheon Builders', pp. 160–92.
73. Walter Scheidel, 'Real Wages in Early Economies: Evidence for Living Standards from 1800 BCE to 1300 CE', *Journal of the Economic and Social History of the Orient*, Vol. 53, No. 3 (2010), pp. 425–62.
74. Ibid., Table 4.
75. DeLaine, 'The Baths of Caracalla', p. 209.
76. Ibid., p. 193.
77. Ibid., pp. 129–30.

78. Özlem Aslan Özkaya and Hasan Böke, 'Properties of Roman Bricks and Mortars Used in Serapis Temple in the City of Pergamon', *Materials Characterization*, Vol. 60 (2009), pp. 995–1,000 (p. 997).
79. DeLaine, 'The Baths of Caracalla', pp. 129–30.
80. Janet DeLaine, 'Strategies and Technologies of Environmental Manipulation in the Roman World: The Thermal Economy of Baths', 'Nachhaltigkeit in der Antike: Diskurse, Praktiken, Perspektiven', *Geographica Historica*, Vol. 42 (2020), pp. 75–93 (my thanks to Professor DeLaine for letting me see it pre-publication).
81. He Que Li and Feng Yun Yu, 'On the Tilt of the Kaifeng Iron Pagoda and Rectification Measures', *Applied Mechanics and Materials*, Vols. 488–9 (2014), pp. 625–8.
82. Drechsler, 'Wang Anshi and the Origins of Modern Public Management in Song Dynasty China', p. 355.
83. Yuheng Bao, Ben Liao and Letitia Lane, *Renaissance in China: The Culture and Art of the Song Dynasty 907–1279* (Edwin Mellen Press, 2006), p. 114.
84. Qinghua Guo, 'Tile and Brick Making in China: A Study of the *Yingzao Fashi*', *Construction History*, Vol. 16 (2000), pp. 3–11 (p. 7).
85. Ibid., p. 9.
86. The academic literature in English on Chinese architecture is considerably less extensive than that of some other world regions, so the very unusual practice of building at architectural scale in iron this early is not much discussed. It is documented in passing by Needham (Joseph Needham and Colin A. Ronan, *The Shorter Science and Civilisation in China*, Vol. 5 (Cambridge University Press, 1978), p. 109) and Steinhardt, *Chinese Architecture* (p. 178), in their much wider texts. Slightly more information is available through this government-backed website giving information on protected buildings in China: http://en.chinaculture.org/library/2008-02/15/content_32459.htm (accessed 27/01/2020).
87. This estimate is based on the figures for American pig iron production in 1810, where the national annual production of pig iron was 50,000 tonnes, for which around 2,600 km² of woodland had to be felled to meet its charcoal requirements (Vaclav Smil, 'Land Requirements of Energy Systems', in Cutler J. Cleveland (ed.), *Encyclopedia of Energy* (Elsevier, 2004), Vol. 3, p. 614).
88. Barker, 'The Origin and Spread of Early-ripening Champa Rice', p. 185.
89. Downdraught kilns, Guo, 'Tile and Brick Making in China', p. 9.
90. Elizabeth Marlowe, 'Framing the Sun: The Arch of Constantine and the Roman Cityscape', *Art Bulletin*, Vol. 88, No. 2 (June 2006), pp. 223–42 (p. 240, n. 52);

David Karmon, *The Ruin of the Eternal City: Antiquity and Preservation in Renaissance Rome* (Oxford University Press, 2011), p. 123. Recent archaeological investigations have uncovered evidence of substantial settlement in the Colosseum: publication will be forthcoming in late 2020.

91. Robert Edward Coates-Stephens, 'Building in Early Medieval Rome: AD 500–1000', unpublished PhD thesis, University College London, 1995, pp. 44–5.
92. Janet DeLaine, 'The Lintel Arch, Corbel and Tie in Western Roman Architecture', *World Archaeology*, Vol. 21 (February 1990), pp. 407–24 (p. 422).

CHAPTER 5 – 'A PROPORTIONAL INDICATOR OF POWER'?: TRADITION, ENERGY AND MOSQUES

1. Jeremy Jones, 'The House of the Prophet and the Concept of the Mosque', in Julian Raby and Jeremy Johns (eds.), *Bayt al-Maqdis: Jerusalem and Early Islam* (Oxford University Press, 1999), pp. 59–113 (p. 110).
2. Jonathan M. Bloom and Sheila S. Blair (eds.), *Grove Encyclopedia of Islamic Art and Architecture* (Oxford University Press, 2009), Vol. 2, 'Mosque', p. 549.
3. G. R. Hawting, *The First Dynasty of Islam: The Umayyad Caliphate AD 661–750* (2nd edn, Routledge, 2000), p. 38.
4. Finbarr Barry Flood, *The Great Mosque of Damascus: Studies on the Makings of an Umayyad Visual Culture* (Brill, 2001), p. 1.
5. Hawting, *The First Dynasty of Islam*, p. 4.
6. Ibid., pp. 63–5.
7. Alan Walmsley, 'Economic Developments and the Nature of Settlement in the Towns and Countryside of Syria-Palestine, ca. 565–800', *Dumbarton Oaks Papers*, Vol. 61 (2007), pp. 319–52 (pp. 350–51).
8. Ibid., pp. 344–5.
9. Bloom and Blair (eds.), *Grove Encyclopedia of Islamic Art and Architecture*, Vol. 2, 'Damascus', p. 515.
10. Flood, *The Great Mosque of Damascus*, p. 2.
11. Ibid., p. 3.
12. Bloom and Blair (eds.), *Grove Encyclopedia of Islamic Art and Architecture*, Vol. 2, 'Damascus', p. 515.
13. Flood, *The Great Mosque of Damascus*, p. 8.
14. Bloom and Blair (eds.), *Grove Encyclopedia of Islamic Art and Architecture*, Vol. 2, 'Damascus', p. 516.

15. Flood, *The Great Mosque of Damascus*, p. 2.
16. Ibid., p. 5.
17. U. Rebstock, 'West Africa and Its Early Empires', in M. Fierro (ed.), *The New Cambridge History of Islam* (Cambridge University Press, 2010), pp. 144–58 (p. 152).
18. John Hunwick, *West Africa, Islam, and the Arab World: Studies in Honor of Basil Davidson* (Wiener, 2006), p. 32.
19. The building we see today is the result of substantial reconstruction and restoration in the sixteenth and nineteenth centuries: Bloom and Blair (eds.), *Grove Encyclopedia of Islamic Art and Architecture*, Vol. 3, 'Timbuktu', p. 328.
20. Ali Ould Sidi, 'Monuments and Traditional Know-how: The Example of Mosques in Timbuktu', *Museum International*, Vol. 58 (2006), pp. 49–58 (p. 50).
21. Marq de Villiers and Sheila Hirtle, *Timbuktu: The Sahara's Fabled City of Gold* (Walker, 2007), p. 160.
22. Sharon E. Nicholson, 'The Nature of Rainfall Variability over Africa on Time Scales of Decades to Millennia', *Global and Planetary Change*, Vol. 26 (2000), pp. 137–58; James C. McCann, 'Climate and Causation in African History', *International Journal of African Historical Studies*, Vol. 32 (1999), pp. 261–79.
23. Shoichiro Takezawa and Mamadou Cisse, 'Découverte du premier palais royal à Gao et ses implications pour l'histoire de l'Afrique de l'Ouest', *Cahier d'Etudes Africaines*, Vol. 208 (2012), pp. 813–44.
24. McCann, 'Climate and Causation in African History', p. 267.
25. Ibid.
26. Rebstock, 'West Africa and Its Early Empires', p. 152, suggests a 12 per cent drop.
27. Stephen A. Dueppen, 'The Archaeology of West Africa, ca. 800 BCE to 1500 CE', *History Compass*, Vol. 14 (2016), pp. 247–63 (pp. 256–7).
28. The rituals are described in Sidi, 'Monuments and Traditional Know-how', pp. 55–6.
29. Eric Fernie, *The Architecture of the Anglo-Saxons* (Batsford, 1983), p. 8.
30. Michael Greenhalgh, *Islam and Marble: From the Origins to Saddam Hussein* (Canberra: Centre for Arab and Islamic Studies, 2006), p. 13.
31. R. Savory, *Iran under the Safavids* (Cambridge University Press, 1980), p. 155.
32. Ibid., pp. 155–6.
33. Charles Melville, 'New Light on Shah 'Abbas and the Construction of Isfahan', *Muqarnas*, Vol. 33 (Brill, 2016), pp. 155–76 (p. 167).
34. Savory, *Iran under the Safavids*, pp. 170–72.

35. Melville, 'New Light on Shah 'Abbas and the Construction of Isfahan', p. 155.
36. Savory, *Iran under the Safavids*, p. 160.
37. Ali Uzay Peker, 'The Monumental Iwan: A Symbolic Space or a Functional Device?', Middle Eastern Technical University, *Journal of the Faculty of Architecture*, Vol. 11 (1991), pp. 5–19.
38. An origin for *muqarnas* in Baghdad is proposed by Yasser Tabbaa in 'The Muqarnas Dome: Its Origin and Meaning', *Muqarnas*, Vol. 3 (1985), pp. 61–74.
39. Rudi Matthee, Willem Floor and Patrick Clawson, *The Monetary History of Iran: From the Safavids to the Qajars* (I.B. Tauris, 2013), p. 79.
40. L. V. Golombek, 'Anatomy of a Mosque: The Masjid-I Shāh of Isfahān', in Charles J. Adams (ed.), *Iranian Civilization and Culture: Essays in Honour of the 2,500th Anniversary of the Founding of the Persian Empire* (McGill, 1973), pp. 5–14 (p. 11).
41. Bloom and Blair (eds.), *Grove Encyclopedia of Islamic Art and Architecture*, Vol. 2, 'Isfahan', p. 297.
42. Gülru Necipoğlu, *The Age of Sinan: Architectural Culture in the Ottoman Empire* (Princeton University Press, 2007), p. 142.
43. Rosa Bacile, 'A Porphyry Workshop in Norman Palermo', in Rosa Bacile and John McNeill (eds.), *Romanesque and the Mediterranean* (Routledge, 2015), pp. 129–49.
44. Greenhalgh, *Islam and Marble*, p. 12.
45. Size of columns scaled from Gülru Necipoğlu, 'Creation of a National Genius: Sinan and the Historiography of "Classical" Ottoman Architecture', *Muqarnas*, Vol. 24 (2007), pp. 141–83 (Fig. 5, p. 150).
46. Necipoğlu, *The Age of Sinan*, p. 142.
47. Greenhalgh, *Islam and Marble*, p. 12.
48. David Peacock and Valerie Maxfield, *The Roman Imperial Quarries: Survey and Excavation at Mons Porphyrites 1994–1998*, Vol. 2, *The Excavations* (London, Egypt Exploration Society, 2007), abstract.
49. Ibid., p. 180.
50. Ibid., passim.
51. Ibid., p. 184.
52. Ibid., pp. 142–3 for *spolia;* p. 179 for other materials.
53. Ibid., p. 180.
54. Ibid., p. 179.
55. Ibid., p. 180.

56. Ünver Rüstem, 'The Spectacle of Legitimacy: The Dome-closing Ceremony of the Sultan Ahmed Mosque', *Muqarnas*, Vol. 33 (2016), pp. 253–343 (pp. 253–4).
57. Halil Inalcik with Donald Quataert (eds.), *An Economic and Social History of the Ottoman Empire, 1300–1914* (Cambridge University Press, 1994), p. 442.
58. Ibid., p. 461.
59. Rüstem, 'The Spectacle of Legitimacy', p. 260.
60. Ibid., p. 258.
61. Ibid., passim.
62. Ibid., p. 255.

CHAPTER 6 – PLAGUE AND PROSPERITY: MEDIEVAL AND EARLY MODERN EUROPE

1. 'If a pig or a chicken, or any other type of animal should eat from the body of a man, or drink his blood, the animal should be killed and thrown to the dogs': Penitential of Ecgbert, Archbishop of York, Book IV, 57, in *Ancient Laws and Institutes of England* (1840), p. 221. My thanks to Dr Ken Lawson for finding the source of this, decades after having taught me about it in school history classes.
2. Michel Audouy, Brian Dix and David Parsons, 'The Tower of All Saints' Church, Earls Barton, Northamptonshire: Its Construction and Context', *Archaeological Journal*, Vol. 152 (1995), pp. 73–94.
3. Ibid., pp. 83 and 89, favours the comparison with timber structure, including Roman timber buildings as depicted on Trajan's Column. Eric Fernie, *The Architecture of the Anglo-Saxons* (Batsford, 1983), pp. 144–5, argues that it is based on Roman 'long-and-short' stonework.
4. Audouy, Dix and Parsons, 'The Tower of All Saints' Church, Earls Barton, Northamptonshire', p. 85.
5. Ibid., p. 73.
6. Paul Frodsham, 'The Stronghold of Its Own Native Past: Some Thoughts on the Past in the Past at Yeavering', in Paul Frodsham and Colm O'Brien (eds.), *Yeavering: People, Power and Place* (Tempus Publishing, 2005), p. 25. Frodsham reports that Brian Hope-Taylor, responsible for the major twentieth-century excavations at Yeavering, noted that the royal site at Yeavering had been destroyed by fire twice, possibly in the mid-600s. Hope-Taylor dated the eventual abandonment of the site to 685 CE.

7. Michael McCormick, Ulf Büntgen, Mark A. Cane, Edward R. Cook, Kyle Harper, Peter Huybers, Thomas Litt, Sturt W. Manning, Paul Andrew Mayewski, Alexander F. M. More, Kurt Nicolussi, and Willy Tegel, 'Climate Change during and after the Roman Empire: Reconstructing the Past from Scientifc and Historical Evidence', *Journal of Interdisciplinary History*, Vol. 43 (Autumn 2012), pp. 169–220 9 (pp. 174–5).
8. Astrid Kander, Paolo Malanima and Paul Warde, *Power to the People: Energy in Europe over the Last Five Centuries* (Princeton University Press, 2014), p. 72, and McCormick et al., 'Climate Change during and after the Roman Empire', pp. 185–99; ibid., p. 190.
9. St John Simpson, *Scythians: Warriors of Ancient Siberia* (British Museum, 2017).
10. Mark J. Johnson, 'Toward a History of Theoderic's Building Program', *Dumbarton Oaks Papers*, Vol. 42 (1988), pp. 73–96 (p. 93); the weight estimate is given in Robert Heizer, 'Ancient Heavy Transport, Methods and Achievements', *Science*, Vol. 153 (19 August 1966), p. 823.
11. Beat Brenk, 'Spolia from Constantine to Charlemagne: Aesthetics versus Ideology', *Dumbarton Oaks Papers*, Vol. 41 (1987), pp. 103–9 (p. 108).
12. For climate, see McCormick et al., 'Climate Change during and after the Roman Empire', pp. 191–9; for rodent origin of plague see David M. Wagner et al., '*Yersinia pestis* and the Plague of Justinian AD 541–543: A Genomic Analysis', *The Lancet Infectious Diseases*, Vol. 14 (1 April 2014), pp. 319–26.
13. George Grantham, 'France', in Harry Kitsikopoulos (ed.), *Agrarian Change and Crisis in Europe, 1200–1500* (Routledge, 2012), pp. 57–92 (p. 64).
14. John Hatcher, *The History of the British Coal Industry*, Vol. 1, *Before 1700: Towards the Age of Coal* (Oxford University Press, 1993), p. 18.
15. Sophie Blain et al., 'Early Medieval Brickmaking: A Cross-Channel Perspective Based on Recent Luminescence and Archaeomagnetic Dating Results', in Tanja Ratilainen, Rivo Bernotas and Christofer Herrmann (eds.), *Fresh Approaches to Brick Production and Use in the Middle Ages: Proceedings of the Session 'Utilization of Brick in the Medieval Period – Production, Construction, Destruction'*, held at the European Association of Archaeologists (EAA) Meeting, 29 August to 1 September 2012 in Helsinki, pp. 1–9 (p. 4).
16. Elisabeth Carpentier and Michel Le Mené, *La France du XIe au XVe siècle: Population, société, économie* (Presses Universitaires de France, 1996), pp. 156–9.
17. Dieter Kimpel, 'Le développement de la taille en série dans l'architecture médiévale et son rôle dans l'histoire économique', *Bulletin Monumental*, Vol. 135 (1977), pp. 195–222 (p. 215).

18. Robert Branner, *The Cathedral of Bourges and Its Place in Gothic Architecture* (MIT Press, 1989), p. 53.
19. Ibid., p. 19.
20. Ibid., pp. 20–22.
21. Eddie Sinclair, 'The Polychromy of Exeter and Salisbury Cathedrals: A Preliminary Comparison', in Arie Wallert, Erma Hermens and Marja Peek (eds.), *Historical Painting Techniques, Materials, and Studio Practice* (The Getty Conservation Institute, 1995), pp. 105–10.
22. Carpentier and Le Mené, *La France du XIe au XVe siècle*, p. 170.
23. Ibid., p. 171.
24. Robert A. Scott, *The Gothic Enterprise: A Guide to Understanding the Medieval Cathedral* (University of California Press, 2003), pp. 31–2.
25. T. F. Thomson, 'Salisbury: The Cathedral Close', *Town Planning Review*, Vol. 17 (1937), pp. 130–33.
26. Stephanie Leroy et al., 'Consolidation or Initial Design? Radiocarbon Dating of Ancient Iron Alloys Sheds Light on the Reinforcements of French Gothic Cathedrals', *Journal of Archaeological Science*, Vol. 53 (2015), pp. 190–201 (pp. 195–9).
27. Branner, *The Cathedral of Bourges and Its Place in Gothic Architecture*, p. 16, n. 88.
28. L. F. Salzman, *Building in England down to 1540: A Documentary History* (Clarendon Press, 1952), p. 119.
29. Stone was shipped in 1287 from Caen to Great Yarmouth as facings for Norwich Cathedral, the voyage of over 200 kilometres costing less than twice the quarry-gate price of the stone: ibid., p. 119.
30. Janet E. Snyder, 'Standardization and Innovation in Design: Limestone Architectural Sculpture in Twelfth-century France', pp. 113–27 in Robert Bork, William W. Clark and Abby McGehee (eds.), *New Approaches to Medieval Architecture*, AVISTA Studies in the History of Medieval Technology, Science and Art, Vol. 8 (Ashgate, 2011).
31. Kimpel, 'Le développement de la taille en série dans l'architecture médiévale et son rôle dans l'histoire économique', p. 215.
32. Ibid., passim.
33. Ibid., p. 215.
34. Salzman, *Building in England down to 1540*, p. 178.
35. David Martlew, 'History and Development of Glass', in Michael Tutton and Elizabeth Hirst (eds.), Jill Pearce (managing editor) and Hentie Louw (consultant), *Windows: History, Repair and Conservation* (Donhead, 2007), pp. 121–58 (p. 122).

36. Salzman, *Building in England down to 1540*, p. 180.
37. Ibid., p. 179.
38. Ibid., p. 176: these figures come from glazing work at the Tower of London in 1287.
39. Contemporary stained glass costs around £130 per square foot if 'of average complexity, without any painting etc.', according to stained-glass producer Pickwick Glass: http://www.pickwickglass.co.uk/Stained%20Glass%20FAQ.html (retrieved 27 June 2019); 4mm clear float glass is £3.51 per square foot at Collis DIY: https://www.collisdiy.co.uk/float-clear-4mm (retrieved 27 June 2019).
40. Salzman, *Building in England down to 1540*, p. 174. The example he gives is St George's Chapel, Windsor. For a royal chapel to be economizing in this way underlines the high cost of glass.
41. L. W. Adlington et al., 'Regional Patterns in Medieval European Glass Composition as a Provenancing Tool', *Journal of Archaeological Science*, Vol. 110 (2019), pp. 1–13.
42. Wim Vroom, *Financing Cathedral Building in the Middle Ages: The Generosity of the Faithful* (Amsterdam, 2010), p. 461.
43. The figure is for Westminster Abbey in 1253, for which particularly full accounts survive. Scott, *The Gothic Enterprise*, p. 33.
44. Hatcher, *The History of the British Coal Industry*, p. 22.
45. Carpentier and Le Mené, *La France du XIe au XVe siècle*, p. 161; Paul Erdkamp, *The Grain Market in the Roman Empire: A Social, Political and Economic Study* (Cambridge University Press, 2005), pp. 34–7.
46. Although a popular story claims that the sculptures represent oxen which assisted in the transportation of stone for the cathedral.
47. Kimpel, 'Le développement de la taille en série dans l'architecture médiévale et son rôle dans l'histoire économique', pp. 214–15.
48. Scott, *The Gothic Enterprise*, pp. 29–31.
49. Grantham, 'France', p. 57.
50. Hatcher, *The History of the British Coal Industry*, pp. 24–5.
51. The economics and demographics of fertility rates in relationship to poverty and child health are discussed well in *Our World in Data*, https://ourworldindata.org/fertility-rate (retrieved 20 February 2020).
52. Suzanne Alston Alchon, *A Pest in the Land: New World Epidemics in a Global Perspective* (University of New Mexico Press, 2003), p. 21. In *The Great Leveler: Violence and the History of Inequality from the Stone Age to the Twenty-first Century* (Princeton University Press, 2017), pp. 296–7, Walter Scheidel offers slightly more conservative estimates of plague losses of 25–45 per

cent of Europe's population, with population levels falling from 94 million to 68 million. He notes that the population loss in England and Wales may have been 50 per cent.

53. Henry Knighton, Augustinian Prior of Leicester, writing in 1390, quoted ibid., p. 300.
54. Edward Miller and Joan Thirsk (eds.,) *The Agrarian History of England and Wales*, Vol. 3, *1348–1500* (Cambridge University Press, 1991), p. 338.
55. Paolo Malanima, 'Italy', in Harry Kitsikopoulos (ed.), *Agrarian Change and Crisis in Europe, 1200–1500* (Routledge, 2012), pp. 93–127 (p. 114).
56. Richard A. Goldthwaite, *The Building of Renaissance Florence* (Johns Hopkins University Press, 1980), p. 50.
57. Ibid., p. 60.
58. Ibid., pp. 105–8.
59. Ibid., pp. 13, 98.
60. Adrian Forty, *Words and Buildings: A Vocabulary of Modern Architecture* (Thames & Hudson, 2000), p. 30.
61. Richard T. Rapp, 'Real Estate and Rational Investment in Early Modern Venice', *Journal of European Economic History*, Vol. 8 (1979), pp. 269–90.
62. A great deal is published on Palladio, from first-rate scholarship to picture books. A good starting point is Guido Beltramini and Howard Burns, *Palladio* (Royal Academy, 2008).
63. Dean Hawkes, *Architecture and Climate: An Environmental History of British Architecture 1600–2000* (Routledge, 2012), p. 5.
64. Loren Partridge, *The Renaissance in Rome 1400–1600* (Everyman, 1996), p. 15.
65. Frederic J. Baumgartner, 'Julius II: Prince, Patron, Pastor', in James Corkery and Thomas Worcester (eds.), *The Papacy since 1500: From Italian Prince to Universal Pastor* (Cambridge University Press, 2010), pp. 12–28 (p. 18).
66. Partridge, *The Renaissance in Rome 1400–1600*, p. 21.
67. Ibid., p. 21.
68. Christof Thoenes, 'Renaissance St Peter's', in William Tronzo (ed.), *St Peter's in the Vatican* (Cambridge University Press, 2005), pp. 64–92 (p. 74).
69. Ibid., p. 75.
70. Ibid., p. 78.
71. Partridge, *The Renaissance in Rome 1400–1600*, p. 52.
72. Henry A. Millon, 'Michelangelo to Marchionni, 1546–1784', in Tronzo (ed.), *St Peter's in the Vatican*, pp. 93–110 (p. 94).
73. Ibid., p. 96.
74. Vroom, *Financing Cathedral Building in the Middle Ages*, p. 456.

75. Partridge, *The Renaissance in Rome 1400–1600*, p. 15, says that the agricultural sector was well developed and successful, whereas an opposite picture of primitive methods and poor productivity, with grain imports needed to subsist, is given by Peter Partner, *The Lands of St Peter: The Papal State in the Middle Ages and the Early Renaissance* (University of California Press, 1972), p. 425. This kind of lack of clarity is problematically widespread in the evidence on agriculture in pre-modern times.
76. Alberto Cassone and Carla Marchese, 'The Economics of Religious Indulgences', *Journal of Institutional and Theoretical Economics*, Vol. 155 (1999), pp. 429–42 (p. 439).
77. Oval churches began in the sixteenth century, before Kepler had published his findings, but they were well known by the time Neumann was active: Sylvie Duvernoy, 'Baroque Oval Churches: Innovative Geometrical Patterns in Early Modern Sacred Architecture', *Nexus Network Journal*, Vol. 17 (2015), pp. 425–56.

CHAPTER 7 – THE MARCH OF BRICKS AND MORTAR: COAL AND THE CITY

1. Ole Didrik Laerum, 'The Blackhouse and *Røykstova:* A Common North Sea Tradition', *Northern Studies*, Vol. 41 (2010), pp. 1–12 (p. 3). In fact the peat bogs of Scotland are a source of considerable fuel, but the low food yields of cool climate and poor soils kept much of rural Scotland's population small and impoverished. The Netherlands, by contrast, achieved a major urban boom, often referred to as 'the Dutch Golden Age', in a context where food yields from reclaimed land were much higher, by burning around 7 billion kWh of peat per year, 1600–1700: J. W. de Zeeuw, 'Peat and the Dutch Golden Age: The Historical Meaning of Energy-attainability', *AAG Bijdragen*, Vol. 21 (1978), pp. 3–38.
2. Janet Rudge, 'Coal Fires, Fresh Air and the Hardy British: A Historical View of Domestic Energy Efficiency and Thermal Comfort in Britain', *Energy Policy*, Vol. 49 (October 2012), pp. 6–11.
3. Peter Brimblecombe, 'Air Pollution and Health History', in Stephen T. Holgate, Jonathan M. Samet, Hillel S. Koren and Robert L. Maynard (eds.), *Air Pollution and Health* (Academic Press, 1999), pp. 5–18 (p. 6).
4. Virginia Smith, *Clean: A History of Personal Hygiene and Purity* (Oxford University Press, 2007), pp. 158–60.

5. John Schofield and Alan Vince, *Medieval Towns: The Archaeology of British Towns in Their European Setting* (Continuum, 2003), p. 232.
6. Kathy L. Pearson, 'Nutrition and the Early-Medieval Diet', *Speculum*, Vol. 72 (January 1997), pp. 1–32 (pp. 4–5).
7. Robert A. Scott, *The Gothic Enterprise: A Guide to Understanding the Medieval Cathedral* (University of California Press, 2003), p. 35.
8. Dean Hawkes, *Architecture and Climate: An Environmental History of British Architecture 1600-2000* (Routledge, 2012), pp. 42–9; Dean Hawkes and Ranald Lawrence are currently conducting quantitative research on Elizabethan climate adaptations in the architecture of Hardwick Hall: R. Lawrence and D. Hawkes, 'Describing the Historic Indoor Climate: Thermal Monitoring at Hardwick Hall', *Architectural Science Review* (forthcoming).
9. A. Kander, P. Malanima and P. Warde, *Power to the People: Energy in Europe over the Last Five Centuries* (Princeton University Press, 2013), pp. 39–41.
10. Jane Humphries and Jacob Weisdorf, 'The Wages of Women in England, 1260–1850', *Journal of Economic History*, Vol. 75 (June 2015), pp. 405–47.
11. The height of the potato's influence on European growth was in the eighteenth and nineteenth centuries, when its calories may have accounted for as much as 47 per cent of urban growth: Nathan Nunn and Nancy Qian, 'The Potato's Contribution to Population and Urbanization: Evidence from an Historical Experiment', National Bureau of Economic Research Working Paper 15157, July 2009, https://www.nber.org/papers/w15157.pdf (retrieved 20 February 2020).
12. Robert C. Allen, 'Tracking the Agricultural Revolution in England', *Economic History Review*, Vol. 52 (May 1999), pp. 209–35 (p. 222).
13. Robert J. Bennett, 'SN 7154 – Urban Population Database, 1801–1911', UK Data Service, Table 1, http://doc.ukdataservice.ac.uk/doc/7154/mrdoc/pdf/guide.pdf (retrieved 11 February 2020).
14. Kander, Malanima and Warde, *Power to the People*, p. 58.
15. Ibid.
16. This estimate is based on the low-confidence assumption that the proportion of firewood required per head for industry is about the same in a city of 10,000 and a city of 100,000.
17. Kander, Malanima and Warde, *Power to the People*, p. 56.
18. John Hatcher, *The History of the British Coal Industry*, Vol. 1, *Before 1700: Towards the Age of Coal* (Oxford University Press, 1993), pp. 24–5.
19. Ibid., p. 25.
20. Ibid., p. 23.
21. Kander, Malanima and Warde, *Power to the People*, pp. 60–61.

22. Hatcher, *The History of the British Coal Industry*, p. 35; the ban on wood for glassworks is likely to have been partly a move to damage existing glass producers in favour of a new businessman who had worked out a method of glass production using coal, and paid the crown for the ban which created an effective monopoly for him: Jose Bellido (ed.), *Landmark Cases in Intellectual Property Law* (Hart, 2017).
23. David Martlew, 'History and Development of Glass', in Michael Tutton and Elizabeth Hirst (eds.), Jill Pearce (managing editor) and Hentie Louw (consultant), *Windows: History, Repair and Conservation* (Donhead, 2007), pp. 121–58 (pp. 124–5).
24. Hentie Louw, 'Window-glass Making in Britain c.1660–c.1860', *Construction History*, Vol. 7 (1991), pp. 47–68 (p. 48).
25. Venice was a surprising place for a glass industry, with the materials and fuel coming tens or in some cases thousands of kilometres to be worked there: David Jacoby, 'Raw Materials for the Glass Industries of Venice and the Terraferma, about 1370–about 1460', *Journal of Glass Studies*, Vol. 35 (1993), pp. 65–90.
26. Louw, 'Window-glass Making in Britain c.1660–c.1860', p. 50.
27. Ibid., p. 47.
28. The tax was instituted in England in 1696 and not fully repealed until 1851. For the effect of the tax in England see Wallace E. Oates and Robert M. Schwab, 'The Window Tax: A Case Study in Excess Burden', *Journal of Economic Perspectives*, Vol. 29 (2015), pp. 163–79. In France, the tax was instituted in 1794 and remained in effect until 1917: see Edgar Kiser and Joshua Kane, 'Revolution and State Structure: The Bureaucratization of Tax Administration in Early Modern England and France', *American Journal of Sociology*, Vol. 107 (2001), pp. 183–223 (p. 214).
29. Hentie Louw and Robert Crayford, 'A Constructional History of the Sash-Window, c.1670–c.1725 (Part 2)', *Architectural History*, Vol. 42 (1999), pp. 173–239 (p. 173).
30. Martlew, 'History and Development of Glass', pp. 127–31.
31. Louw and Crayford, 'A Constructional History of the Sash-Window, c.1670–c.1725 (Part 2)', p. 185.
32. Jim Bennett, 'Instruments and Ingenuity', in Michael Cooper and Michael Hunter (eds.), *Robert Hooke: Tercentennial Studies* (Ashgate, 2006), pp. 65–76.
33. Lisa Jardine, *The Curious Life of Robert Hooke: The Man Who Measured London* (HarperCollins, 2003), p. 142.

34. Hentie Louw and Robert Crayford, 'A Constructional History of the Sash-Window, c.1670–c.1725 (Part 1)', *Architectural History*, Vol. 41 (1998), pp. 82–130.
35. Ibid., p. 95.
36. Frank Spellman, *Environmental Engineering Dictionary* (Rowman and Littlefield, 2018), p. 184.
37. Steven Parissien, *The Georgian House* (Aurum, 1995), p. 220; John Dean and Nick Hill, 'Burn Marks on Buildings: Accidental or Deliberate?', *Vernacular Architecture*, Vol. 45 (2014), pp. 1–15 (p. 6).
38. Taking a candle to produce 12.57 lumens and a 100w tungsten bulb to produce 1,600 lumens.
39. The *OED* gives the first instance of the phrase in English as 1699.
40. Matias D. Cattaneo, Sebastian Galiani, Paul J. Gertler, Sebastian Martinez and Rocio Titiunik, 'Housing, Health, and Happiness', *American Economic Journal: Economic Policy*, Vol. 1 (February 2009), pp. 75–105.
41. Gill Newton, 'Infant Mortality Variations, Feeding Practices and Social Status in London between 1550 and 1750', *Social History of Medicine*, Vol. 24 (August 2011), pp. 260–80 (Fig. 3).
42. Frank Sharman, 'Fires and Fire Laws up to the Middle of the Eighteenth Century', *Cambrian Law Review*, Vol. 22 (1991), p. 43.
43. Sarah Rees Jones, 'Building Domesticity in the City: English Urban Housing before the Black Death', in Maryanne Kowaleski and P. J. P. Goldberg (eds.), *Medieval Domesticity: Home, Housing and Household in Medieval England* (Cambridge University Press, 2008), pp. 66–91 (p. 67).
44. Elizabeth McKellar, *The Birth of Modern London: The Development and Design of the City 1660–1720* (Manchester University Press, 1999), p. 159.
45. Ibid., p. 71; quotes from Richard Neve, *The City and Countrey Purchaser and Builder's Dictionary* (London, 1703), p. 71: 'The greatest objection against London-houses (being for the most part Brick) is their slightness, occasioned by the Fines exacted by Landlords. So that few houses at the common rate of Building last longer than the Ground-lease, and that is about fifty or sixty years . . . And this way of Building is wonderful beneficial to Trades relating to it, for they never want Work in so great a City, where Houses here and there are always Repairing, or Building up again.'
46. McKellar, *The Birth of Modern London*, p. 72.
47. Ibid., p. 74.
48. James Ayres, *Building the Georgian City* (Yale University Press, 1998), p. 61.
49. William M. Cavert, 'Industrial Coal Consumption in Early Modern London', *Urban History*, Vol. 44 (2017), pp. 424–43 (pp. 433–4).

50. Lime energy cost ibid., p. 435; glass and coal costs p. 437.
51. Ibid., p. 438.
52. McKellar, *The Birth of Modern London*, p. 12.
53. Ibid., p. 14.
54. Rising rickets recorded in Brimblecombe, 'Air Pollution and Health History', pp. 9–10.
55. McKellar, *The Birth of Modern London*, p. 26.
56. Cavert, 'Industrial Coal Consumption in Early Modern London', p. 427.
57. Sharman, 'Fires and Fire Laws up to the Middle of the Eighteenth Century', p. 47.
58. Nicholas Barbon, *An Apology for the Builder, or, A Discourse Shewing the Cause and Effects of the Increase of Building* (London: Cave Pullen, 1685).
59. Kander, Malanima and Warde, *Power to the People*, p. 193.
60. Barbon quoted by Roger North, in McKellar, *The Birth of Modern London*, p. 42.
61. Ibid., p. 28.
62. Ibid., p. 43.
63. Ibid., p. 74.
64. Ibid., p. 45.
65. Ibid., pp. 78–9.
66. Ibid., p. 43.
67. Ibid., pp. 38–9.
68. Peter Razzell and Christine Spence, 'The History of Infant, Child and Adult Mortality in London, 1550–1850', *London Journal*, Vol. 32 (2007), pp. 271–92.
69. Quoted in Elizabeth McKellar, *Landscapes of London: The City, the Country and the Suburbs, 1660–1840* (Yale University Press, 2014), p. 20.
70. William J. Hausman, 'Public Policy and the Supply of Coal to London, 1700–1770: A Summary', *Journal of Economic History*, Vol. 37 (1977), pp. 252–4.

CHAPTER 8 – 'THAT WHICH ALL THE WORLD DESIRES': VICTORIAN LIVERPOOL

1. James Boswell, *The Life of Samuel Johnson* (1791), 22 March 1776.
2. Benoit Cushman-Roisin and Bruna Tanaka Cremonini, *Useful Numbers for Environmental Studies and Meaningful Comparisons*, https://www.dartmouth.edu/~cushman/books/Numbers/Chap1-Materials.pdf (retrieved 24 December 2019).

3. Vaclav Smil, 'World history and energy', *Encyclopedia of Energy* (Elsevier, 2004), pp. 549–61 (p. 554).
4. Astrid Kander, Paolo Malanima and Paul Warde, *Power to the People: Energy in Europe over the Last Five Centuries* (Princeton University Press, 2014), p. 174.
5. 20,000 tonnes of pig iron produced annually in England in the 1720s represented more than 1,100 km^2 of coppiced growth (Smil, 'World history and energy', p. 554). So 2,250,000 tonnes (1850s annual production of Britain) produced with charcoal to 1720s English ratios, would have required the output of 123,750 km^2 of coppiced wood for charcoal. England is 130,395 km^2.
6. Sarah Brown and Peter de Figueiredo, *Religion and Place: Liverpool's Historic Places of Worship* (English Heritage, 2008), pp. 16–20.
7. Kander, Malanima and Warde, *Power to the People*, Table 3.1, p. 38.
8. 1901 census recorded fractionally under 37 million people in England, Wales and Scotland.
9. https://ourworldindata.org/grapher/long-term-energy-transitions?country=England%20%26%20Wales (retrieved 22 January 2020).
10. Kander, Malanima and Warde, *Power to the People*, p. 194.
11. Henry P. Colburn, 'Connectivity and Communication in the Achaemenid Empire', *Journal of the Economic and Social History of the Orient*, Vol. 56 (2013), pp. 29–52.
12. Nancy Ritchie-Noakes, 'The Construction of Albert Dock and Its Warehouses', in Adrian Jarvis and Kenneth Smith (eds.), *Albert Dock, Trade and Technology* (National Museums and Galleries on Merseyside, 1999), pp. 35–42 (p. 37).
13. Fifty-two weeks of labour in 1851 given as £42 18s in Peter H. Lindert and Jeffrey G. Williamson, 'English Workers' Living Standards during the Industrial Revolution: A New Look', *Economic History Review*, Vol. 36 (February 1983), Table 3, p. 7.
14. Ian Weir and Adrian Jarvis, 'The Development of Fireproof Construction, with Special Reference to Liverpool's Dock Warehouses, 1825 to 1875', in Jarvis and Smith (eds.), *Albert Dock*, pp. 69–76 (p. 71).
15. Ibid., pp. 71, 74.
16. For example, an arson attack against two rooms in the south-west corner in 1883, which did only £1,300 worth of damage to the building thanks to fire-proofing and containment: ibid., p. 73.
17. Smith and Weir, 'The Engineering History of Liverpool's Old Masonry Dock Walls', in Jarvis and Smith (eds.), *Albert Dock*, p. 46; Joseph Sharples, *Liverpool* (Yale University Press, 2004), p. 96.

18. Smith and Weir, 'The Engineering History of Liverpool's old Masonry Dock Walls', in Jarvis and Smith (eds.), *Albert Dock*, p. 43.
19. Joseph Sharples, 'Thomas Rickman's Liverpool', in Megan Aldrich and Alexandra Buchanan (eds.), *Thomas Rickman and the Victorians* (Victorian Society, 2019), p. 147.
20. Ritchie-Noakes, 'The Construction of Albert Dock and Its Warehouses', in Jarvis and Smith (eds.), *Albert Dock*, pp. 35–42 (pp. 37–8).
21. Ibid., p. 38.
22. Albert Dock handled around 5.8 per cent of the tonnage of goods that came through Liverpool's port, but 9.1 per cent of its revenue: Graeme J. Milne and Graham Tonks, 'Specialised Port Facilities on Trial: Liverpool's Albert Dock in the Nineteenth Century', in Jarvis and Smith (eds.), *Albert Dock*, pp. 97–106 (pp. 98, 100).
23. Henry Booth, *An Account of the Liverpool and Manchester Railway* (Wales & Baines, 1830), pp. 11–12.
24. 'Trial of Locomotive Carriages', *The Times*, 12 October 1829.
25. Kander, Malanima and Warde, *Power to the People*, pp. 192–3.
26. Jack Simmons, *St Pancras Station* (Historical Publications, 3rd edn, revised and expanded by Robert Thorne, 2012), p. 3.
27. Sharples, *Liverpool*, p. 185.
28. Simmons, *St Pancras Station*, p. 15.
29. Ibid., p. 21. Price equivalent: 116,500 times the annual 1851 pay of £49 10s given in Lindert and Williamson, 'English Workers' Living Standards during the Industrial Revolution', since when pay had risen to some extent.
30. Simmons, *St Pancras Station*, pp. 26–7.
31. Thomas Hardy, 'The Levelled Churchyard'.
32. Simmons, *St Pancras Station*, p. 40.
33. Ibid., p. 46.
34. Ibid., p. 33.
35. Ibid., pp. 51–2.
36. Gavin Stamp, *Gothic for the Steam Age: An Illustrated Biography of George Gilbert Scott* (Aurum, 2015), p. 18.
37. Sir George Gilbert Scott, *Personal and Professional Recollections* (1879; facsimile 1995), p. 271.
38. Sharples, *Liverpool*, pp. 245–9.
39. Ibid., p. 247.
40. School for the Blind, now a restaurant, discussed ibid., pp. 233–4.
41. Sally Sheard, 'James Newlands and the Origins of the Municipal Engineer', *Engineering History and Heritage*, Vol. 168 (May 2015), pp. 83–9 (p. 83).

42. *Dictionary of National Biography.*
43. An excellent account of the origin and problems of the courts and cellars of Liverpool is given in I. C. Taylor, 'The Court and Cellar Dwelling: The Eighteenth Century Origin of the Liverpool Slum', *Transactions of the Historic Society of Lancashire and Cheshire*, Vol. 122 (1970), pp. 67–90.
44. Stephen Halliday, 'Duncan of Liverpool: Britain's First Medical Officer', *Journal of Medical Biography*, Vol. 11 (August 2003), pp. 142–9 (pp. 144–5).
45. Ibid., p. 145.
46. Colum Giles, *Building a Better Society: Liverpool's Historic Institutional Buildings* (English Heritage, 2008), p. 53.
47. Dictionary of National Biography.
48. Mark Swenarton, *Homes Fit for Heroes: The Politics and Architecture of Early State Housing in Britain* (Heinemann, 1981), pp. 12–13.
49. Kander, Malanima and Warde, *Power to the People*, p. 265.
50. Colum Giles and Bob Hawkins, *Storehouses of Empire: Liverpool's Historic Warehouses* (English Heritage, 2004), pp. 14–15.
51. Ibid., pp. 15–16.
52. Maureen Dillon, *Artificial Sunshine: A Social History of Domestic Lighting* (National Trust, 2002), pp. 14–15.
53. Historic England listing entry, 'No. 1 Gasholder, Kennington Lane Gasholder Station', https://historicengland.org.uk/listing/the-list/list-entry/1427396 (retrieved 20 February 2020).
54. Adam Menuge, *Ordinary Landscapes, Special Places: Anfield, Breckfield and the Growth of Liverpool's Suburbs* (English Heritage, 2008), pp. 7–26.
55. Ibid., p. 29.
56. Sharples, *Liverpool*, pp. 281–3.
57. Joseph Sharples, 'The Residential Development of Sefton Park, Liverpool, c.1872–c.1900', *Transactions of the Historic Society of Lancashire and Cheshire*, Vol. 165 (2016), pp. 57–78.
58. My thanks to Richard Dod for showing me round an example that has been preserved in remarkable original condition thanks to his efforts.
59. Katy Layton-Jones and Robert Lee, *Places of Health and Amusement: Liverpool's Historic Parks and Gardens* (English Heritage, 2008), p. 38.
60. Andrew Saint, *Richard Norman Shaw* (rev. edn, Yale University Press, 2009), p. 289.
61. Ibid., p. 82.
62. Sharples, 'Thomas Rickman's Liverpool', in Aldrich and Buchanan (eds.), *Thomas Rickman and the Victorians*, pp. 147–8.
63. Saint, *Richard Norman Shaw*, p. 288.

64. Ibid., p. 291.
65. 'Artibus Legibus Consiliis Locum Municipes Constituerunt Anno Domini MDCCCXLI'.
66. Joseph Sharples, personal communication, December 2019.
67. Sharples, 'Thomas Rickman's Liverpool', in Aldrich and Buchanan (eds.), *Thomas Rickman and the Victorians*, pp. 143–4.
68. Sharples, *Liverpool*, p. 50; Janet DeLaine, 'The Pantheon Builders: Estimating Manpower for Construction', in Tod A. Marder and Mark Wilson Jones (eds.), *The Pantheon: From Antiquity to the Present* (Cambridge University Press, 2015), pp. 160–92.
69. D. Jaggar and R. R. Morton, *Design and the Economics of Building* (Taylor and Francis, 2003), p. 103.
70. Robert Bruegmann, 'Central Heating and Forced Ventilation: Origins and Effects on Architectural Design', *Journal of the Society of Architectural Historians*, Vol. 37 (October 1978), pp. 143–60 (p. 152).
71. Pugin, *Contrasts*, p. 3.
72. Ruskin, *Stones of Venice*, Vol. 3, Chapter II, XXXII.
73. Pugin, *Contrasts*, plates.
74. James W. P. Campbell, *Brick: A World History* (Thames & Hudson, 2003), p. 208.
75. Paul Joseph and Svetlana Tretsiakova-McNally, 'Sustainable Non-metallic Building Materials', *Sustainability*, Vol. 2 (2010), p. 403, Table 1. Labour calculated at 0.075kW: see Introduction, above.
76. Sharples, *Liverpool*, p. 223.
77. See, for example, M. H. Port, *Imperial London: Civil Government Building in London, 1851–1915* (Yale University Press, 1995).
78. Saint, *Richard Norman Shaw*, p. 209.
79. John Ruskin, 'Traffic', in *The Crown of Wild Olive* (George Allen, 1882), pp. 90–91.
80. See, for example, Lauren S. Weingarden, 'Aesthetics Politicized: William Morris to the Bauhaus', *Journal of Architectural Education*, Vol. 38 (1985), pp. 8–13 (pp. 8–10). Morris's medievalism can be seen clearly in his *News from Nowhere: An Epoch of Rest* (1890). The post-industrial society described in this utopian tract is explicitly modelled on medieval England, with repeated references to the fourteenth century throughout the text.
81. Peter Davey, *Arts and Crafts Architecture* (Phaidon, 1997), p. 174.
82. Baillie Scott, *Houses and Gardens* (1906), quoted in Dean Hawkes, *Architecture and Climate: An Environmental History of British Architecture 1600–2000* (Routledge, 2012), p. 165.

83. Mary Myers, 'England's Domestic Dream', *Country Life*, Vol. 205 (31 August 2011), pp. 82–3.
84. Charles Harvey and Jon Press, 'John Ruskin and the Ethical Foundations of Morris & Company, 1861–96', *Journal of Business Ethics*, Vol. 14 (1995), pp. 181–94 (pp. 185, 188).
85. Elizabeth Carolyn Miller, 'William Morris, Extraction Capitalism, and the Aesthetics of Surface', *Victorian Studies*, Vol. 57 (2015), pp. 395–404 (n. 2). My thanks are due to Ewan Harrison for extensive help with my research on the relationship between the Arts and Crafts movement and industry.
86. G. A. Bremner, 'Material, Movement and Memory: Some Thoughts on Architecture and Experience in the Age of Mechanisation', in E. Gillin and H. Joyce (eds.), *Experiencing Architecture in the Nineteenth Century: Buildings and Society in the Modern Age* (Bloomsbury, 2018), pp. 175–91 (p. 185).
87. Stamp, *Gothic for the Steam Age*, p. 59.
88. J. G. Kohl, quoted in Joseph Sharples and John Stonard, *Built on Commerce: Liverpool's Central Business District* (English Heritage, 2015), p. 1.
89. Joseph Sharples, '"The visible embodiment of modern commerce": Speculative Office Buildings in Liverpool, c.1780–1870', *Architectural History*, Vol. 61 (2018), pp. 131–73 (p. 133).
90. Sharples and Stonard, *Built on Commerce*, p. 33.
91. Quoted ibid., p. 50.
92. Quoted ibid., p. 36.
93. For cotton's role see Giles and Hawkins, *Storehouses of Empire*, p. 5.
94. Sharples and Stonard, *Built on Commerce*, pp. 8–10.
95. Ibid., p. 10.
96. Ibid., p. 42.
97. Ibid., p. 41.
98. Ibid., pp. 41–2.
99. Quoted ibid., p. 50.
100. G. A. Bremner, '"In bright tints ... nature's own formation": The Uses and Meaning of Marble in Victorian Building Culture', in N. J. Napoli and W. Tronzo (eds.), *Radical Marble: Architectural Innovation from Antiquity to the Present* (Routledge, 2018), pp. 72–92 (p. 78).
101. Ibid., p. 79.
102. Quoted in Sharples and Stonard, *Built on Commerce*, p. 52.
103. A fascinating analysis of the design challenges of lighting, ventilation and architectural expression in Victorian architecture is Ranald Lawrence, *The Victorian Art School: Architecture, History, Environment* (Routledge, 2020).
104. Ibid., pp. 51–2.

105. 'Architecture in Liverpool', *The Builder* (20 January 1866), pp. 40–41.
106. Sharples and Stonard, *Built on Commerce*, p. 41.
107. Ibid., p. 61.
108. B. Pugh, *The Hydraulic Age: Public Power Supplies before Electricity* (Mechanical Engineering Publications, 1980), pp. 113–14.
109. Sharples, *Liverpool*, pp. 68, 70–71.
110. https://transportgeography.org/?page_id=2135 (retrieved 4 August 2019).

CHAPTER 9 – FORM FOLLOWS FUEL: INDUSTRY AND CONSTRUCTION IN THE USA, 1850–1920

1. John Belchem, *Irish, Catholic and Scouse: The History of the Liverpool Irish, 1800–1939* (Oxford University Press, 2007), p. 2.
2. Edward Cheshire, 'The Results of the Census of Great Britain in 1851, with a Description of the Machinery and Processes Employed to Obtain the Returns', *Journal of the Statistical Society of London*, Vol. 17 (March 1854), pp. 45–72 (p. 49); Statista, 'Population Density of the United States from 1790 to 2019 in Residents per Square Mile of Land Area', https://www.statista.com/statistics/183475/united-states-population-density/ (retrieved 24 February 2020).
3. Alan Krell, *The Devil's Rope: A Cultural History of Barbed Wire* (Reaktion, 2002).
4. Although in some areas of the USA the abundance of forest and limited local access to coal gave rise to wood-burning steam locomotives.
5. Ryan E. Smith, 'History of Prefabrication: A Cultural Survey', *Proceedings of the Third International Congress on Construction History*, Cottbus, May 2009.
6. P. Vervoort, 'Lakehead Terminal Elevators: Aspects of Their Engineering History', *Canadian Journal of Civil Engineering*, Vol. 17, No. 3 (1990), pp. 404–12.
7. Stuart Banner, *American Property: A History of How, Why and What We Own* (Harvard University Press, 2011), p. 182.
8. Cecil D. Elliott, *Technics and Architecture: The Development of Materials and Systems for Buildings* (MIT Press, 1992), pp. 7, 18; Paul E. Sprague, 'The Origin of Balloon Framing', *Journal of the Society of Architectural Historians*, Vol. 40 (December 1981), pp. 311–19 (p. 311).
9. Frank Lloyd Wright, 'The Art and Craft of the Machine, 1901, 1930', in Leland M. Roth (ed.), *America Builds: Source Documents in American Architecture and Planning* (Harper & Row, 1983), pp. 364–76 (p. 372).

10. Real Estate Record Association, *A History of Real Estate, Building and Architecture in New York City during the Last Quarter of a Century* (Record and Guide, 1898), p. 352.
11. Ibid., p. 355.
12. Cass Gilbert, 'The Financial Importance of Rapid Building', *Engineering Record*, Vol. 41 (30 June 1900), p. 624, quoted in Michael Holleran, '"The Machine That Makes the Land Pay": Recent Skyscraper Scholarship', *Journal of Urban History*, Vol. 25, Issue 6 (1 September 1999), pp. 860–67 (p. 861).
13. Real Estate Record Association, *A History of Real Estate*, p. 367.
14. Ibid., p. 370.
15. Ibid.
16. Ibid., p. 373.
17. Gail Fenske, *The Skyscraper and the City: The Woolworth Building and the Making of Modern New York* (University of Chicago Press, 2008), p. 191.
18. Thomas J. Misa, *A Nation of Steel: The Making of Modern America, 1865–1925* (Johns Hopkins University Press, 1999), Chapter 1.
19. Fenske, *The Skyscraper and the City*, p. 167.
20. Misa, *A Nation of Steel*, pp. 68–9.
21. Real Estate Record Association, *A History of Real Estate*, p. 417–19.
22. Ibid., p. 415.
23. Carol Willis, *Form Follows Finance: Skyscrapers and Skylines in New York and Chicago* (Princeton Architectural Press, 1995), p. 45.
24. Thomas Parke Hughes, 'British Electrical Industry Lag: 1882–1888', *Technology and Culture*, Vol. 3 (Winter 1962), pp. 27–44 (p. 27).
25. Thomas P. Hughes, *Networks of Power: Electrification in Western Society, 1880–1930* (Johns Hopkins University Press, 1983).
26. Willis, *Form Follows Finance*, pp. 24–5.
27. Ibid., p. 26.
28. Q. Collette, I. Wouters and L. Lauriks, 'Evolution of Historical Riveted Connections: Joining Typologies, Installation Techniques and Calculation Methods', *Structural Repairs and Maintenance of Heritage Architecture*, Vol. 118 (2011), pp. 295–306 (p. 296).
29. A 4mm arc welding electrode has a power of 3.840kW, http://www.arcraftplasma.com/welding/weldingdata/powersources.htm (retrieved 12 March 2020). A labourer has a power of 0.075kW: see Introduction, above.
30. Fenske, *The Skyscraper and the City*, pp. 193–4.
31. Ibid., pp. 182–4.
32. Ibid., pp. 193–4.

33. Willis, *Form Follows Finance*, p. 32.
34. The shape of the Empire State Building, for example, was largely determined by financial considerations before the appointment of the architects as the optimum way to maximize profit, and an extra ten storeys were then added to make it a mean-minded sixty centimetres taller than the slightly earlier Chrysler Building, earning it the marketable title of tallest building in the world (Willis, *Form Follows Finance*, pp. 95–9).
35. Fenske, *The Skyscraper and the City*, p. 43.
36. George H. Douglas, *Skyscrapers: A Social History of the Very Tall Building in America* (McFarland, 1996), p. 54.
37. Ibid., p. 53.
38. Ibid., p. 56.
39. Fenske, *The Skyscraper and the City*, p. 11.
40. Ibid., pp. 261–4 for fire, p. 258 for sewerage.
41. Ibid., p. 258.
42. Ibid., p. 213.
43. Sarah Watts, 'Built Languages of Class: Skyscrapers and Labor Protest in Victorian Public Space', in Roberta Moudry (ed.), *The American Skyscraper: Cultural Histories* (Cambridge University Press, 2005), pp. 185–200 (p. 191).
44. Fenske, *The Skyscraper and the City*, p. 215.
45. Willis, *Form Follows Finance*, p. 45.
46. Fenske, *The Skyscraper and the City*, p. 213.
47. Ibid., p. 166.
48. Ibid., p. 71.
49. Edwin A. Cochrane, *The Cathedral of Commerce* (Broadway Park Place, 1918), p. 26.
50. Ernest Flagg, quoted in Gail Fenske, 'The Beaux-Arts Architect and the Skyscraper', in Moudry (ed.), *The American Skyscraper*, pp. 19–37 (p. 22).
51. 'Form ever follows function' is used repeatedly in Louis H. Sullivan, 'The Tall Office Building Artistically Considered', *Lippincott's Magazine* (March 1896), pp. 403–9 (pp. 408–9).
52. Frank Lloyd Wright, *Studies and Executed Buildings* (Wasmuth, 1910), p. 12.
53. Fenske, *The Skyscraper and the City*, p. 192.
54. David Edgerton, *The Shock of the Old: Technology and Global History since 1900* (Oxford, 2006), pp. 34–5.

CHAPTER 10 – 'THE BEAUTY OF SPEED': THE RISE OF OIL AND ELECTRICITY, 1914–39

1. Sara Reguer, 'Persian Oil and the First Lord: A Chapter in the Career of Winston Churchill', *Military Affairs*, Vol. 46 (1982), p. 134.
2. For dangerous races see James J. Flink, *The Automobile Age* (MIT Press, 1990), p. 31.
3. Filippo Tommaso Marinetti, 'Le Futurisme', *Le Figaro* (20 February 1909). It is available in lots of translations. The one I have quoted from is at http://www.ubu.com/papers/marinetti_futurist-manifesto.html (retrieved 10 March 2020).
4. Sant'Elia's drawings are reproduced in many places, including a good selection in Viviana Birolli, 'Antonio Sant'Elia et *La Città Nuova*: représenter la ville moderne', *Livraisons d'histoire de l'architecture*, Vol. 32 (2016), pp. 89–104, available free online, https://journals.openedition.org/lha/641.
5. Ibid., p. 89.
6. Prices in 1910 dollars, from David E. Nye, *America's Assembly Line* (MIT Press, 2013), p. 32.
7. Ibid., p. 18.
8. J. F. Crowley, 'The Use and Advantages of Electric Power in the Factory, as Illustrated by Its Application to the Jute Industry (Concluded)', *Journal of the Royal Society of Arts*, Vol. 70, No. 3,640 (1922), pp. 691–8 (p. 692).
9. Nye, *America's Assembly Line*, p. 25.
10. https://www.thehenryford.org/collections-and-research/digital-collections/artifact/167313.
11. Richard B. Du Boff, 'The Introduction of Electric Power in American Manufacturing', *Economic History Review*, Vol. 20 (December 1967), pp. 509–18 (p. 515).
12. Nye, *America's Assembly Line*, pp. 19, 27.
13. Cedric Price, 'The City as an Egg', c. 2001, Cedric Price Fonds, Canadian Center for Architecture, DR2004:1520:001.
14. Frank Lloyd Wright, *The Disappearing City* (William Farquhar Payson, 1932) and *Frank Lloyd Wright: An Autobiography* (Faber and Faber, 1945), pp. 289–91.
15. Charles K. Hyde, 'Assembly-line Architecture: Albert Kahn and the Evolution of the U.S. Auto Factory, 1905–1940', *Journal of the Society for Industrial Archeology*, Vol. 22, No. 2 (1996), pp. 5–24 (p. 5).
16. Henry Ford, *My Life and Work* (Doubleday, Page and Co., 1922), p. 72.
17. It is often referred to by the title given it by its first English translator, Frederick Etchells: *Towards a New Architecture* (Brewer and Warren, 1927).

18. Ibid., pp. 134–5.
19. 'Plan Voisin', in Le Corbusier and Pierre Jeanneret, *Oeuvre complete*, Vol. 1, *1910–1929*.
20. Le Corbusier, *Towards a New Architecture*, p. 233.
21. Ibid., p. 142.
22. Walter Gropius, *The New Architecture and the Bauhaus* (1925), English translation by P. Morton Shand (MIT Press, 1965), p. 25.
23. In 1967, a barrel of Portland cement cost 1,526,498 Btu: Mike Jackson, 'Embodied Energy and Historic Preservation: A Needed Reassessment,' *APT Bulletin: The Journal of Preservation Technology* 36 (2005), pp. 47–52 (p. 48).
24. See 'Introduction' above.
25. 1870, 29 GJ per capita; 1910, 54 GJ per capita, figures from Astrid Kander, Paolo Malanima and Paul Warde, *Power to the People: Energy in Europe over the Last Five Centuries* (Princeton University Press, 2014), p. 254.
26. Rosa Urbano Gutiérrez, 'Le pan de verre scientifique: Le Corbusier and the Saint-Gobain Glass Laboratory Experiments (1931–32)', *Architectural Research Quarterly*, Vol. 17, No. 1 (March 2013) pp. 63–71.
27. Le Corbusier, in a 1929 lecture in Buenos Aires, quoted in Dean Hawkes, *The Environmental Imagination: Technics and Poetics of the Architectural Environment* (Routledge, 2007), p. 33.
28. Ibid., p. 64.
29. Jiat-Hwee Chang, 'Thermal Comfort and Climatic Design in the Tropics: An Historical Critique', *Journal of Architecture*, Vol. 21, No. 8 (2016), pp. 1,171–202.
30. Urbano Gutiérrez, 'Le pan de verre scientifique', p. 65.
31. Rosa Urbano Gutiérrez, '"Pierre, revoir tout le système fenêtres": Le Corbusier and the Development of Glazing and Air-conditioning Technology with the Mur Neutralisant (1928–1933)', *Construction History*, Vol. 27 (2012), pp. 107–28 (pp. 116–20).
32. Jacques Sbriglio, *Le Corbusier: The Villa Savoye* (Birkhäuser, 1999), p. 134.
33. Le Corbusier, *Vers une architecture* (G. Crès et Cie, 1923), p. 73.
34. Tim Benton, *The Villas of Le Corbusier, 1920–1930* (Yale University Press, 1987), pp. 65–73.
35. Ibid., p. 65.
36. Ibid., p. 73.
37. Ibid., p. 72.
38. George Orwell, *Down and Out in Paris and London* (1933).
39. Sbriglio, *Le Corbusier*, p. 145.
40. Ibid., pp. 193–5.

41. Ibid., pp. 109–10.
42. Thomas P. Hughes, *Networks of Power: Electrification in Western Society, 1880–1930* (Johns Hopkins University Press, 1983), p. 364.
43. Sbriglio, *Le Corbusier*, p. 138.
44. Dean Hawkes, *The Environmental Imagination: Technics and Poetics of the Architectural Environment* (Routledge, 2007), p. 39.
45. Ibid., p. 145.
46. The French reads '*ceci pourra côuter quelque argent*', but the understatement translates as meaning it will cost a lot of money: ibid., p. 146.
47. J. B. Williams, *The Electric Century: How the Taming of Lightning Shaped the Modern World* (Springer, 2017), p. 54.
48. Sbriglio, *Le Corbusier*, p. 108.
49. England and Wales went from 134.4GJ per person in 1880 to 152.8GJ per person in 1910, Germany from 43.2GJ to 86.3GJ per person over the same period: Kander, Malanima and Warde, *Power to the People*, p. 243, Table 7.13.
50. Reginald Isaacs, *Gropius: An Illustrated Biography of the Creator of the Bauhaus* (Bulfinch, 1991), p. 25. My thanks to Patrick Zamarian, who drew my attention to this.
51. Daniel A. Barber, 'Heating the Bauhaus: Understanding the History of Architecture in the Context of Energy Policy and Energy Transition', report published by Kleinman Center for Energy Policy, October 2019, available at https://kleinmanenergy.upenn.edu/sites/default/files/proceedingsreports/KCEP-Heating-the-Bauhaus-Singles.pdf (retrieved 27 February 2020). It is an excellent, so far rare example of energy history and architectural history being investigated together.
52. Ibid., p. 14.
53. Paul Scheerbart, *Glass Architecture* (1914), translated and quoted by Detlef Mertins, 'The Enticing and Threatening Face of Prehistory: Walter Benjamin and the Utopia of Glass', *Assemblage*, Vol. 29 (April 1996), pp. 6–23 (p. 10).
54. Pamela Kachurin, 'Working (for) the State: Vladimir Tatlin's Career in Early Soviet Russia and the Origins of *The Monument to the Third International*', *Modernism/modernity*, Vol. 19 (January 2012), pp. 19–41 (p. 25).
55. Daniel Barber has uncovered fascinating examples of a period either side of the Second World War in which major modernist architects discussed and experimented with designing comfortable conditions without air conditioning: *Modern Architecture and Climate: Design before Air Conditioning* (Princeton University Press, 2020).
56. A portion of the essay is reproduced at https://www.archdaily.com/798529/the-longish-read-ornament-and-crime-adolf-loos (retrieved 5 February 2020).

57. Jean-Louis Cohen, *Architecture in Uniform: Designing and Building for World War II* (Yale University Press, 2011), pp. 88–9.
58. Ibid., pp. 88–9.
59. Joseph Siry, 'Air-conditioning Comes to the Nation's Capital, 1928–60', *Journal of the Society of Architectural Historians* (2018), pp. 448–72 (p. 466).

CHAPTER 11 – 'TOO CHEAP TO METER': THE POST-WAR BOOM, 1939–90

1. Thomas Wellock, '"Too cheap to meter": A History of the Phrase', United States Nuclear Regulatory Commission blog, https://public-blog.nrc-gateway.gov/2016/06/03/too-cheap-to-meter-a-history-of-the-phrase/ (retrieved 6 February 2020).
2. For London's power consumption see Anthony Price, 'UK Power: An Ever Changing Challenge for Civil Engineers', *Proceedings of the Institution of Civil Engineers – Civil Engineering*, Vol. 158 (November 2005), pp. 4–11; for Battersea's coal consumption see Felix Lowe, 'Battersea Power Station: A Timeline', *Telegraph*, 20 June 2008, https://www.telegraph.co.uk/finance/newsbysector/constructionandproperty/2791940/Battersea-Power-Station-A-timeline.html (retrieved 20 February 2020).
3. Stuart R. J. Walker, 'Barriers to the Deployment of a 100MW Tidal Energy Array in the UK', *International Journal of Energy Engineering*, Vol. 3 (June 2013), pp. 80–92.
4. Otto Saumarez Smith, 'Cooling Towers are a Powerful Presence in the Landscape – and Deserve to be Saved', *Apollo*, 24 February 2020.
5. Energy figures for 1965 from https://ourworldindata.org/energy. 40,650 million kWh, divided by 0.075kWh to get hours of manual labour equivalence, gives 540,000 million hours.
6. 'Extract from Notes from the Advisory Panel on Home Affairs on Reconstruction Problems: The Five Giants on the Road', 25 June 1942, National Archives (T 161/1165), https://www.nationalarchives.gov.uk/education/resources/attlees-britain/five-giants/ (retrieved 26 February 2020).
7. Andrew Saint, *Towards a Social Architecture: The Role of School Building in Post-war England* (Yale University Press, 1987).
8. Philip Johnson, 'School at Hunstanton: Comment by Philip Johnson as an American Follower of Mies van der Rohe', *Architectural Review* (August 1954), pp. 148–62.

9. Thomas D. Kelly and Grecia R. Matos, 'Historical Statistics for Mineral and Material Commodities in the United States', National Minerals Information Center, 2014 version, https://www.usgs.gov/centers/nmic/historical-statistics-mineral-and-material-commodities-united-states (retrieved 10 September 2019).
10. In 1967 a barrel of Portland cement cost 1,526,498 Btu (447 kWh), not including delivery: Mike Jackson, 'Embodied Energy and Historic Preservation: A Needed Reassessment', *APT Bulletin: The Journal of Preservation Technology*, Vol. 36 (2005), pp. 47–52 (p. 48). A US barrel of cement is 376 lbs (171 kg), https://www.eia.gov/energyexplained/units-and-calculators/. So a tonne of cement in 1967 in the USA cost 2,614 kWh of energy. In 1973 in the USA 81,941,000 metric tonnes of cement were consumed. 'Supply-demand' statistics from https://www.usgs.gov/centers/nmic/historical-statistics-mineral-and-material-commodities-united-states. So total energy cost of that cement, at 1967 rates, not including delivery, was: 214,193,774,000 kWh. This equates to 244,513,440,639 m^2 of a year's charcoal coppice (244,513 km^2) at 0.876 kWh/m^2: see Introduction, above. Michigan is 250,487 km^2, the eleventh-biggest state. In terms of labour, the same energy use equates to 2,855,916,986,666 hours of human labour at 0.075 kW, 1,144,197,510 years of fifty-two working weeks of six eight-hour days. There were 211,900,000 people in the USA in 1973. So 5.4 times the population of the USA, working a full adult male labouring year, would provide equivalent energy to the level of energy inputs required for the country's cement consumption in 1973.
11. James Ashby, 'The Aluminium Legacy: The History of the Metal and Its Role in Architecture', *Construction History*, Vol. 15 (1999), pp. 79–90 (p. 81).
12. Benjamin M. Gimarc, 'Friedrich Wöhler's Lost Aluminum', *Chem Matters* (October 1990), pp. 14–15.
13. Aluminium cost 279 MJ per kg to manufacture from ore (although recycled scrap costs under 10 per cent of the amount of energy) in 1979. Steel costs 25–28 MJ per kg to manufacture from ore. Benoit Cushman-Roisin and Bruna Tanaka Cremonini, *Useful Numbers for Environmental Studies and Meaningful Comparisons*, available at http://www.dartmouth.edu/~cushman/books/Numbers/Chap1-Materials.pdf (retrieved 24 August 2019), pp. 1, 3. Jackson, 'Embodied Energy and Historic Preservation', gives various figures for different types of aluminium and steel which suggest that aluminium costs between five and seven times more energy per pound for primary production of aluminium.

14. Meredith L. Clausen, 'Belluschi and the Equitable Building in History', *Journal of the Society of Architectural Historians*, Vol. 50 (June 1991), pp. 109–29 (p. 112).
15. 'National Museum, Accra, Ghana, publicity, March 1957 to January 1987', Denys Lasdun Papers, RIBA Archives Collection, LaD/11/1.
16. Ashby, 'The Aluminium Legacy', p. 86.
17. Ibid.
18. Primary plastic production (as opposed to recycling, which saves more than half the energy) varies according to plastic type, with PVC being at the lower end, at 56–62 MJ/kg and epoxy at the upper end at 127–140 MJ/kg, Cushman-Roisin and Cremonini, *Useful Numbers*, p. 6.
19. Mark Girouard, *Big Jim: The Life and Work of James Stirling* (Chatto and Windus, 1998), pp. 145–6.
20. 'Plastics: Architects of Modern and Sustainable Buildings', https://www.plasticseurope.org/application/files/2115/3897/8310/Final_BC_brochure_111212_web_version_UPD2018.pdf (retrieved 20 February 2020).
21. Peter Apps, 'Fact Check: How Many People Live in Buildings with Dangerous Cladding?' *Inside Housing*, 30 June 2020, https://www.insidehousing.co.uk/insight/insight/fact-check-how-many-people-live-in-buildings-with-dangerous-cladding-67000, retrieved 21 September 2020.
22. Clausen, 'Belluschi and the Equitable Building in History', p. 118.
23. David Arnold, 'Air Conditioning in Office Buildings after World War II', *ASHRAE Journal* (July 1999), p. 35. My thanks to Professor Daniel Barber, who drew my attention to this article.
24. 'Construction: Details: Curtain Walling on the Lever Building, New York', *Architects' Journal* (16 June 1955), pp. 827–30 (p. 827).
25. Arnold, 'Air Conditioning in Office Buildings after World War II', 1999, p. 35.
26. The best source on the Barbican and its architects is Elain Harwood, *Chamberlin, Powell & Bon* (RIBA, 2011).
27. Barnabas Calder, *Raw Concrete: The Beauty of Brutalism* (William Heinemann, 2016), pp. 11–12.
28. Stefan Muthesius and Miles Glendinning, *Towerblock: Modern Public Housing in England, Scotland, Wales and Northern Ireland* (Yale University Press, 1994), p. 319.
29. UK 1,345 TWh of oil energy in 1973, Japan 3,111 TWh, but note that the USA remained by a long way the world's most oil-hungry economy, consuming 9,929 TWh that same year: Our World in Data, https://ourworldindata.org/grapher/oil-consumption-by-country?tab=chart&country=GBR+JPN+USA+RUS (retrieved 10 February 2020).

30. Caroline S. Tauxe, 'Mystics, Modernists, and Constructions of Brasilia', *Ecumene*, Vol. 3 (January 1996), pp. 43–61 (pp. 49–50).
31. The developer was Harry Hyams. Oliver Marriott, *Property Boom* (Hamish Hamilton, 1967), p. 117. Average house price of £3,735 from Nationwide Building Society, 'UK house prices since 1952', https://www.nationwide.co.uk/-/media/MainSite/documents/about/house-price-index/downloads/uk-house-price-since-1952.xls (retrieved 24 August 2019).
32. Ewan Harrison, 'R. Seifert & Partners: Architecture for Profit and the Post-war State' (unpublished doctoral thesis, University of Liverpool, 2020).
33. Elain Harwood and Alan Powers (eds.), *The Heroic Period of Conservation* (Twentieth Century Society, 2006).
34. 1950 world tourism receipts c. $2.1 billion, scheduled flight passengers 31 million; 1984 $70 billion, passengers 832 million: Peter J. Lyth and Marc L. J. Dierikx, 'From Privilege to Popularity: The Growth of Leisure and Air Travel since 1945', *Journal of Transport History*, Vol. 15 (1994), pp. 97–116.
35. Eeva-Liisa Pelkonen, *Eero Saarinen: Shaping the Future* (Yale University Press, 2006), p. 301.
36. Nathalie Roseau, 'Airports as Urban Narratives: Toward a Cultural History of the Global Infrastructures', *Transfers: Interdisciplinary Journal of Mobility Studies*, Vol. 2 (Spring 2012), pp. 32–54 (p. 45).
37. Sarah Nichols, 'Opération Béton: Constructing Concrete in Switzerland, 1880–1972', unpublished doctoral thesis submitted to ETH Zurich, 2020. Nichols has uncovered discussions as early as 1913 that show expert commentators, including the leading Swiss concrete engineer Robert Maillart, arguing that the difficulty of adapting or dismantling it made monolithic concrete inappropriate for factory buildings. My warm thanks to Sarah Nichols for sharing material from her unpublished doctoral research with me.
38. Jonathan Glancey, 'Sufferin' Satellites! We've Built the Future!' *Guardian* 28 April 2008, https://www.theguardian.com/artanddesign/2008/apr/28/architecture (retrieved 27 February 2020). My thanks to Otto Saumarez Smith, who drew my attention to it.
39. Warren Chalk, Frank Brian Harvey and Ron Herron, 'Project No. 64: Walking City', in Kester Rattenbury and Murray Fraser (eds.), *The Archigram Archival Project* (University of Westminster), http://archigram.westminster.ac.uk/project.php?id=60 (retrieved 27 February 2020).
40. The Buchanan Report, discussed in Otto Saumarez Smith, *Boom Cities: Architect Planners and the Politics of Radical Urban Renewal in 1960s Britain* (Oxford University Press, 2019), p. 171. Depressingly, as Saumarez Smith points out, the same influential report regarded cycling as a disappearing and obsolete

form of transport, a prophecy which looked for a while like becoming self-fulfilling before the more recent renaissance of cycle transport for health and sustainability.

41. Cedric Price, 'Fun Palace', *New Scientist* (14 May 1964), p. 433; the best history of the project is Stanley Mathews, *From Agit-Prop to Free Space: The Architecture of Cedric Price* (Black Dog, 2007).
42. Quoted in Robert Ouellette, 'The Manulife Centre is a Brute', in Michael McClelland and Graeme Stewart (eds.), *Concrete Toronto: A Guide to Concrete Architecture from the Fifties to the Seventies* (Coach House, 2004), p. 194.
43. Barnabas Calder, 'The Limits of 1960s Radicalism: The Fun Palace versus the National Theatre', in Alistair Fair (ed.), *Setting the Scene: Perspectives on Twentieth-Century Theatre Architecture* (Routledge, 2015), pp. 163–78 (p. 172).
44. Reyner Banham, 'A Home is Not a House', *Art in America* (April 1965).
45. Giovanna Borasi and Mirko Zardini (eds.), *Sorry, Out of Gas: Architecture's Response to the 1973 Oil Crisis* (Canadian Center for Architecture, 2007), pp. 50ff.
46. Leonard R. Bachman, *Integrated Buildings: The Systems Basis of Architecture* (Wiley, 2003), pp. 373–9.
47. Muthesius and Glendinning, *Towerblock*, pp. 322–3.
48. Good recent publications on British postmodernism include Elain Harwood and Geraint Franklin, *Post-Modern Buildings in Britain* (Twentieth Century Society, 2017), and Terry Farrell and Adam Nathaniel Furman, *Revisiting Postmodernism* (RIBA, 2017).
49. Robert Venturi, *Complexity and Contradiction in Architecture* (1966), pp. 16–17.
50. Borasi and Zirdini (eds.), *Sorry, Out of Gas*.
51. United Nations Report of the World Commission on Environment and Development, 'Our Common Future', Chapter 2, I.1.

CHAPTER 12 – TODAY'S GREAT ENERGY REVOLUTION

1. UN Environment and International Energy Agency, *Global Status Report 2017: Towards a Zero-emission, Efficient, and Resilient Buildings and Construction Sector* (2017), p. 6.
2. Ibid., p. 2.

3. Andrew Saint, *Architect and Engineer: A Study in Sibling Rivalry* (Yale University Press, 2008). Saint gives examples of the collaboration between Auguste Perret and the engineering firm Latron et Vincent in the design of a famous early concrete frame building at Rue Franklin as being decisive in the Perrets' understanding of concrete construction (pp. 232–3) and details the relationship between Frank Lloyd Wright and the trained engineer-cum-construction manager Paul Mueller (pp. 249–57), for example.
4. Zofia Trafas White, 'Computers and the Sydney Opera House', Victoria and Albert Museum website, https://www.vam.ac.uk/articles/computers-and-the-sydney-opera-house (retrieved 26 February 2020).
5. Graham Kendall, 'Apollo 11 Anniversary: Could an iPhone Fly Me to the Moon?' *Independent*, 9 July 2019.
6. Alexander J. Hahn and Alexander J. J. Hahn, *Mathematical Excursions to the World's Great Buildings* (Princeton University Press, 2012), p. 255.
7. Ibid., p. 253.
8. Andrew Saint, 'Thoughts from Walsall', in Anthony Alofsin (ed.), *A Modernist Museum in Perspective: The East Building, National Gallery of Art* (Yale University Press, 2009), pp. 171–83.
9. https://www.architectmagazine.com/design/buildings/guangzhou-opera-house_o
10. Barnabas Calder, '"Never so serious": Venturi's Sainsbury Wing at the National Gallery, London', in Alofsin (ed.), *A Modernist Museum in Perspective*, pp. 183–99.
11. An excellent social media campaign was led by the architect and artist Adam Nathaniel Furman in 2019, under the hashtags #arch4all and #archishame. It is discussed in Warwick Mihaly, 'The Architecture of Exploitation', *Architecture AU* (9 September 2019), https://architectureau.com/articles/the-architecture-of-exploitation/ (retrieved 26 February 2020).
12. Osamu Ishiyama, quoted in Thomas Daniell, *An Anatomy of Influence* (Architectural Association, 2018), p. 200.
13. John Joseph Burns, 'The Chengzhongcun: Collectivity through Individuality', in Austin Williams and Theodoros Dounas (eds.), *Masterplanning the Future: Modernism: East, West & Across the World* (XJTLU, 2012), pp. 117–25.
14. Mee Kam Ng, 'City Profile: Shenzhen', *Cities*, Vol. 20 (2003), pp. 429–41 (p. 431).
15. https://www.washingtonpost.com/news/wonk/wp/2015/03/24/how-china-used-more-cement-in-3-years-than-the-u-s-did-in-the-entire-20th-century/
16. Michael J. Gibbs, Peter Soyka and David Conneely, 'CO_2 Emissions from Cement Production', IPPC Background Paper, *Good Practice Guidance and Uncertainty Management in National Greenhouse Gas Inventories* (IPCC, 2000), p. 177.

17. For cement and steel contributions to industrial emissions see Michael Pooler, 'Cleaning up Steel is Key to Tackling Climate Change Emissions', *Financial Times* (1 January 2019); 51 per cent of steel is used for architecture and infrastructure according to the World Steel Association, https://www.worldsteel.org/steel-by-topic/steel-markets.html (retrieved 11 March 2020).
18. World cement production in 2019 was 4,200,000,000 tonnes: https://www.statista.com/statistics/1087115/global-cement-production-volume/. Early 2000s cement cost 5.8MJ per kg: Shahrir Shams, Kashif Mahmud and Md. Al-Amin, 'A Comparative Analysis of Building Materials for Sustainable Construction with Emphasis on CO_2 Reduction', *International Journal of Environment and Sustainable Development*, Vol. 10 (2011), pp. 364–74, Table 1. So total 2019 world cement production may have required 24,360,000,000,000MJ, or 6,766,666,666,667kWh. Which is the annual production of 7,724,505,327,245m² of charcoal coppice (7,724,505km²) at 0.876kWh/m². Australia is 7,692,000km². In terms of human labour, 6,766,666,666,667kWh is 90,222,222,222,226 hours of labour at 0.075kW, 36,146,723,647 years of fifty-two six-day weeks, each day being eight hours of labour, by 4.7 times the population of the earth.
19. Fiona Harvey, 'China pledges to become carbon neutral before 2060', *Guardian* 22 September 2020.
20. *BP Statistical Review – 2019: China's Energy Market in 2018*, https://www.bp.com/content/dam/bp/business-sites/en/global/corporate/pdfs/energy-economics/statistical-review/bp-stats-review-2019-china-insights.pdf (accessed 5 December 2019).
21. *BP Statistical Review – 2019*, https://www.bp.com/content/dam/bp/business-sites/en/global/corporate/pdfs/energy-economics/statistical-review/bp-stats-review-2019-full-report.pdf (accessed 5 December 2019).
22. https://www.architectsdeclare.com/ (figure for 22 February 2021).
23. https://ourworldindata.org/emissions-by-sector (retrieved 23 September 2020).
24. Intergovernmental Panel on Climate Change Special Report, 'Global Warming of 1.5°C', https://www.ipcc.ch/sr15/ (accessed 26 February 2020).
25. John Bambridge, 'Bloomberg London HQ Rated World's Most Sustainable Office', *Construction Week*, 2 October 2017, https://www.constructionweekonline.com/article-46569-bloomberg-london-hq-rated-worlds-most-sustainable-office (retrieved 26 February 2020).
26. Bloomberg, 'Bloomberg's New European Headquarters Rated World's Most Sustainable Office Building', 1 October 2017, https://www.bloomberg.com/company/press/bloomberg-most-sustainable-office-building/ (retrieved 26 February 2020).

27. A UK campaign to promote retrofitting and reuse of existing buildings wherever possible is being led by Will Hurst and *Architects' Journal*: Will Hurst, 'Introducing RetroFirst: A New *AJ* Campaign Championing Reuse in the Built Environment', *Architects' Journal* (12 September 2019).
28. Simon Sturgis, 'RIBA Stirling Prize 2018: The Most Sustainable Shortlist Ever?', *Architects' Journal* (3 October 2018), https://www.architectsjournal.co.uk/opinion/riba-stirling-prize-2018-the-most-sustainable-shortlist-ever/10035714.article (retrieved 26 February 2020).
29. John H. Schofield, 'Efficacy of LEED-certification in Reducing Energy Consumption and Greenhouse Gas Emission for Large New York City Office Buildings', *Energy and Buildings*, Vol. 67 (December 2013), pp. 517–24; 'In U.S. Building Industry, is It Too Easy to be Green?', *USA Today*, 13 June 2020, https://eu.usatoday.com/story/news/nation/2012/10/24/green-building-leed-certification/1650517/ (retrieved 26 February 2020).
30. Hattie Harman, 'Cork House: Matthew Barnett Howland with Dido Milne and Oliver Wilton', *Architects' Journal* (26 September 2019). The research was conducted in collaboration with, among others, the Bartlett School of Architecture UCL, the University of Bath, Sturgis Carbon Profiling LLP and Arup.
31. Department for Environment, Food and Rural Affairs report, 'UK Statistics on Waste', 7 March 2019, p. 8, https://assets.publishing.service.gov.uk/government/uploads/system/uploads/attachment_data/file/784263/UK_Statistics_on_Waste_statistical_notice_March_2019_rev_FINAL.pdf.
32. Andrea Di Maria, Johan Eyckmans and Karel Van Acker, 'Downcycling versus Recycling of Construction and Demolition Waste: Combining LCA and LCC to Support Sustainable Policy Making', *Waste Management*, Vol. 75 (May 2018), pp. 3–21.

CONCLUSION

1. Phyllis Lambert recalling a conversation with Mies, in '"Joan of Architecture" and the Difficulty of Simplicity', *CTBUH Journal* (2013, Issue IV), pp. 46–8 (p. 47).
2. Architectural history and energy is a huge topic, so far largely untouched by historians, though interest is just starting to grow, with a session at the Society of Architectural Historians (USA) Annual Meeting at Montreal, 2021, on the theme, and the Society of Architectural Historians annual

symposium, 2021, on architecture, energy and environment. We are developing a research network forming of academics working on different aspects of architectural history, heritage and energy. I would welcome enquiries from prospective research PhD students wanting to work on aspects of the history of energy and architecture in any period and region.

3. Steve Webb, Zoom call with the author, 9 October 2020.
4. Gilles Perraudin, paper given at Building Centre webinar, 'Contemporary stone architecture: the art and science of building in stone', 17 September 2020.
5. Wolfgang Feist, University of Innsbruck, commentary on Rachel Mitchell and Sukumar Natarajan, 'UK Passive House and the Energy Performance Gap', *Energy and Buildings*, Vol. 224, 1 October 2020. Commentary available here: https://www.researchgate.net/publication/342644074_UK_Passivhaus_and_the_energy_performance_gap, retrieved 24 September 2020.
6. 'How much heating energy do you use?', Ovo Energy, https://www.ovoenergy.com/guides/energy-guides/how-much-heating-energy-do-you-use.html retrieved 23 September 2020.
7. There is extensive and fast-growing literature on the technical challenges of improving environmental performance in older houses. It is worth working with a specialist consultancy as they know how to avoid potential pitfalls that can be damaging to either the environmental performance or the fabric of your house, and consultancy fees are a modest proportion of the cost of the project. Information and practices in your area can be found at passivhaustrust.org.uk/.

Index

A

D

E

F

I

J

R

S

T

X

Y

Z

1 **Economics: The User's Guide**
Ha-Joon Chang

2 **Human Evolution**
Robin Dunbar

3 **Revolutionary Russia: 1891–1991**
Orlando Figes

4 **The Domesticated Brain**
Bruce Hood

5 **Greek and Roman Political Ideas**
Melissa Lane

6 **Classical Literature**
Richard Jenkyns

7 **Who Governs Britain?**
Anthony King

8 **How to See the World**
Nicholas Mirzoeff

9 **The Meaning of Science**
Tim Lewens

10 **Social Class in the 21st Century**
Mike Savage

11 **The European Union: A Citizen's Guide**
Chris Bickerton

12 **The Caliphate**
Hugh Kennedy

13 **Islam: The Essentials**
Tariq Ramadan

---

14 **Basic Income: And How We Can Make It Happen**
Guy Standing

---

15 **Think Like an Anthropologist**
Matthew Engelke

---

16 **Hermeneutics: Facts and Interpretation in the Age of Information**
John D. Caputo

---

17 **Being Ecological**
Timothy Morton

---

18 **Object-Oriented Ontology: A New Theory of Everything**
Graham Harman

---

19 **Marx and Marxism**
Gregory Claeys

---

20 **The Human Planet: How We Created the Anthropocene**
Simon L. Lewis and Mark A. Maslin

---

21 **Think Again: How to Reason and Argue**
Walter Sinnott-Armstrong

---

22 **Parenting the First Twelve Years: What the Evidence Tells Us**
Victoria L. Cooper, Heather Montgomery, Kieron Sheehy

---

23 **Social Mobility: And Its Enemies**
Lee Elliot Major and Stephen Machin

---

24 **National Populism: The Revolt Against Liberal Democracy**
Roger Eatwell and Matthew Goodwin

---

25 **What We Really Do All Day: Insights from the Centre for Time Use Research**
Jonathan Gershuny and Oriel Sullivan

---

26 **A Short History of Brexit**
Kevin O'Rourke

---

27 **A Political History of the World**
Jonathan Holslag

---

28 **Our Universe: An Astronomer's Guide**
Jo Dunkley

---

29 **Chinese Thought: From Confucius to Cook Ding**
Roel Sterckx

---

30 **The Art of Statistics: Learning from Data**
David Spiegelhalter

---

31 **This is Shakespeare**
Emma Smith

---

32 **The Government of No One: The Theory and Practice of Anarchism**
Ruth Kinna

---

33 **The Plunder of the Commons: A Manifesto for Sharing Public Wealth**
Guy Standing

---

34 **Artificial Intelligence: A Guide for Thinking Humans**
Melanie Mitchell

---

35 **Can We Be Happier? Evidence and Ethics**
Richard Layard

---

36 **Feeding Britain: Our Food Problems and How to Fix Them**
Tim Lang

---

37 **The Road to Conscious Machines: The Story of AI**
Michael Wooldridge

---

38 **Feminisms: A Global History**
Lucy Delap
